COMPOSITIONAL ANALYSIS
OF POLYMERS
An Engineering Approach

COMPOSITIONAL ANALYSIS OF POLYMERS
An Engineering Approach

Edited by

Aleksandr M. Kochnev, PhD

Oleg V. Stoyanov, DSc

Gennady E. Zaikov, DSc, and

Renat M. Akhmetkhanov, DSc

Apple Academic Press Inc. | Apple Academic Press Inc.
3333 Mistwell Crescent | 9 Spinnaker Way
Oakville, ON L6L 0A2 | Waretown, NJ 08758
Canada | USA

© 2016 by Apple Academic Press, Inc.

First issued in paperback 2021

Exclusive worldwide distribution by CRC Press, a member of Taylor & Francis Group

No claim to original U.S. Government works

ISBN-13: 978-1-77463-562-9 (pbk)
ISBN-13: 978-1-77188-148-7 (hbk)

Library and Archives Canada Cataloguing in Publication

Compositional analysis of polymers : an engineering approach / edited by Aleksandr M. Kochnev, PhD, Oleg V. Stoyanov, DSc, Gennady E. Zaikov, DSc, and Renat M. Akhmetkhanov, DSc.

Includes bibliographical references and index.
Issued in print and electronic formats.
ISBN 978-1-77188-148-7 (hardcover).--ISBN 978-1-77188-288-0 (pdf)
1. Polymers. 2. Polymers--Analysis. 3. Composite materials. 4. Polymerization.
I. Zaikov, G. E. (Gennadiĭ Efremovich), 1935- author, editor II. Stoyanov, Oleg V., editor
III. Kochnev, Aleksandr M., editor IV. Akhmetkhanov, Renat M., editor

QD381.C68 2016   547'.7   C2016-900203-9   C2016-900204-7

Library of Congress Cataloging-in-Publication Data

Names: Kochnev, Aleksandr M. | Stoyanov, Oleg V. | Zaikov, G. E. (Gennadiĭ Efremovich), 1935- | Akhmetkhanov, Renat M.
Title: Compositional analysis of polymers : an engineering approach / [edited by] Aleksandr M. Kochnev, PhD, Oleg V. Stoyanov, DSc, Gennady E. Zaikov, DSc, and Renat M. Akhmetkhanov, DSc.
Description: Toronto : Apple Academic Press, 2016. | Includes bibliographical references and index.
Identifiers: LCCN 2016000579 (print) | LCCN 2016002235 (ebook) | ISBN 9781771881487 (hardcover : alk. paper) | ISBN 9781771882880 ()
Subjects: LCSH: Polymers. | Polymerization. | Copolymers.
Classification: LCC QD381 .C6588 2016 (print) | LCC QD381 (ebook) | DDC 620.1/92--dc23
LC record available at http://lccn.loc.gov/2016000579

Apple Academic Press also publishes its books in a variety of electronic formats. Some content that appears in print may not be available in electronic format. For information about Apple Academic Press products, visit our website at **www.appleacademicpress.com** and the CRC Press website at **www.crcpress.com**

ABOUT THE EDITORS

Aleksandr M. Kochnev, PhD
Aleksandr M. Kochnev is Deputy of Director of Kazan National Research Technological University. He is a specialist in the chemistry, physics, and mechanics of polymers, composites, and nanocomposites. He is also very active in synthetic organic chemistry, physical chemistry, and chemical physics. He has published 10 books and about 100 original papers.

Oleg V. Stoyanov, DSc
Oleg V. Stoyanov is Head of the Chair of Industrial Safety and Professor and Chair of Technology of Processing Plastic and Composite Materials (TPPCM) at Kazan State Technological University in Russia. He is a world-renowned scientist in the field of chemistry and the physics of oligomers, polymers, composites, and nanocomposites.

Gennady E. Zaikov, DSc
Gennady E. Zaikov is Head of the Polymer Division at the N. M. Emanuel Institute of Biochemical Physics, Russian Academy of Sciences, Moscow, Russia, and professor at Moscow State Academy of Fine Chemical Technology, Russia, as well as Professor at Kazan National Research Technological University, Kazan, Russia. He is also a prolific author, researcher, and lecturer. He has received several awards for his work, including the Russian Federation Scholarship for Outstanding Scientists. He has been a member of many professional organizations and on the editorial boards of many international science journals.

Renat M. Akhmetkhanov, DSc
Renat M. Akhmetkhanov is Dean of Chemical Faculty, Bashkir State University, Ufa, Bashkirian Republic, Russian Federation).

He is an expert in the fields of chemical kinetics, chemistry and physics of polymers, composites, and nanocomposites as well as synthesis and modification of polymers (including degradation and stabilization of polymers and composites). He has published about 500 original papers and reviews as well as several monographs.

CONTENTS

LIST OF CONTRIBUTORS

M. I. Abdullin
Bashkir State University, 100 Mingazhev Str., Ufa 450074, Russia,
E-mail: koksharova.yulya@yandex.ru

O. M. Alekseeva
Emanuel Institute of Biochemical Physics, Russian Academy of Sciences, Moscow, Russia

G. Yu. Amantaeva
Faculty of Chemistry, Bashkir State University, 32 ZakiValidi Str., Ufa 450043, Russian Federation

N. A. Amineva
Faculty of Chemistry, Bashkir State University, 32 ZakiValidi Str., Ufa 450043, Russian Federation

G. F. Aminova
Federal State Educational Institution of Higher Professional Education, Ufa State Petroleum Technological University, Kosmonavtov Str. 1, Ufa 450062, Republic of Bashkortostan, Russian Federation

Yu. O. Andriasyan
Emanuel Institute of Biochemical Physics of Russian Academy of Sciences (IBCP RAS), Moscow119991, Kosygina Str. 4, Russia

M. S. Babaev
Institute of Organic Chemistry, Ufa Scientific Centre of RAS, pr. Oktyabrya 71, Ufa, Russia. E-mail: b.marat.c@mail.ru

T. S. Babicheva
Educational and Research Institute of Nanostructures and Biosystems, Saratov State University, 83 Astrakhanskaya Str., Saratov 410012, Russian Federation. E-mail: Tatyana.babicheva.1993@mail.ru

A. A. Basyrov
Bashkir State University, 100 Mingazhev Str., Ufa 450074, Russia, E-mail: koksharova.yulya@yandex.ru

T. V. Berestova
Faculty of Chemistry, Bashkir State University, 32 ZakiValidi Str., Ufa 450043, Russian Federation. E-mail: berestovatv@gmail.com

A. O. Burakova
Kazan National Research Technological University, 68 Karl Marx Str., Kazan 420015, Republic of Tatarstan, Russian Federation

V. V. Chernova
Bashkir State University, Republic of Bashkortostan, Ufa 450074, ul.ZakiValidi 32, Russia

R. R. Daminev
Federal State Educational Institution of Higher Professional Education Ufa State Petroleum Technological University, Prospect Str. 2, Sterlitamak 453118, Sterlitamak, Republic of Bashkortostan, Russian Federation. E-mail: Taiffa27@mail.ru

R. J. Deberdeev
Federal State Educational Institution of Higher Professional Education Kazan State Technological University, Karl Marx Str. 68, Kazan 420015, Republic of Tatarstan, Russian Federation

T. R. Deberdeev
Federal State Educational Institution of Higher Professional Education Kazan State Technological University, Karl Marx Str. 68, Kazan 420015, Republic of Tatarstan, Russian Federation

R. N. Fatkullin
Federal State Educational Institution of Higher Professional Education, Ufa State Petroleum Technological University,Republic of Bashkortostan, 453118 Sterlitamak, Prospect Str. 2, Russian Federation. E-mail: fatkullin.rna@kaus.ru

A. K. Friesen
Institute of Organic Chemistry, Ufa Scientific Center, Russian Academy of Sciences, pr. Oktyabrya 71, Ufa 450054, Russia

A. S. Gadeev
Bashkir State University, 100 Mingazhev Str., Ufa 450074, Russia, E-mail: koksharova.yulya@yandex.ru

N.O. Gegel
Educational and Research Institute of Nanostructures and Biosystems, Saratov State University, 83 Astrakhanskaya Str., Saratov 410012, Russian Federation. E-mail: GegelNO@yandex.ru
K. Z. Gumargalieva
N.N. Semenov Institute of Chemical Physics, Russian Academy of Sciences, Moscow, Russia

A. L. Iordanskii
Semenov Institute of Chemical Physics, Russian Academy of Sciences, KosyginStr. 4, Moscow 119991, Russia

I. G. Kalinina
N. N. Semenov Institute of Chemical Physics, Russian Academy of Sciences, Moscow, Russia

A. V. Khvatov
Emanuel Institute of Biochemical Physics of Russian Academy of Sciences, Kosigina Str. 4, Moscow, Russia

Yu. A. Kim
Institute of Cell Biophysics, Russian Academy of Sciences, Pushchino, Moscow, Russia. E-mail: olgavek@yandex.ru

Yu. A. Koksharova
Bashkir State University, 100 Mingazhev Str., Ufa 450074, Russia. E-mail: koksharova.yulya@yandex.ru

N. N. Kolesnikova
Emanuel Institute of Biochemical Physics of Russian Academy of Sciences, Kosigina Str. 4, Moscow, Russia. E-mail: kolesnikova@sky.chph.ras.ru

S. V. Kolesov
Institute of Organic Chemistry, Ufa Scientific Center, Russian Academy of Sciences, Ufa 450054, Republic of Bashkortostan, Russian Federation. E-mail: kolesov@anrb.ru

N. V. Koltaev
Bashkir State University, 100 Mingazhev Str., Ufa 450074, Russia.
 E-mail: koksharova.yulya@yandex.ru

A. V. Krementsova
Emanuel Institute of Biochemical Physics, Russian Academy of Sciences, Moscow, Russia

A. V. Krivandin
Emanuel Institute of Biochemical Physics, Russian Academy of Sciences, Moscow, Russia

E. I. Kulish
Bashkir State University, Republic of Bashkortostan, Ufa 450074, ul.ZakiValidi 32, Russia. E-mail: alenakulish@rambler.ru

L. G. Kuzina
Faculty of Chemistry, Bashkir State University, 32 ZakiValidi Str., Ufa 450043, Russian Federation
Yu. K. Lukanina
Emanuel Institute of Biochemical Physics of Russian Academy of Sciences, Kosigina Str. 4, Moscow, Russia

A. R. Maskova
Federal State Educational Institution of Higher Professional Education Ufa State Petroleum Technological University, Kosmonavtov Str. 1, Ufa 450062, Republic of Bashkortostan, Russian Federation

I. A. Massalimov
Faculty of Chemistry, Bashkir State University, 32 ZakiValidi Str., Ufa 450043, Russian Federation

A. K. Mazitova
Federal State Educational Institution of Higher Professional Education Ufa State Petroleum Technological University, Kosmonavtov Str. 1, Ufa 450062, Republic of Bashkortostan, Russia

E. N. Miftakhova
Department of Physics and Mathematics, Ufa State Aviation Technical University, Ishimbay Branch, Republic of Bashkortostan, Ishimbay City, Gubkina Str. 15, 453200, Russia

G. V. Miftakhova
Faculty of Chemistry, Bashkir State University, 32 ZakiValidi Str., Ufa 450043, Russian Federation

T. A. Mikhailova
Department of Mathematical Modelling, Sterlitamak Branch of Bashkir State University, Republic of Bashkortostan, Sterlitamak City, Lenin Avenue 37, 453103 Russia. E-mail: T.A.Mikhailova@yandex.ru

I. A. Mikhaylov
Plekhanov Russian University of Economics (PRUE), Moscow117997, Stremyanny per. 36, Russia

S.A Mustafina
Department of Mathematical Modelling, Sterlitamak Branch of Bashkir State University, Republic of Bashkortostan, Sterlitamak City, Lenin Avenue 37, 453103 Russia. E-mail: Mustafina_SA@mail.ru

R.F. Nafikova
Federal State Educational Institution of Higher Professional Education, Ufa State Petroleum Technological University, Prospect Str. 2, Sterlitamak 453118, Sterlitamak, Republic of Bashkortostan, Russian Federation. E-mail: nafikova.rf@kaus.ru

I. I. Nasibullin
Institute of Organic Chemistry, Ufa Scientific Center, Russian Academy of Sciences, pr. Oktyabrya 71, Ufa 450054, Russia

S. N. Nikolaev
Bashkir State University, 100 Mingazhev Str., Ufa 450074, Russia,
E-mail: koksharova.yulya@yandex.ru

A. A. Ol`khov
Plekhanov Russian University of Economics, Stremyanny per. 36, Moscow117997, Russia. E-mail: aolkhov72@yandex.ru

V. Pankov
Voronezh State University, Voronezh 394006, Universitetskayapl No.1, Russian Federation

A. A. Popov
Emanuel Institute of Biochemical Physics of Russian Academy of Sciences, Kosigina Str. 4, Moscow, Russia

S. A. Semenov
N.N. Semenov Institute of Chemical Physics, Russian Academy of Sciences, Moscow, Russia

O. V. Shatalova
Emanuel Institute of Biochemical Physics, Russian Academy of Sciences, Moscow, Russia

A. B. Shipovskaya
Educational and Research Institute of Nanostructures and Biosystems, Saratov State University, 83 Astrakhanskaya Str., Saratov 410012, Russian Federation. E-mail: ShipovskayaAB@rambler.ru

E. B. Shirokikh
Kazan National Research Technological University, 68 Karl Marx Str., Kazan 420015, Republic of Tatarstan, Russian Federation

N.A. Shkeneva
Bashkir Soda Company JSC, Tekhnicheskaya Str. 32, The Republic of Bashkortostan, Sterlitamak, The Russian Federation. E-mail: shkeneva.natalia@gmail.com

D. A. Shiyan
Kazan National Research Technological University, 68 Karl Marx Str., Kazan 420015, Republic of Tatarstan, Russian Federation

A. S. Shurshina
Bashkir State University, Republic of Bashkortostan, Ufa 450074, ul.ZakiValidi 32, Russia

N. N. Sigaeva
Institute of Organic Chemistry, Ufa Scientific Center, Russian Academy of Sciences, pr. Oktyabrya 71, Ufa 450054, Russia

L.B. Stepanova
Federal State Educational Institution of Higher Professional Education Kazan State Technological University, Karl Marx Str. 68, Kazan 420015, Republic of Tatarstan, Russian Federation. E-mail: Lenadez@mail.ru

A. A. Sultanova
Ufa State University of Economics and Service, Chernyshevsky Str. 145, 450078 Ufa, Russia

K. A. Tereshchenco
Kazan National Research Technological University, 68 Karl Marx Str., Kazan 420015, Republic of Tatarstan, Russian Federation

I. F. Tuktarova
Bashkir State University, Republic of Bashkortostan, Ufa 450074, ul.ZakiValidi 32, Russia

N. V. Ulitin
Kazan National Research Technological University, 68 Karl Marx Str., Kazan 420015, Republic of Tatarstan, Russian Federation. E-mail: n.v.ulitin@mail.ru

R. R. Usmanova
Ufa State Technical University of Aviation, 12 Karl Marks Str., Ufa 450000, Bashkortostan, Russia. E-mail: usmanovarr@mail.ru

V. N. Verezhnikov
Voronezh State University, Voronezh 394006, Universitetskayapl No.1, Russian Federation

V. M. Yanborisov
Ufa State University of Economics and Service, 24 Aknazarova Str.,Ufa 450022, Republic of Bashkortostan, Russian Federation. E-mail: YanborisovVM@mail.ru

V. P. Yudin
Voronezh Affiliated Societies of Science Researching Synthetic Rubber, Voronezh 394014, Mendeleeva Str. No.3, Building B, Russian Federation. E-mail: skilful25@mail.ru

G. E. Zaikov
N.M. Emanuel Institute of Biochemical Physics, Russian Academy of Sciences, 4 Kosygin Str., Moscow 119334, Russia. E-mail:chembio@sky.chph.ras.ru

V. P. Zakharov
Bashkir State University, 32 Validy Str., Ufa 450076, Republic of Bashkortostan, Russian Federation. E-mail: zaharovvp@mail.ru

LIST OF ABBREVIATIONS

AMCh	Amikacin chloride
AMS	Amikacinsulphate
AIBN	Azoisobutyronitrile
BPO	Benzoyl peroxides
BP	Benzoyl peroxide
bPP	Propylene-based block copolymer
BSF	Besieged silica filler
CaO	Calcium oxide
CCD	Charge-coupled device
CEPDM	Chlorinated ethylene–propylene–dienecaoutchoucs
ChT	Chitosan
ChTA	Chitosan acetate
CMC	Critical micelle concentration
CPhZ	Cefazolin sodium salt
CSR/CSE	Complex stabilizer
DAP	Dialkyl phthalate
DASM-4	Differential adiabatic scanning microcalorimeter
DBP	Dibutyl phthalate
DINP	Diisononylphthalate
DMEM	Dulbecco's Modified Eagle Medium
DMPC	Dimyristoilphosphatidylcholine
DOP	Dioctyl phthalate
DOTF	Dioctylphthalate
DPET	Dipenterythritol
DSC	Differential scanning calorimeter
DTA	Deferential thermal analysis
ENB	Ethylidenenorbornene
EPDM	Ethylene–propylene–dienecaoutchoucs
EPR	Electron paramagnetic resonance
ESO	Epoxidized soybean oil
FTIR	Fourier transform infrared spectroscopy
FLWIP	First layer of the water-insoluble polymer
GA	Glycolic acid
GE	Grafting efficiency
GF	Grafting frequency

GMS	Gentamicin sulfate
HCl	Hydrogen chloride
HET	Hexachloroendomethylenetetrahydrophthalic
HIPS	High-impact polystyrenes
IR	Infrared
KOH	Potassium hydroxide
KPD	Polydispersity coefficient
MC	Monte Carlo method
MFI	Melt flow index parameter
MgO	Magnesium oxide
MMA	Methyl methacrylate
MMD	Molecular mass distribution
MM	Molecular mass
Mn	Number average
mPP	Maleic anhydride grafted PP
Mw	Molecular weight
NaCl	Sodium chloride
NaOH	Sodium hydroxide
NIR	Near infrared
NMR	Nuclear magnetic resonance
OIT	2-n octyl-4-isothiazolin-3-one
PAM	Polyacrylamide
PB-PS-Li	Polybutadienyl–polystyryl lithium
PDADMAC	Poly-N,N-diallyl-N,N-dimethylammonium chloride
PE	Polyethylene
PE	polyelectrolyte
PEEK	Polyether ketones
PEKKA	Polyether ketone ketone
PET	Polyethylene terephthalate
PF	Phenol formaldehyde resins
PHB	Poly(3-hydroxybutyrate)
PIB	Polyisobutylene
PLP	Pulsed laser polymerization
PMMA	Polymethyl methacrylate
PNB	Pyrolytic boron nitride
PP	Polypropylene
PPO	Poly(phenylene oxide)
PSC	Polyelectrolyte–surfactant complexes
PVAC	Polyvinyl acetate
PVC	Polyvinyl chloride

PVDF	Polyvinylidene fluoride
PVE	Polyvinyl ether
PVK	Poly(n-vinylcarbazol)
PVOH	Polyvinyl alcohol
PVP	Poly(N-vinyl-pyrrolidone)
PVPP	Polyvinyl pyrrolidone
rPP	Propylene-based random copolymers
SAXS	Small-angle X-ray scattering
SDBS	Sodium dodecylbenzenesulfonate
SDS	Sodium dodecyl sulfate
SEM	Scanning electron microscope
SF	Surfactants
SIP	Spatially discontinuous polymerization
SLWIP	Second layer of the water-insoluble polymer
S-SBR	Styrene-butadiene rubber
St	Styrene
TC	Technical carbon
TCNE	Tetracyanoethylene
TEA	Triethanolamine
TESPT	Bis-(triethoxysilylpropyl)-tetrasulfide
TGA	Thermogravimetric analysis
TNF	Trinitrofluorene
USD	US dollar
UV	Ultraviolet
UV–vis	Ultraviolet–visible
XRD	X-ray diffraction
ZnO	Zinc oxide

PREFACE

Technical and technological development demands the creation of new materials that are stronger, more reliable, and more durable, i.e., materials with new properties. Up-to-date projects in the creation of new materials go along the way of nanotechnology. This new book covers a broad range of polymeric materials and provides researchers in polymer science and technology with an important resource covering many aspects involved in the functional materials production chain.

The aim of this new volume is to provide a solid understanding of the recent developments in advanced polymeric materials from macro- to nano-length scales. Composites are becoming more and more important as they can help to improve our quality of life.This volume presents the latest developments and trends in advanced polymer materials and structures. It discusses the developments of advanced polymers and respective tools to characterize and predict the material properties and behavior. This book has an important role in advancing polymer materials in macro- and nanoscale. Its aim is to provides original, theoretical, and important experimental results that use non-routine methodologies. It also includes chapters on novel applications of more familiar experimental techniques and analyses of composite problems that indicate the need for new experimental approaches.

This new book:

- is a collection of articles that highlight some important areas of current interest in key polymeric materials and technology
- gives an up-to-date and thorough exposition of the present state of the art of key polymeric materials and technology
- describes the types of techniques now available to the engineers and technicians and discusses their capabilities, limitations, and applications
- provides a balance between materials science and chemical aspects, basic and applied research
- focuses on topics with more advanced methods
- emphasizes on precise mathematical development and actual experimental details
- explains modification methods for changing of different materials properties.

CHAPTER 1

MORPHOLOGY OF POLYPROPYLENES OF DIFFERENT CHEMICAL STRUCTURE

N. N. KOLESNIKOVA, YU. K. LUKANINA, A. V. KHVATOV, and A. A. POPOV

Emanuel Institute of Biochemical Physics of Russian Academy of Sciences, Kosigina Str. 4, Moscow, Russia,, E-mail: kolesnikova@sky.chph.ras.ru

CONTENTS

ABSTRACT

The melting behavior of polypropylenes of different chemical structures (isotactic homopolypropylene, propylene-based block and random copolymers, and maleic anhydride grafted polypropylene) was studied by differential scanning calorimeter (DSC) and optical microscopy. Melting behavior and the crystal structure of polypropylene and its copolymers were observed depending on the crystallization rate, chemical nature of comonomer units, and regularity of comonomer units' arrangement in the polypropylene main chain.

1.1 INTRODUCTION

Polypropylene and its copolymers are the most important commodity thermoplastic polymers which are found in a wide variety of applications due to its excellent strength, toughness, high chemical resistance, and high melting point. Polypropylene versatility along with its low cost and technological potentials has rendered it as one of the most used materials worldwide.[1–5]

Isotactic homopolypropylene (PP) is a crystalline polymer whose properties are dependent on the degree of crystallinity. It exhibits three crystalline forms: the monoclinic α-phase, the hexagonal β-phase, and the orthorhombic γ-phase.[6–10] When PP is formed, its crystal formation changes according to the heat treatment temperature and the conditions of cooling process. These changes create differences in strength, heat resistance, and pressure bonding properties. Accordingly, it is interesting and important to find out how to manage the structure of the polymer by varying the crystallization conditions during processing.

Propylene-based block copolymer (bPP)—the chain of molecules of propylene gap in the chain of ethylene copolymer. Propylene block copolymers produced in the form of homogeneous color of the granules with high impact strength (at low temperatures) and high flexibility; increased long-term thermal stability; resistance to oxidative degradation during production, and processing of polypropylene as well as the operation of the product out of it. Owing to the crystalline structure of the polypropylene block copolymer is a thermoplastic structural sufficiently economical polymer.[11–12]

Propylene-based random copolymers (rPP) are made by copolymerizing propylene and small amounts of ethylene (usually up to 7 wt.%) and their structure is similar to isotactic polypropylene, but the regular repeating of propylene units along the macromolecular chains is randomly disrupted by the presence of the comonomer ones. The presence of ethylene units reduces the melting point and crystallinity by introducing irregularities into the main polymeric chain. The advantages of this class of polymers are improved transparency, relative softness, lower sealing temperature, and moderate low-temperature impact strength due to the lowered glass-transition temperature.[13–15]

Maleic anhydride grafted PP (mPP) is the most common functional adhesion promoter. It was proved to be effective functional molecule for the reactive compatibilization.[1,16–18]

It is generally known that the properties of semicrystalline polymers depend strongly on the size and shape of their supramolecular structure.[1,16,19,20] The purpose of this chapter is to present new results concerning the melting behavior and the crystal structure of polypropylene and its copolymers which depends on the crystallization rates, chemical nature of comonomer units, and regularity of comonomer units arrangement in the polypropylene main chain.

1.2   EXPERIMENTAL

1.1.1   MATERIALS AND SAMPLE PREPARATIONS

All polymers used in this study are commercially available. The isotactic homo-polypropylene iPP—Caplen 01030, melt index 1.2 g/10 min; the propylene-based random copolymer rPP—SEETEC R3400, content of ethylene groups—3 wt.%, melt index 8 g/10 min; the propylene-based block copolymer bPP—SEETEC M1400, content of ethylene groups—9 wt.%, melt index 8 g/10 min, the maleic anhydride grafted polypropylene mPP—Polybond 3002, content of polar groups—0.6 wt.%, melt index 12 g/10 min.

1.1.2   DSC MEASUREMENTS

DSC measurements were made on DSC thermal system Microcalorimeter DSM-10ma. Its temperature scale was calibrated from the melting characteristics of indium. The experiments were conducted in nonisothermal mode. The sample was first heated up to 200 °C and maintained at this temperature for 10 min in order to erase any previous morphological history which the sample might be carrying. The sample was then non-isothermally crystallized when it was cooled down to room temperature at different cooling rates (1, 2, 4, 8, 16, 32, 64 °C/min). It was subsequently heated at a heating rate of 8 °C/min. The sample then repeatedly non-isothermally crystallized with the same cooling rates to permit structure microphotographs. The sample was approximately 7 mg. All curves were normalized to the unit weight of the sample.

The percent crystallinity, X_c, of the samples is calculated by[21]:

$$X_c = \Delta H_m / [\Delta H_{m0}]$$

where ΔH_m and ΔH_{m0} are melting enthalpy of sample and 100% crystalline PP (147 J/g[22]), respectively.

1.1.3 MICROSCOPY

The Axio Imager microscopes Z2m with transmitted-light differential interference contrast (TL DIC) was used to obtained microphotographs.

1.3 RESULTS AND DISCUSSION

The influence of crystallization rate on melting behavior of PP and copolymers were investigated. The melting endotherms of PP and its copolymers after various crystallization rates are presented in Figure 1.1. Henceforth, for all investigated polymers, the peak at the lower temperature is called peak-1, and at the higher temperature is called peak-2. The values of melting temperature (T_m) and crystallinity (X_c) are listed in Table 1.1.

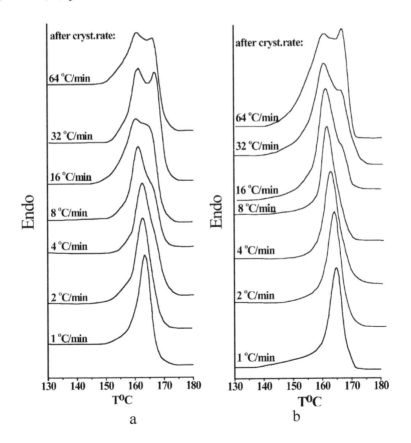

a

b

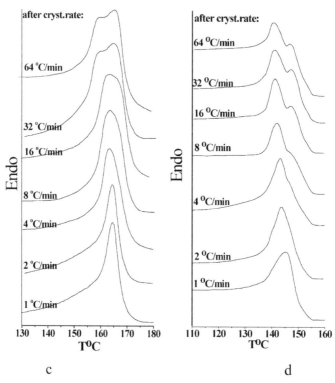

FIGURE 1.1 DSC thermograms of samples at 8 °C/min heating: (a) PP, (b) mPP, (c) bPP, and (d) rPP after crystallization with different rates.

TABLE 1.1 Thermal Parameters of Polymers Samples during Melting Process after Different Crystallization Rates.

Sample	Crystalliza-tion rate (°C/min)	Tm (°C)	Xc (%)	Sample	Crystalli-zation rate (°C/min)	Tm (°C)	Xc (%)
	64	161, 166	50		64	160, 166	63
	32	161, 166	49		32	160, 166	62
	16	160, 165	46		16	161, 166	49
PP	8	161	47	mPP	8	162	54
	4	162	49		4	163	53
	2	163	54		2	164	55
	1	163	58		1	165	63

TABLE 1.1   (*Continued*)

	64	160, 166	62		64	140, 147	31
	32	160, 165	60		32	140, 147	29
	16	163, 166	59		16	141, 148	29
bPP	8	163	60	rPP	8	142, 147	29
	4	163	61		4	143, 147	25
	2	164	63		2	144	25
	1	165	63		1	145	26

Figure 1.1a shows the corresponding DSC thermographs of the sample PP recorded after cooling with different rates. From Figure 1.1a, it can be seen that melting behavior depends remarkably on the cooling rate. At lower crystalliza-tion rates < 8 °C/min, the general feature of the DSC curves is the single melt-ing peak which localized mainly at a temperature about 163 °C. With increase in cooling rate, one melting peak transformed into double melting peaks. At crystallization rate 8 °C/min on the DSC melting curves appears shoulder peak at about 165 °C, which grows into a complete peak-2 at the crystallization tem-perature 32 and 64 °C/min, whereas the position of the peak-1 displaced at the location of lower temperature 161 °C. Such dependence is explained by the formation of more advanced and stable crystal structures at low cooling rate, whereas with increase in cooling rate increases supercooling and a large amount of defective crystals exposed to reorganization and recrystallization during heat-ing process.

In Figure 1.1b, mPP samples shows melting endotherms after different crys-tallization rates. In addition to pure PP, the DSC curve for mPP shows single melting peak under low cooling rate. It should be noted that appearance and increases of shoulder peak for such copolymer observed at higher cooling rates—16 °C/min. Another distinguishing feature of mPP melting after different crystallization rates is the significant decrease in melting temperature of peak-1 (from 165 °C after cooling rate 1 °C/min to 160 °C after cooling rate 64 °C/min).

Figure 1.1c gives the heating DSC thermographs for the bPP samples pre-pared also at various cooling rates. Curves have the similar character described above by maintaining the main polypropylene chain under the block introduc-tion of ethylene units. This later transition from a single peak to a double peak will be explained in terms of polypropylene chemical structural change by the introduction of functional groups. Since the regular introduction of functional groups decreases the mobility of the system and increases the viscosity of the system at low cooling rates which leads to the rapid formation of crystallization centers around them to form more perfect crystals than in pure PP. This displayed the higher melting temperatures of mPP and bPP. In the transition to higher cooling rates of up to 16 °C/min, at which the viscosity increases, the rearrangement of copolymer molecules is difficult, which in turn reduces the rate of crystallization, resulting in the formation of qualitatively less perfect package. The melting temperature T_m decreases to 5 °C. A further increase in the material crystallization rate contributes to recrystallization and formation of amid bulk imperfect crystals a high melting crystalline structure.

The statistical introduction of ethylene units into the main polypropylene chain affects greatly the crystalline structure of rPP in Figure 1.1d. Firstly, a decrease in the melting temperature of the polymer matrix of about 20 °C is observed. Secondly, the peak-2 as a shoulder peak appears at crystallization rate of 4 °C/min. For the propylene-based random copolymer the ethylene monomers are often considered as a defect points in polypropylene matrix. This induces structure heterogeneity in the long chains of polypropylene namely short propylene sequences leading to a decrease in crystallizable sequence length. Even small comonomer content (3%), as in our case, results in short crystalline sequence.

From Table 1.1, we can observe the relationship between the crystallization rate and crystallinity of polymer samples. Comparing the crystalline degree of the samples at a rate of crystallization (e.g., 1 °C/min), one can note that the regular introduction of the comonomer units of any chemical nature (mPP and bPP) leads to an increase in the polypropylene crystalline degree by 5%. Such incorporated co-units are not defect points; it leads to the facilitating folding crystalline structure. Due to above-mentioned statistical distribution of ethylene co-units, defect points in the propylene random copolymer rPP crystallinity decreases by 30%. On the other hand, crystallinity of PP, mPP, and bPP is characterized by decrease with increased crystallization rate, followed by increase in crystallization at higher rates (32 and 64 °C/min). Samples of rPP are characterized by the constancy of crystalline degree at low cooling rates and its growth of 4–5% even at a cooling rate of 8 °C/min. These dependences as well as the formation of shoulder and peak-2 on the DSC thermograms are explainable in terms of reorganization and recrystallization of the polypropylene supramolecular structure.

The above-mentioned crystalline structures of PP, mPP, bPP, and rPP obtained after different crystallization rates were visualized by microscopy. Figure 1.2 shows microphotographs of PP, mPP, bPP, and rPP obtained after different cooling rates.

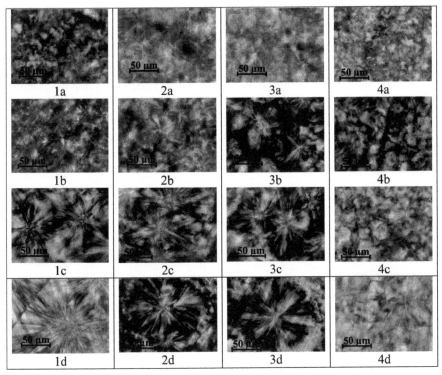

FIGURE 1.2 Microscope photographs of (1) PP, (2) mPP, (3) bPP, and (4) rPP obtained after different crystallization rates: (a) 64 °C/min, (b) 16 °C/min, (c) 4 °C/min, and (d) 1 °C/min.

For all polymer samples in the train of crystallization the line radial bundles emanating from a single point—polypropylene crystalline structures are formed. The crystalline sizes of the samples vary widely depending on the polymer crystallization rate and its chemical structure. At higher cooling rates and structural inhomogeneity of the propylene, long chains in random copolymer formed submicroscopic fine-grain-like crystal structures. Crystallization at a lower cooling rates, leads to the formation of crystals in diameter of about 100 microns. It is necessary to note that the size and shape of the supramolecular structures have great influence on the mechanical properties of the polymer. Samples with fine-grain-like structure have high strength and have good elastic properties. Samples with large crystals destroyed fragile. Elasticity loss of crystalline polymers is manifested in the appearance of cracks and breaks at the crystal interface. An increase in its size leads to increased fragility and decreased strength.

1.4 CONCLUSION

The melting behavior of four polymers: polypropylene, bPP, propylene-based random copolymer, and mPP were studied with DSC and microscopy. Obtained results indicate that both chemical structure of polypropylene chain and crystallization rate show great influence on the crystalline structure of polypropylene. It was found that the regular introduction of the comonomer units of any chemical nature (mPP and bPP) and higher crystallization rates lead to the formation of large crystals. On the other hand, structural inhomogeneity of polypropylene chain and increase in crystallization rates promote the formation of polymer fine-grain-like structure.

KEYWORDS

- **polypropylene**
- **melting**
- **crystal structure**
- **DSC**
- **microscopy**

REFERENCES

1. *Polypropylene*; Dogan, F., Ed.; Intech: Croatia, 2012.
2. Zhu, M. F.; Yang, H. H.; Polypropylene Fibers. In *Handbook of Fiber Chemistry*; Lewin, M., Ed.; CRC Press, 2007, 139–260.
3. Papageorgiou, G. Z.; Achilias, D. S.; Bikiaris, D. N. et al. *Thermochim Acta* **2005**, *427*, 117.
4. Seki, M.; Nakano, H.; Yamauchi, S. et al. *Macromolecules* **1999**, *32*, 3227.
5. Cai, H. J.; Luo, X. L.; Ma, D. Z. et al. *J. Appl. Polym. Sci.* **1999**, *71*, 93.
6. Zia, Q.; Radusch, H. J.; Androsch, R. Microscopy: Science, Technology, Application and Education. In *Formatex;* Mendez-Vilas, A., Diaz, J., Eds.; 2010; 1940–1950.
7. Androsch, R.; Di Lorenzo, M. L.; Schick, C. et al. *Polymer* **2010**, *51*, 4639.
8. Cho, K; Li, F. *Macromolecules* **1998**, *31*, 7495.
9. Holsti-Miettinen, R.; Seppala, J.; Ikkala, O. T. *Polym. Eng. Sci.* **1992**, *32*, 868.
10. Nandi, S.; Ghosh, A. K. *J. Polym. Res.* **2007**, *14*, 387.
11. Jacoby, P.; Bersted, B. H.; Kissel, W. J. et al. *J. Polym. Sci., Polym Phys.* **1986**, *24*, 461.
12. Lovinger, A. J. *J Polym. Sci., Polym. Phys.*, **1983**, *21*, 97.
13. Purez, E.; Zucchi, D.; Sacchi, M. C. et al. *Polymer*, **1999**, *40*, 675.
14. Gou, Q. Q.; Li, H. H.; Yu, Z. Q. et al. *Chin. Sci. Bull.* **2008**, *53*, 1804.
15. Papageorgious, D. G.; Papageorgious, G. Z.; Bikiaris, D. N. et al. *Europ. Polym. J.*, **2013**, *49*, 1577.
16. Papageorgiou, D. G.; Bikiaris, D. N.; Chrissafis, K. *Thermochim. Acta* **2012**, *543*, 288.
17. Cho, K.; Li, F.; Choi, J. *Polymer,* **1999**, *40*, 1719.

18. Salmah, H.; Ruzaidi, C. M. and Supri, A. G. *J. Phys. Sci.,* **2009**, *20,* 99.
19. Lukanina, Yu. K.; Kolesnikova, N. N.; Khvatov, A. V. et al. *J. Balk. Tribol. Assoc.* **2012**, *18,* 142.
20. Nguen, S.; Perez, C. J.; Desimone, M., et al. *Int. J. Adhesion Adhesives* **2013**, *46,* 44.
21. Bershteyn, V. A.; Egorov, V. M. *Differential Scanning Calorimetry for Physicochemistry of Polymers*; Khimia: Leningrad, 1990.
22. Ambrogi, I. *Polypropylene*; Khimia: Leningrad, 1967.

CHAPTER 2

THERMODYNAMIC AND STRUCTURAL PARAMETERS OF PHOSPHOLIPID MEMBRANES

O. M. ALEKSEEVA[1], A. V. KRIVANDIN[1], O. V. SHATALOVA[1], A. V. KREMENTSOVA[1], and YU. A. KIM[2]

[1]Emanuel Institute of Biochemical Physics, Russian Academy of Sciences, Moscow, Russia
[2]Institute of Cell Biophysics, Russian Academy of Sciences, Pushchino, Moscow, Russia, E-mail: olgavek@yandex.ru

CONTENTS

The action of aqueous solutions of melamine salt of bis(oximethyl)phosphinic acid (Melafen), used for a crop production, on the structural and thermodynamic properties of phospholipid membranes in multilammelar liposomes formed from individual synthetic dimyristoilphosphatidylcholine (DMPC) or natural phospholipid mixture (egg lecithin) was studied by differential scanning microcalorymetry (DSC) and small-angle X-ray scattering (SAXS).

The method of DSC was used to appraise the influence of Melafen aqueous solutions (10^{-17} to 10^{-3} M) on thermodynamic parameters of lipid melting (temperature, enthalpy, and cooperativity of phase transition) in suspension of liposomes formed from DMPC. On the basis of DSC data, it was concluded that the microdomain structural organization of DMPC membranes was altered by Melafen aqueous solutions as polymodal type.

We did not reveal any noticeable integral structural changes in membranes of egg lecithin multilammelar liposomes as well as in their stacking parameters (period and degree of order) in Melafen aqueous solutions (10^{-21} to 10^{-6} M) by the SAXS method. This range of Melafen concentrations is wider than the one used for a crop production.

Relying on the results of this work it can be inferred that microdomain structure of membranes composed of individual neutral phospholipids is changed in Melafen aqueous solutions at the wide concentration range of Melafen. At the same time, Melafen does not affect the overall structure of lipid membranes composed of a natural phospholipid mixture.

2.1 INTRODUCTION

In most of the cases, the primary targets for biologically active substances (BAS) are the cell membranes. In connection with this fact, the actions of Melafen aqueous solutions to the structural components of membranes have been examined. Considering that Melafen is used as the plant growth regulator (it inhibits the development of seeds when large concentrations and activates when low and ultra-small), we held the complex examination by applying the aqueous solutions of Melafen over a wide range of concentrations from 10^{-21} up to 10^{-3} M.

It is known that the main structural bases of membranes are the lipids. In our examination for formation of model experimental objects, the lipids of one of the eight most important classes of living organisms, the glycerophospholipids (the phospholipids),[1] were used. Most of the phospholipid molecules consist of hydrophilic polar head with charged (or neutral) phosphate group, and hydrophobic nonpolar part. Thus, the molecule may be neutral or negatively charged. The hydrophobic part consists of fatty acid residues. Such architecture of phospholipid molecules permits phospholipids to form bilayer or the hexagonal structure in water media. This fact is as the subject to stereo-specific parameters and size relation of polar head and nonpolar fatty acid residues. Bilayer or the hexagonal lipid structures are typical for cellular compartment, and were being discovered in vivo.

Different biologically active substances may exert the significant influence on the lipid bilayer. The structure of lipid bilayer is possible to study, watching changes of phase transitions of lipids and so to gain information about place when the material localizes in bilayer, and about subtle mechanisms cooperation with lipids. The phase's transitions into lipid membranes have been studied with sufficient details, and it is available to discover the reliable information on how the molecules localize in lipid bilayer. Localization was mirrored in lipids thermodynamic characteristics.

The lipid bilayer exists in two main phase states—crystal (*gel Lβ phase*) and liquid-crystal (*Lα phase)* with temperature dependence. The bilayer transition from crystal into liquid crystalline state (and back) occurs at strictly specified temperature, characteristic data for lipid (T_m). Differential scanning calorimetric (DSC) study determines the heat capacity of lipids or membranes at constant pressure in suspension from temperature.

The low molecular weight compounds may shift the phase transition temperature. Any causes of that phenomenon are the preferential interaction of these materials with solid or liquid states of membrane. The most of the learned materials (having, e.g., anesthetic activity) decrease the temperature of membrane-phase main transition. These substances connect with outer layer on the surface of membrane, primarily with the liquid phase. These relationships stabilize any areas of liquid phase, and the temperature of membrane melting decreases. BAS, which are able to shift the temperature of phase transition, increase the transition duration that are mirrored as a width of endothermic peak in the thermograms curve.

With engagement of model experimental objects, it has managed to define any targets for influence of Melafen aqueous solution on cell membrane structures.

In our experiments, the multilayered liposomes, formed from phospholipids, were used as model membranes. Thermodynamic parameters of lipid melting were determined by the DSC method in the presence of biologically active substances tested.

The method is based on the measurement of power, which was brought to cells. Cells that contained control and experimental objects are warmed up with strictly equal speed. This makes it possible to estimate the thermodynamic parameters of explored melting systems: enthalpy, temperature, and half of width of main phase transition. But in diluted solutions, the thermal effects, conditioned by macromolecules, are extremely small (the biopolymers heat capacity in 0.3% solution is only thousandth quota from common heat capacity of dissolvent). Furthermore, only measurements by means of very sensitive differential adiabatic scanning microcalorimeter (DASM-4)[2] allowed estimating of melting parameters. The melting parameters are correlated with the structural changes in organization of lipid microdomains in bilayer in the presence of BAS.

Performance characteristics of DASM-4[2] are as follows: temperature range, 0–100 °C; warm-up speed, 0.2–2.0 grad/min; workers volume of cells, 1.0 mL; sensibility on heat capacity, 1.6×10^{-5} J grad^{-1} mL^{-1}; determination accuracy of relative heat capacity, 8×10^{-5} J grad^{-1} mL^{-1}; record accuracy of temperature, 0.1 grad.

The sensibility of DASM-4 is determined by three construction principles of device: differential measurement circuit, continual warm-up with exactly designated speed, and complete adiabatic of measuring cells. When feeding the electric current in heaters of microcalorymetry cells, these cells rise in temperature. Simultaneously, proportionally integrated controllers automatically provide, with a high degree of accuracy, the temperature equalization of thermal shields with cell's temperature. The cameras happen to be in conditions, close to adiabatic, when the camera heat-exchange with environment is virtually absent. These conditions are implemented in wide temperature range when cells were heated with a constant rate.[2]

The synthetic individual glycerophospholipid dimyristoilphosphatidylcholine (DMPC) was chosen in our work as material for formation of liposome. As hydrocarbon chains are two residues of saturated myristic acid. The molecules of DMPC have cylindrical shape. Because the size of phosphate head is small; that is why from such phospholipid it is possible to form of bilayer liposomes with short crook of membrane.[3]

When liposomes were melted, two thermally-induced endothermic transitions were discovered: first—pretransition and second—main transition (Fig. 2.1 (1)).

The interpretation of calorimetric curve that describes the main transition for individual substance is most simple. The peak, which is not clearly expressed in the area of 14–15 °C, reflects the pretransition, when the restructuring ranked from order phospholipids packing (3) as "gel-phase"(6) to "ripple-phase" (4), (7). Then the main endothermic transition occurs into less rank order condition "liquid crystal" (5), (8). The "ripple-phase" presence was observed for certain phosphatidylcholines membrane by electron microscopy method. In particular, it was seen that the corrugation period is different and typical for each kind of lipids. The corrugation periodicity and the quantities were changed by many factors affecting the bilayer structure.[4]

For exploration of membrane structure of bilayer membrane, composed of DMPC, the multilamellar liposomes were used (Fig. 2.2 (2)). Such liposomes were formed when thin films were acted by hydration in phosphate buffer, when pH was neutral, and when temperature was above phase transition. Subtle films on the walls of flask were acquired when DMPC was dried under vacuum. These multilayer films were acted by hydration consistently as the stratum for stratum. Thus, such method allowed to study multilayer vesicular structures, composed of bilayer membranes. Obtained liposomes have the multilamellar structure by the size up to 2000 Å from data of electron microscopy.[5]

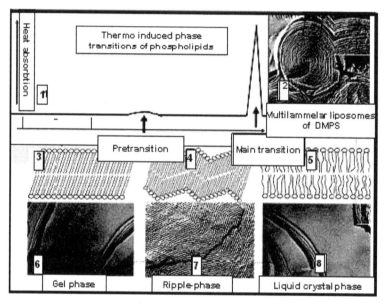

FIGURE 2.1 Modified scheme[4] of thermally induced endothermic phase transitions of DMPC in bilayer of multilamellar liposomes when melting by the DSC method. (1) Thermograms of endothermic phase transitions that were registered on DASM-4 when thermally induced phase transitions of DMPC: pre-transition and main transition; (2) electron microscopic photo of multilamellar liposomes, obtained by Tarahovsky by means of crio-transmission microscopy[4]; (3, 4, 5) diagrams of phospholipid DMPC molecules locations when three phase states: "gel-phase" (3), "ripple-phase" (4), "liquid-crystal phase" (5); (6, 7, 8) electron micrographs when freeze spelling of phospholipid bilayer is able to "gel-phase" (6), "ripple-phase" (7), and "liquid-crystal phase" (8).[4]

The proposed approach with aid of DSC method for inspection of properties of membranes is fairly correct. Since thermally induced phase main transition, that is, the transition from solid–gel state in rank order to the liquid state–liquid crystal, plays a major role in membranes of living organism. When in liquid phase, all conformational rearrangements of proteins occur easily. The oligomers formation, and the lipids flip-flop moving, lateral shifting occurs easily too. In solid environment, membranes of all these species of biological mobility have been hampered. And, respectively, all structural and functional activities are inhibited. The membrane structures remained, when liquid crystal state of membrane exists. And the braking of all conformation processes is lacking.[4] This is why examining of BAS action on thermally induced transitions of lipid phase is the most important link in circuit test of biological effects of BAS.

Melafen is a material applied in agriculture as plant growth regulator. Subject to concentrations, it can operate as the stimulator (10^{-18} to 10^{-13} M) or as the inhibitor (10^{-9} to 10^{-3} M) of the developments of plant seeds and body. When studying its effects on the experimental objects, we used the aqueous solutions of Melafen from large concentrations up to ultrasmall ones (10^{-3} to 10^{-21} M).

Taking into account the close interdependence of vegetation and animal bodies in nature, it was necessary to study the action of plant growth regulator to objects of animal origin. As a simple model of animal cellular biomembranes, being the primary target for biologically active substances, the phospholipid multilamellar liposomes, formed from DMPC, were used.

For second part of this work, the method of small-angle X-ray scattering (SAXS) was used. As experimental object the egg lecithin liposomes were formed.

These two methods, applauded for testing of Melafen–lipids relationships with two membranes models, formed from synthetic individual phospholipid and natural mixture of egg lecithin, were used for solving the primary aim of these investigation—evaluations of the impact of plant growth regulator on membranes of animal origin.

2.2 RESULTS AND DISCUSSION

2.2.1 DIFFERENTIAL SCANNING MICROCALORIMETY

In first part of this work, we investigated the action of Melafen aqueous solution under the wide range of concentrations (10^{-17} to 10^{-3} M) to the phase state of model phospholipid membranes. Data that obtained by the DSC method, are submitted: thermodynamic parameters of liposomes DMPC melting in the presence of aqueous solutions of Melafen over a wide range of concentrations are indicated in Table 2.1. The results were obtained by means of program "Microcal Origin 5.0."[5]

The temperature, in which maximum increasing DMPC heat capacity is seen for main endothermic phase transition (T_{max}), occurs at 24.1–24.3 °C. The Melafen aqueous solutions were added to liposomes right before melting endoterm registration. Solutions were used over a wide range of Melafen concentrations (10^{-17} to 10^{-3} M). Thermograms of endothermic thermally induced main transition of DMPC in the presence of aqueous solutions of Melafen when large concentrations are submitted at Figure 2.2. On the assumption of data, that were shown in Table 2.1, one can infer that Melafen at low and ultrasmall concentrations in aqueous solutions affected the fine structure of membranes in liposomes from DMPC. Melafen at concentrations above 10^{-17} M enhanced the enthalpy by 1.6–8%. The cooperatives of thermally induced transitions was subtracted, as demonstrated by half width increasing on half altitude transition, up to 17%. And insignificantly (in limits of accuracy of method), T_{max} was shifted to higher temperatures at 0.1 °C.

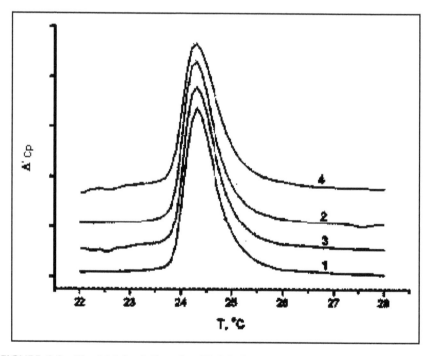

FIGURE 2.2 The Melafen influencing DMPC liposomes melting. The peaks of main endothermic transitions were shown. Thermograms of melting of multilamellar liposomes formed from DMPC, and melting in the presence of large concentrations of aqueous solution of Melafen. (1) Control 0.15 mg/mL DMPC, (2) DMPC + 10^{-5} M Melafen, (3) DMPC + 10^{-3} M Melafen, and (4) DMPC + 10^{-2} M Melafen.

TABLE 2.1 Effect of Melafen Aqueous Solutions Under Various Concentrations on Parameters of Thermally Induced Endothermic DMPC Main Phase Transition.

Melafen concentration in DMPC suspension (M)	T_{max} (°C)	Cooperativity of main transition (arb.un.)	Enthalpy of main transition (arb.un.)
Without Melafen	24.1	0.6	358.1
10^{-21} M	24.1	0.7	351.3
10^{-19} M	24.1	0.7	347.1
10^{-17} M	24.2	0.7	363.9
10^{-15} M	24.2	0.6	379.5
10^{-13} M	24.2	0.6	370.1
10^{-11} M	24.2	0.7	375.2
10^{-7} M	24.2	0.7	366.4
10^{-5} M	24.2	0.7	372.6
10^{-3} M	24.2	0.8	385.4

The standard DSC conditions of liposomes melting had not shown any important changes for DMPC liposomes, when Melafen aqueous solution concentrations were added. T_{max} of main endothermic peak, which corresponded to thermally induced transition, insignificantly were changed. The enthalpy and the width of polymodal transitions (the reciprocal value of the transition cooperatives) have undergone changes.

As a next step of investigation of Melafen influencing the DMPC liposomes, we complicated the experimental model study by the special modification. In order to educe the Melafen solution actions to the phospholipid membrane structures the specific approach of different speed of heat supplied (1 degree/min has been used; 0.5 degree/min; 0.25 degree/min; 0.125 degree/min) to cells with control and experimental samples was used by us. Such approach use in order to educe some eventual restructuring of additive cooperative units in membrane microdomains.

From data of thermograms registered when different melting rates of membranes, the concentration dependences of melting enthalpy have been built (the data are introduced in Fig. 2.3). T_{max} and the half width on half altitude of main transition under the wide range of the Melafen concentration strongly marked the extreme, which have been observed in the area of Melafen concentrations 10^{-14}–10^{-10} M, when all examined melting rates (data were not shown). The extremes and dose-depending of T_{max} and half width on half altitude of transition were appeared in the same Melafen concentration range, when such method was applauded for DMPC liposomes melting.

It is known that the deceleration of heat supplied to cells with control and experimental samples changes parameter of thermally induced transitions. Evidently, the relaxation processes in bilayer of phospholipid membranes were so much different, when such approach was occurred. Indeed, in control samples the changes have been noted too. The deceleration of heat supplied to cells considerably reduces the main phase transition T_{max} and the enthalpy, but the cooperatives were changed so small. The complex polymodal picture of action of Melafen solutions, when large and medium concentrations, was found, when addition of Melafen aqueous solutions in wide concentration range was produced. It should be noted that enthalpy changes were great in control samples. It was great when ultrasmall 10^{-17} and 10^{-16}, and large 10^{-6} and 10^{-5} M of Melafen concentrations (Fig. 2.3). Melafen concentration 10^{-13} and 10^{-12} M "hindered" enthalpy changes. And for changing T_{max} and for width of transitions, the same regularity was observed.

So, it has been ascertained that additions of Melafen aqueous solutions to DMPC liposomes exerted the significant complex dose-dependence effects on lipid bilayer structure. It should be noted that Melafen the hydrophilic substance, deprived of any possibility to be incorporated into lipid bilayer. However, it is known that bilayer melting of the growth of membrane permeability was observed. This phenomenon may be linked with the processes of short-lived nano-size pour forming.[6] The bilayer melting was held when temperatures were physiologically appropriate.

As considering the multilamellar liposomes as defects of bilayer, and some pores, which are permeable for water,[6] it can be assumed that Melafen aqueous solutions operate to all liposomes in multilayer. However, all of Melafen effects may take place only on the surface of each bilayer. Evidently, Melafen may act as transducer of water structures near the surface of bilayer, or it directly influences on DMPC phospholipid heads. It was disclosed that in concentration range 10^{-14}–10^{-10} M, Melafen exerts the maximum impact on thermally induced transition parameters: temperature, in which maximum increase in heat capacity is seen (T_{max}), enthalpy, and cooperatives. Obtained data by model membranes appears to and on biomembranes, once again indicated that are likely, affect may be determined by formation in water media the supramolecular complexes Melafen aqueous in vitro.[7]

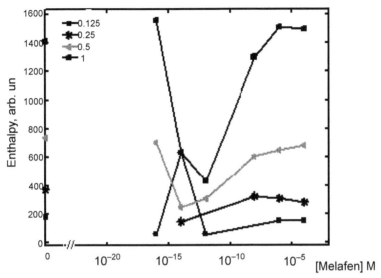

FIGURE 2.3 The of Melafen aqueous solutions with various concentrations effect to enthalpy of main thermally induced DMPC transition when different speeds of heat supplied to DSC cells with control and experimental samples. Different speed of heat supplied (1 degree/min has been used; 0.5 degree/min; 0.25 degree/min; 0.125 degree/min) to cells with control and experimental samples (Melafen concentration was given in logarithmic scale. The point "0" was conditional).

As was mentioned above, low and ultra-small Melafen concentrations do not exert destructive action on model membranes that are formed from individual neutral phospholipid. But restructuring the lipid microdomains occurred in bilayer. Large, middle, and ultrasmall Melafen concentration changed the intramembranous organization microdomains structures of lipids in liposomes formed from individual neutral phospholipid. The polymodal changes of dose-dependent parameters of thermally induced transition may take place due to certain relaxation conditions of structural elements of bilayer by means of certain changing melting rates. The experimental model object that were formed from individual neutral phospholipid DMPC, is likely, more vulnerable for Melafen actions compared to liposomes that were composed of neutral and charged phospholipids mixture. Mixtures fairly close simulate the nature stable membranes. Precisely equilibrium distribution of phospholipids provides the bilayer stability. For example, it is known that the destruction of plasma membrane of cells is associated with the phospholipid redistribution while programmed cell death (apoptosis). So while apoptosis, in starting equable on charges and topography of native membrane was occurred, the phosphatidylserine moving to the exterior leaf of bilayer was occurred too. These phosphatidylserine moving causes the charge destabilization and leads to destruction of the packing regularity in bilayer. The structure breaks down, that was followed by membrane fragmentation, and the cell death occurs.[8]

That is why the following part of our work was the study on the effect of Melafen aqueous solutions over a wide range of concentrations to the structural organization of lipids in membranes that were composed from mixture of nature phospholipids. The multilamellar liposomes formed from egg lecithin were used as experimental object. Such multilamellar liposomes formed from egg lecithin are the convenient models, which may be approximated compositionally and structured to natural membranes.

The locations of membranes change in the cell interior due to cellular life activity. Some membranes are coming together. The structures of most membranes are sequenced, disordered, or removed. The actions of the biologically active substances may be lacking certain target of its effects. However, much of BAS touch the whole processes. It may influence the structural orders of different organizational levels from the regularity degree of membranes and bilayer thickness up to reciprocal locations of bilayer.

The election of egg lecithin for exploration of Melafen action on membranes, which are composed of nature phospholipids, had been conditioned by the following reasons. The egg lecithin primarily contains phosphatidylcholine and phosphatidylethanolamine, but so less it contains phosphatidylinositol and sfingocholine. The polar groups, such as choline, ethanolamine, and inositol, are entering into the composition of these phospholipid molecules. And fatty acids residues of different quantity and degree of saturation are contained in these molecules. These phospholipids are present in all cell membranes of animal origin. It composes the basis of structural membrane integrity. Phospholipid supports the membrane components functional activity. For active works of integrated and associated membrane proteins (receptors, channels, ionic exchangers, pumps, and enzymes) the certain environment compositions near these proteins are needed. For maintaining integrated and associated membrane proteins activity, the certain types of annular lipids are so important. The proteins primarily interact with zwitterionic lipid phosphatidylcholine. The molecule in general is electroneutral, since it carries positive and negative charges.

The fluidity of membrane lipid phase can control protein conformation fluctuations, which are required for maintaining the corresponding protein activity. Membrane thickness was conditioned by the length of fatty acid residues and sizes of phospholipid heads in phospholipid molecules. Membrane thickness permits to control function intensity of proteins that are built into membrane. For example, the large integral protein Ca^{2+}-ATPase (from family of SERCA2) changes it affinity for ligands in dependence of membrane thickness. So, the operating of depth or outcropping of membranous loops with ligand-binding centers regulates the work activity of this Ca^{2+} pump. Ca^{2+} ATPase of sarcoplasmic reticulum (the main Ca^{2+} store in muscle cells) is surrounded by the phosphatidylethanolamine, which has the cone molecular shape. This type of molecular structure allows lipid molecules to form the hexagonal phase. For example, associated proteins, G-proteins, which are linked and implanted in membrane, are surrounded by phosphatidylethanolamine too.[9] G-proteins conduct the signals from receptors located in the outer surface of plasmalemma to cytoplasm enzymes. Their activities are in great dependence in lipid environment.

In connection with enumerated arguments, we separated the phospholipid models for test action of biologically active substance as experimental models, which are fairly actuating for many cell structural and functional parameters.

Above, it was seen that DSC method, which was based on different heating rate of cell with experimental sample of liposomes DMPC, brought to light the significant influences of Melafen aqueous solution (over a wide range of concentrations) on organization of lipid microdomains in bilayer. The three parameters of thermally induced endothermic main transitions of DMPC were registered by us. It was testified that the microdomains organization of phospholipid in bilayer was subjected to Melafen concentration.

It was known that Melafen is the hydrophilic molecule. But we noted that its action extends to all thickness of multilamellar liposomes. Since the peak of main transition remain clearly discovered. In visible, Melafen enters via defects of bilayer. The studies of nano-pores are possible during phase transitions also. Melafen action shall be exercised on hydrophilic surface of each bilayer. It can be assumed that Melafen changes the water medium around liposomes. It either forms the supramolecular complexes of Melafen-aqua, or it changes the gas solubility in the water solution.

However, in liposomes formed from the egg lecithin, that is, from the natural phospholipid mixture, we could not measure the parameters of microdomain organization of lipid bilayer by the DSC method correctly because the heat capacity peaks for each phospholipid in such mixture were overlapped. Due to this fact, we did not get any clear picture suitable for quantitative parameter estimation of thermally induced transitions in the egg lecithin membranes. Hence, we investigated the structural changes of lipid membranes, composed of natural phospholipid mixture, under the action of Melafen with the aid of another method, namely by the small-angle X-ray scattering technique. Thus, we have studied the next more complicated level of membrane structural organization in multilammelar liposomes.

2.2.2 SMALL-ANGLE X-RAY SCATTERING

In the second part of this work, the action of Melafen on the structure of phospholipid membranes in multilayer liposomes was studied by one of the methods of X-ray diffraction analysis, namely by small-angle X-ray scattering (SAXS).[10] This method permits one of the bases of the analysis of the measured SAXS intensity to get information about low-resolution structure of lipid membranes and to characterize membrane stacking parameters in liposomes.

The SAXS study of liposomes was carried out with an automated small-angle X-ray diffractometer assembled in the Institute of Biochemical Physics RAS. This diffractometer was developed on the basis of the construction of the small-angle X-ray diffractometer AMUR-K elaborated and exploited earlier in the Institute of Crystallography RAS.

Because the diffractometer used in this study is a custom-built device, a general description of its construction is given herein.

The X-ray source in this diffractometer is an X-ray generator IRIS-M (Nauch-Pribor, Orel, Russia) with a fine focus X-ray tube with a copper anode BSV29Cu (Svetlana-Roentgen, St. Petersburg, Russia). The X-ray beam passes through a Ni β-filter, and then is focused with a glass mirror and collimated with tantalum slits. A sample for SAXS study (liposome suspension) is placed in a thin-walled (wall thickness ~10 μm) glass capillary of ~1 mm outer diameter (sample volume ~15 μL).

SAXS patterns are recorded with the gas-filled (85% Xe, 15% Me, 4 atm) linear position-sensitive detector constructed in the Joint Institute for Nuclear Research (Dubna, Russia).[11] The detector has 1024 channels of discretization; the width of each channel is 97 micrometers. For CuKα X-ray radiation (λ_{avr} = 0.1542 nm) the detector has the special resolution of 150–200 micrometers and the efficiency of registration about 75%. SAXS patterns recorded with the detector are transferred to a personal computer for storage and processing.

For an X-ray tube with a copper anode and sample-to-detector distances from 130 to 425 mm accessible in the diffractometer, it is possible to record the intensity of X-ray scattering in the range of the values of diffraction vector module $S = (2\sin\theta)/\lambda$ from ~0.015 to ~3.5 nm^{-1} (λ is the wavelength of X-radiation, θ is a half of a scattering angle). This range of S values corresponds to the range of Wolf–Bragg distances $D = (S)^{-1}$ from ~67 to ~0.286 nm.

Due to the X-ray beam focusing with a mirror and simultaneous registration of X-ray diffraction pattern with a position-sensitive detector, a high intensity of X-ray scattering and high sensitivity of registration are achieved in such a diffractometer. This is very important for efficient SAXS study of liposomes and other biologic objects, which have labile structure and scatter X-rays rather poorly.

Multilayered liposomes for SAXS study were formed by the method of hydration of lipid films. At first, the egg lecithin (Sigma, USA) was dissolved in chloroform in a glass flask. Then chloroform was evaporated from this solution under argon gas flow and a thin lipid film was formed on the walls of the flask. For a complete removal of chloroform from a lipid film it was kept under vacuum not less than 12 h. After this, a small amount of buffer solution (50 mM phosphate buffer, pH 7.5, 200 mM of NaCl) was poured in the flask with a lipid film, the flask was filled with argon gas, heated up to ~45 °C and then was vigorously shaken with an electric shaker until a lipid film was totally dispersed in the buffer solution. The lipid concentration in liposome suspensions used in SAXS study was about 10%. Melafen was dissolved in the same buffer solution and was added to liposome suspensions at ~25 °C in quantities required to achieve Melafen concentration in these suspensions equal to 10^{-21}, 10^{-18}, 10^{-2}, and 10^{-6} M. Prepared liposome suspensions with Melafen were stored in closed vials under argon gas at a temperature of ~5 °C before being used in the SAXS study.

The SAXS study of liposome suspensions was done at the ambient room temperature (~22 °C). The duration of the SAXS intensity measurement for each liposome specimen was 30–50 min.

For background scattering correction the SAXS intensity of the capillary with buffer solution was measured and then subtracted from the SAXS intensity measured for each liposome sample. To take account of lipid concentration variation in liposome samples arising from addition of Melafen solutions in liposome dispersion, the SAXS intensity for all samples was normalized to have the equal peak intensity at $S \approx 0.14$ nm^{-1}. SAXS patterns smoothing and desmearing with method[12] were done with PRO software developed in the Institute of Crystallography RAS (SAXS patterns smearing is intrinsic for diffractometers with slit collimation systems).

SAXS patterns for all liposome samples studied (liposome suspensions containing 10^{-21}, 10^{-18}, 10^{-12}, 10^{-6} M of Melafen and without Melafen) disclose two diffraction peaks (reflections) with maxima at $S \approx 0.14$ and $S \approx 0.29$ nm^{-1}. As an example, two experimental SAXS patterns of liposome suspensions containing 10^{-6} mM of Melafen and without Melafen are shown in Figure 2.4a. In these patterns (Fig. 2.4a), background scattering was subtracted and intensities were normalized. The same SAXS patterns after smoothing and collimation correction are submitted in Figure 2.4b.

The X-ray diffraction peaks on the SAXS patterns for all studied liposome suspensions were equidistant (the abscissa of the second peak is approximately twice the abscissa of the first peak). Consequently, these diffraction peaks can be considered as the first and the second orders of reflection for membrane multilayers in liposomes (lipid lamellar phase).

The normalized SAXS intensities for all liposome samples studied (liposome suspensions containing Melafen and control suspensions without Melafen) were very similar, as illustrated for two samples of liposome suspensions containing 10^{-6} mM of Melafen and without Melafen in Figures 2.4a and 2.4b.

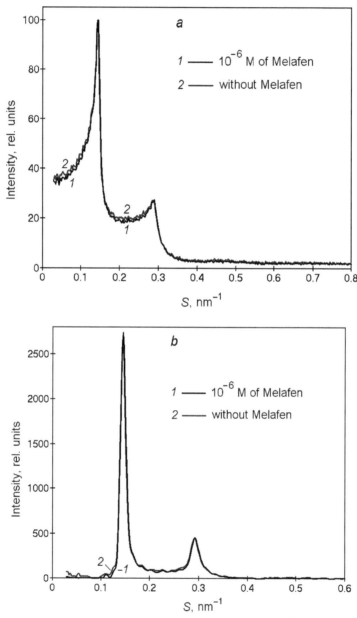

FIGURE 2.4 SAXS patterns of liposome suspension containing 10^{-6} M of Melafen (1) and liposome suspension without Melafen (2): *a*—experimental SAXS patterns after subtraction of background scattering and intensity normalization, *b*—the same SAXS patterns after smoothing and collimation correction. $S = (2\sin\theta)/\lambda$.

This indicates that the structure of lipid membranes and their stacking parameters in liposomes are not noticeably altered by Melafen addition to liposome suspension.

The period D of membrane stacking in liposomes was calculated for liposome samples after collimation correction according to the Wolf–Bragg formula as $D = h(S_h)^{-1}$. In this formula h is the order of reflection and S_h is the abscissa of this reflection determined as the value of S at which the first derivative of SAXS intensity goes to zero.

Results obtained showed that the values of period D of membrane stacking in liposomes at all studied Melafen concentrations and in liposomes without Melafen were the same with the accuracy ± 0.01 nm (Table 2.2). Such accuracy does not exceed the experimental error. So, we can state that within experimental error of the SAXS method Melafen when concentrations 10^{-21}, 10^{-18}, 10^{-12}, and 10^{-6} M has no influence on membrane stacking period in multilammelar liposomes formed from the egg lecithin.

TABLE 2.2 Lipid Membranes Stacking Period D in Liposomes when Various Melafen Concentrations in Buffer Solution. The Values of D were Calculated after Collimation Correction of SAXS Patterns and for Each Pattern were Averaged on Two Orders of Diffraction.

Melafen concentrations	D, nm
Without Melafen (control No. 1)	6.88
10^{-21} M	6.89
10^{-18} M	6.88
10^{-12} M	6.88
10^{-6} M	6.88
Without Melafen (control No. 2)	6.90

The widths of corresponding diffraction peaks were also one and the same for all liposome suspensions studied as illustrated for two samples in Figures 2.4a and 2.4b. This is the evidence for the same size of the regions of coherent scattering and the same degree of membrane stacking order in these regions for membrane multilayers in liposome suspensions containing 10^{-21}, 10^{-18}, 10^{-12}, and 10^{-6} M of Melafen and in liposome suspensions without Melafen.

For further analysis of Melafen action on lipid membrane structure in liposomes the electron density profiles of these membranes when various concentrations of Melafen in buffer solution were calculated on the basis of the SAXS data.

One-dimensional centrosymmetric electron density profiles $\rho(x)$ of membrane multilayers in liposomes were calculated at the relative scale in the normal direction to membrane planes by means of Fourier transformation of collimation corrected SAXS patterns utilizing the formula:

$$\rho(x) = 2 / D \sum_{h=1}^{N} P(h) h \sqrt{I(h)} \cos\left(\frac{2\pi hx}{D}\right),$$ (2.1)

In this formula, D is a stacking period of membranes in liposomes, h is an order of reflection (in our case, this is a number of a diffraction maximum), $P(h)$ is a phase sign (+ or −) for the h order reflection, $I(h)$ is the integrated intensity of the h order reflection, and N is a maximum order of reflections (in our case N = 2). The multiplier h in eq 2.1 is the intensity correction factor for membrane multilayers disorientation in liposome suspension (Lorenz factor).

The electron density profiles of lipid membranes calculated with eq 2.1 and utilizing SAXS patterns depicted in Figure 2.4b for liposome suspensions containing 10^{-6} M of Melafen and without Melafen are shown in Figure 2.5

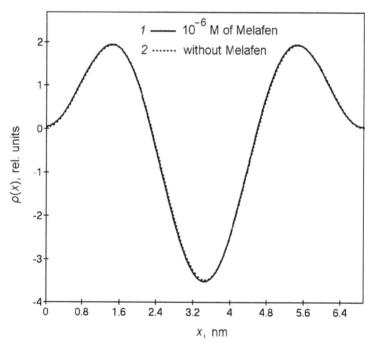

FIGURE 2.5 Electron density profiles of phospholipid membranes in multilamellar liposomes formed from egg lecithin: 1—for liposome suspension containing 10^{-6} M of Melafen; 2—for liposome suspension without Melafen (x is a distance in the direction normal to membrane planes).

The peaks of the electron density on these profiles at $x \approx 1.4$ nm and $x \approx 5.4$ nm can be associated with the position of lipid polar groups in a lipid bilayer, and the electron density minimum at $x \approx 3.4$ nm corresponds to the central hydrophobic part of a lipid bilayer. The thickness of lipid membrane defined as the distance between the peaks of the electron density at $x \approx 1.4$ nm and $x \approx 5.4$ nm is about 4 nm. As is seen from Figure 2.5, the electron density profiles of lipid membranes for liposome suspensions containing 10^{-6} M of Melafen and without Melafen are identical. The same electron density profiles of lipid membranes were obtained when Melafen concentrations in buffer solution equal to 10^{-21}, 10^{-18}, and 10^{-2} M.

So, the results of the SAXS study of liposome suspensions show that Melafen, which presents in wide concentration range (from 10^{-21} up to 10^{-6} M), has no visible effect on the structure and stacking parameters of phospholipid membranes in multilayered liposomes formed from egg lecithin. The absence of destructive changes of phospholipid membranes in the presence of Melafen in solution may be considered as a positive factor for practical applications of this biologically active substance.

2.3   CONCLUSION

These investigations are actual because the cell membranes are the primary targets for exogenous operative factors (biologically active substances in our case) entering into animal body by all ways. Lipids are the major and prevailing components of the animal cellular membranes. That is why phospholipids were chosen for our investigations, as the most widespread components of membrane lipid phase. For differential scanning calorimetric testing of membrane structural properties the synthetic individual neutral phospholipid DMPC was used. When melting, DMPC multilammelar liposomes have simple thermograms with two phase endothermic transitions in physiological temperature range. These thermograms are interpreted easily. Pretransition and the main phase transition in the presence of biologically active substances changed its parameters (enthalpy, T_{max} and cooperativity of the main phase transition). Our formulation of experiment was complicated by the special modification. This modification consisted in four different speeds of heat supplied to DSC cells, which contained control and experimental samples. This type of model study is fairly physiological because restructurings of lipid phase in membranes take place in animal organism constantly. In certain cases (for instance, when hibernation) membrane restructurings are temperature-dependent. And the rates of temperature changing are different sometimes also.

For experiments with the aid of the small-angle X-ray scattering method, the multilamellar liposomes formed from a natural phospholipid mixture were used. The multilamellar liposomes formed from egg lecithin are a difficult object for differential scanning calorimetric test. The phospholipid mixture gives the confluent peaks of heat capacity and the estimation of calorimetric parameters by deconvolution procedure does not allow correct deciphering of the biologically active substances effect on the certain lipid microdomains. Therefore we used the SAXS technique to study the next organizational level of lipid membranes—the lamellar structure and mutual arrangement of lipid membranes in multilayer liposomes.

We did not reveal by the SAXS method any noticeable structural changes of the egg lecithin membranes at Melafen concentrations used for a crop production. The SAXS method disclosed that the bilayer organization of multilamellar liposomes formed from egg lecithin was not changed in the presence of Melafen.

But on the base of differential scanning microcalorymetry data, we discovered that the microdomain structure organization of DMPC liposome membrane was changed by Melafen aqueous solutions in the wide concentration range. The types of these variations were rather complex.

We concluded that Melafen changed the microdomain organization of structure in membrane formed from individual neutral phospholipid and did not affect the membrane structure at the next organizational level. Namely, Melafen did not change the overall size and spacing of bilayers in multilamellar liposomes formed from natural phospholipid mixture. These results were reported shortly elsewhere.[13]

However, we should indicate that the aqueous solutions of Melafen altered the structural properties of labile objects, which can easily change their conformation. Such influence was shown for objects that are not related to plant life, but to animal life, namely for liposomes formed from individual phospholipid[13] and for the water soluble protein (bovine serum albumin).[14] As membrane composition becomes more complicated, the membrane structure becomes more resistant to the bed environment (or biologically active substances as in our case). These phenomena in our works were shown for liposomes formed from a mixture of natural phospholipids and for erythrocyte ghost.[15] The aqueous solutions of Melafen did not influence on these objects. However, when experimental object acquired any functions, the structure of this object began to be exposed to Melafen influence. In this case, the alteration of structural parameters of the model object under the action of biologically active substances may be mediated by the influence of these substances on some functions of this object.

As an example of this phenomenon, one can mention the shape changing of erythrocytes.[16] The aqueous solutions of Melafen exert significant influence on cell function. The purine-dependent calcium signaling in insulated thymocytes, lymphocytes, and cells of ascetic Ehrlich carcinoma decreased to the complete oppression when large concentrations of Melafen.[17]

The main finding following from all our studies[13–17] is the necessity of very strict observance of concentration limitations of Melafen when it will use as the plant growth regulator. This is vitally important for the prevention of Melafen negative influence on bodies of animal origin.

KEYWORDS

- **melafen**
- **multilammelar liposomes**
- **differential microcalorymetry**
- **small-angle X-ray scattering**

REFERENCES

1. Fahy, E.; Subramaniam, S.; Brown, H. A.; Glass, C. K.; Merill, A. H. Jr.; Murphy, R. C.; Raetz, C. R.; Russell, D. W.; Seyama, Y.; Shaw, W.; Shimizu, T.; Spener, F.; van Meer, G.; Van-Nieuwenhze, M. S.; White, S.; Witztum, J. L.; Dennis, E. A. A Comprehensive Classification System for Lipids. *J. Lipid Res.* **2005**, *46*, 839–861.

2. Privalov, P. L.; Plotnikov, V. V. Three Generations of Scanning Microcalorimeters for Liquids. *Therm. Acta.* **1989**, *139*, 257–277.

3. Tharahovsky, Y. S.; Kim, Yu. A.; Abdrasilov, B. S.; Muzafarov, E. N. *Flavonoids: Biochemistry, Biophysicist, Medicine*; Synchrobook: Puschino, 2013; p 308.

4. Tharahovsky, Y. S. *Intellectual Lipid Nanoconteiners in Address Delivery of Drug Substances*; M. Publishing House LKI: Moscow, 2011; p 280.

5. Tharahovsky, Y. S.; Kuznetsov, S. M.; Vasilyev, N. A.; Egorochkin, M. A.; Kim Yu. A. Taxifolin Interaction (dihydrocvercitin) with Multilamellar Liposomes from Dimitristoyl Phosphatidylcholine. *Biophysicist* **2008**, *53* (*1*), 78–84.

6. Antonov, V. F.; Smirnova, E. Yu.; Shevchenko, E. V. Lipid Membrane in Phase Transformations. *M. Sci.*, **1992**, 125.

7. Konovalov, A. I.; Rigkina, I. S.; Fattachov, S-G. G. Supramolecular Structure Based on Hydrophilic Derivative of Melamine and Bis (hydroxymethyl) Phosphinic Acid (Melafen) and Surface-Active Substance Message 1. Structure and Melafen Self-Association in Water and Chloroform. *Rep. Acad. Sci. Series Chem.* **2008**, *6*, 1207–1214.

8. Hampton, M. B.; Vanags, D. M.; Porn-Ares, M. I.; Orrenius, S. Involvement of Extracellular Calcium in Phosphatidylserine Exposure during Apoptosis. *FEBS Lett.* **1996**, *399*, 277–282.

9. Escriba, P. V.; Ozaita, A.; Ribas, C.; Miralles, A.; Fodor, E.; Farkas, T.; Garcia-Sevilla, J. A. Role of Lipid Polymorphism in G Protein-Membrane Interactions: Nonlamellar–Prone Phospholipids and Peripheral Protein Binding to Membranes. Proc. *Natl.Acad. Sci. U.S.A.* **1997**, *94*, 11375–11380.

10. Feigin, L. A.; Svergun, D. I. *Structure Analysis by Small-Angle X-ray and Neutron Scattering*; Plenum Press: New York, 1987.

11. Cheremukina, G. A.; Chernenko, S. P.; Ivanov, A. B.; Pashekhonov, V. D.; Smykov, L. P.; Zanevsky, Yu. V. Automatized One-Dimensional X-Ray Detector. *Isotopenpraxis* **1990**, *26*, 547–549.

12. Shedrin, B. M.; Feigin, L. A. *Collimation Correction for Small-Angle X-Ray Scattering. The Case of Final Dimensions of Slits*; Crystallographia: Moscow, 1966; Vol. 11, pp 159–163 (in Russian).

13. Alekseeva, O. M.; Krivandin, A. V.; Shatalova, O. V.; Rikov, V. A.; Fattacfov, C-G. G.; Burlakova, E. B.; Konovalov, A. I. A study of Melafen Interaction with Phospholipid Membranes. *Doklady Biochem. Biophys*. **2009**, *427*, 218–220.

14. Alekseeva, O. M.; Kim, Yu. A.; Zaikov, G. E. The Interactions of Melafen and Ihfans with Animal's Soluble Protein. *Her. Kazan Technol. Univ.*, **2014**; *17* (*7*), 164–167.

15. Alekseeva, O. M.; Fatkullina, L. D.; Kim, Yu. A.; Zaikov, G. E. The Melafen Influence to the Erythrocyte's Proteins and Lipids. Her. Kazan Technol. Univ., **2014**; *17* (*9*), 176–181.

16. Albantova, A. A.; Binyukov, V. I.; Alekseeva, O. M.; Mill, E. M. The Investigation Influence of Phenozan, ICHPHAN-10 on the Erythrocytes *in vivo* by AFM Method. «Modern Problems in Biochemical Physicsew Horizons;" 2012, Varfolomeev, S. D., Burlakova, E. B., Popov, A. A., Zaikov, G. E., Ed.; Nova Science Publishers: New York, Chapter 5, pp 45–48.

17. Alekseeva, O. M. The Influence of Melafen—Plant Growth Regulator to Some Metabolic Pathways of Animal Cells. *Polym. Res. J.* **2013**, *7* (*1*), 15–23.

CHAPTER 3

UPDATES ON VINYL POLYMERS

G. E. ZAIKOV

Russian Academy of Sciences, Moscow, Russia

CONTENTS

The vinyl group CH2 = CH– is present in ethylene and its derivatives, in which the hydrogen atom has been replaced by an atom or a group of atoms. As a result of polymerization of this type of compounds, called vinyl monomers, the vinyl polymers are formed.

Depending on the type of substituent, the vinyl polymers can be divided into:
- polyolefins, resulting from the polymerization of alkenes (olefins),
- styrene polymers containing aromatic substituents,
- poly(vinyl halides), containing halogen atoms,
- poly(vinyl amines),
- poly(vinyl ethers),
- poly(vinyl ketones),
- poly(vinyl esters),
- acrylic polymers, containing groups CN, $CONH_2$, COOH, and COOR.

3.1 POLYETHYLENE (PE)

Polymerization of ethylene is very difficult, and therefore, the first synthesis of polyethylene was carried out accidentally. During the test of the reactivity of chemical compounds under pressure, a mixture of ethylene and benzoic aldehyde was introduced into the reactor. The reaction product did not contain aromatic groups and proved to be polyethylene. Further tests of the ethylene polymerization under elevated pressure were initially unsuccessful and only the discovery of the reaction initiation by atmospheric oxygen enabled the development of this synthesis on an industrial scale. The small molecular weight of polyethylene and its excellent insulation properties preferred it to be used as a raw material in the manufacture of submarine cable under the La Manche channel.

The first charge of technically produced polyethylene was obtained on 01.09.1939, coinciding with the start of World War II. The lightness and the dielectric properties of polyethylene have enabled the construction of light aircraft radars, which contributed to the sinking of several German submarines, and thereby to break the blockade of the sea around England, influencing in this way the outcome of the war.

The ethylene polymerization process is an exothermic one and the heat of reaction is of 93 [kJ/mol].

The reaction is carried out in adiabatic autoclave or tubular-type reactors. Instead of the previously applied oxygen the hydroperoxides are used as initiators. The conversion degree of ethylene during one cycle is about 23%. The unreacted ethylene is recycled to the circulation, and the molten polyethylene is extruded in strips, cooled, and granulated.

The discovery of organometallic coordination catalysts by Ziegler and Natta allowed to obtain high-density polyethylene, which is characterized by a greater degree of crystallinity, worse transparency, and better mechanical properties in comparison with the high-pressure polyethylene, called a low-density polyethylene.

The high-density polyethylene is also obtained using the technique introduced in the UnitedStates in 1957. The method is called the Philips medium pressure technique. As catalyst, the chromium trioxide, fixed on a support consisting of a mixture of silicon dioxide and aluminum oxide, is used. In this method, a 5–7%solution of ethylene in the aromatic or the aliphatic hydrocarbons from pentane to decane is heated at a temperature of 150 °C and at a pressure of 3.5 MPa. Under these conditions, ethylene polymerizes virtually in a single cycle.

The structure of polyethylene depends on the method used for its production. During the high-pressure free-radical polymerization, the grafting side reactions take place, which cause side branching in polyethylene pmolecules. It makesits crystallization difficult and lowers the density. The products of ethylene polymerization obtained by Ziegler–Natta or Phillips methods are practically free of branching.

Polyethylene dissolves in hot aromatic and halogen derivative aliphatic hydrocarbons and exhibit good chemical resistance, except powerful oxidizing agents and concentrated mineral acids.

3.2 POLYPROPYLENE (PP)

The free-radical polymerization of propylene proceeds unsatisfactory. The cationic polymerization at a temperature of−50 °C and in the presence Friedel–Crafts catalysts leads to the formation of an amorphous polypropylene with molecular weight of 1000–10,000 and with appearance of wax, which can be used to improve the viscosity of oils.

The use of Ziegler–Natta catalysts for propylene polymerization allowed to obtain isotactic polypropylene with superior utility performances, enabling a widespread application of polypropylene products for the manufacturing of construction plastics and synthetic fibers.

The properties of polypropylene depend largely on its molecular weight, degree of polydispersity, its tacticity, and crystallinity.

The average values are as follows:

density [g/cm3]	0.90–0.91
melting temperature of crystallites [°C]	70–75
glass transition temperature [°C]	−35

tensile strength [Pa]	$(1–3.6)\times 10^7$
bursting strength [Pa]	$3–4\times 10^7$
drawing elongation [%]	5–15
ultimate elongation [%]	500–700
compressive strength [Pa]	$(0.7–1)\times 10^8$
specific heat at temperature of 23 °C [J/g K]	1.92
dielectric strength [kV/m]	350–800
surface resistance [cm/cm]	5×10^{13}
oblique resistance [cm $\times$ cm]	$1015–10^{17}$
dielectric loss factor [104Hz]	0.0019
water absorption after 24 h [%]	0.03

Polypropylene is characterized by a high chemical resistance. It shows no sensitivity to water, solutions of strong acids, bases, and inorganic salts. However, it is susceptible to the action of strong oxidants and nonpolar liquids. Polypropylene is a thermoplastic material, which is suitable for injection and extrusion moldings as well as for the production of foils and fibers.

The products made of polypropylene retain the shape up to the temperature of 423 K (150 °C). However, the value of the elasticity modulus at 413 K (140 °C) is only 10% of that at room temperature. Therefore, in practice, polypropylene is used at temperatures below 408 K (135 °C).

3.3 POLYISOBUTYLENE (PIB)

Polyisobutylene is obtained by the cationic polymerization of isobutylene in the presence of Friedel–Craffts catalysts. The most widely used catalyst is the bromine trifluoride and t-butyl alcohol as cocatalyst. The polymerization reaction proceeds according to the following scheme:

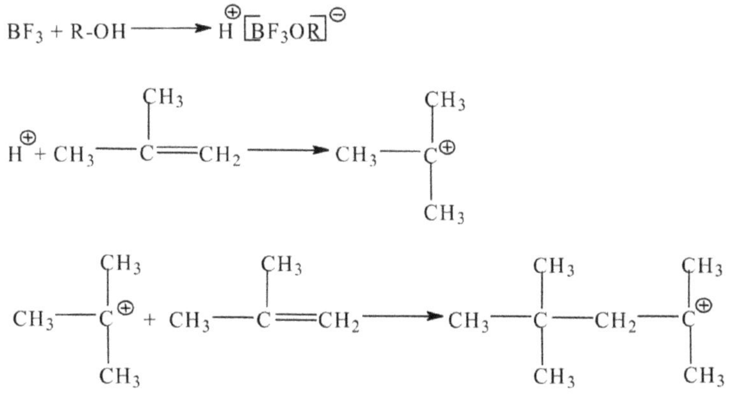

$$BF_3 + R\text{-}OH \longrightarrow \overset{\oplus}{H} \left[BF_3OR\right]^{\ominus}$$

Completion of the chain growth takes place by the transfer of proton to a new monomer molecule or a negative ion.

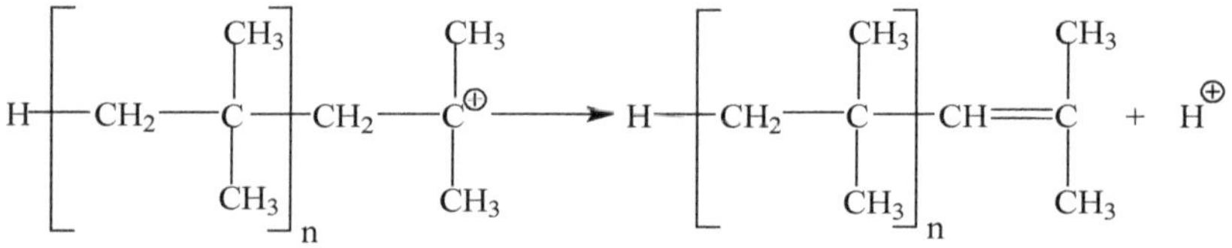

Polymerization inhibitors are sulfur, mercaptans, hydrogen sulfide, hydrogen chloride, etc. The bromine trifluoride is a very energetic catalyst in this process. The reaction is one of the fastest in the chemistry of macromolecular compounds and is highly exothermic (heat emission is of 54.4 kJ/mol). It impedes the process greatly, because it requires rapid removal of the generated heat. For this reason, the isobutylene polymerization reaction is carried out in solution.

The course of polymerization is highly influenced by the way how monomer is mixed with catalyst and by temperature of the conducted process. Products with high molecular weight are obtained by polymerization of isobutylene at low temperatures (below −70 °C), while the increase in temperature leads to the formation of polymers with a low molecular weight.

The properties of polyisobutylene vary depending on its molecular weight. Low molecular weight products (M< 5000) are oils, while polymers with an average molecular weight of 100,000–500,000 are typical elastomers, characterized by a very low glass transition temperature (199 K) and a low tendency to crystallize.

Some properties polyisobutylene are as follows:

density [g/cm^3]	0.91–0.93
glass transition temperature [°C]	−74
brittle point [°C]	−50
bursting strength.[Pa]	$(2–13.5) \times 10^6$
elongation [%]	1000–2000
water absorbability [%]	0–0.05
dielectric breakdown [kV/mm]	23–25
cross-resistance [Ω/cm]	10^{15}–10^{16}
dielectric loss factor [tan δ]	0.0004–0.0008
dielectric constant [at 10^6 Hz]	2.3–2.2

The polyisobutylene is relatively easily soluble in aromatic hydrocarbons, carbon disulfide, andhalogenated derivatives. It is insoluble in alcohols, ketones, esters, and other polar solvents.

Polyisobutylene is characterized by a high chemical resistance and a resistance to the water action. At ambient temperatures polyisobutylene is insensitive to the action of almost all acids, even the nitric acid, and in some cases, resists to the aqua regia too. It is also resistant to bases and to halogens. High chemical resistance of polyisobutylene originates from its saturated character. It belongs to the family of weakly polar polymers. Due to that polyisobutylene is a good dielectric.

In the presence of oxygen and under the UV light polyisobutylene oxidises with formation of oily decay products. The introduction of active fillers (soot, talc, etc.) enhances its resistance to oxygen and to radiation.

Polyisobutylene has found application in various fields of technology and in particular for wire and cable insulation, as an anticorrosion lining as well as for obtaining protective coatings and adhesives.

The polyisobutylene oils have found application as brake fluids and additives for lubricating oils to improve their viscosity characteristics.

3.4 POLY(METHYLPENTENE)

The poly(methylpentene) is formed by the polymerization of 4-methyl-1-pentene on Ziegler–Natta coordination catalysts:

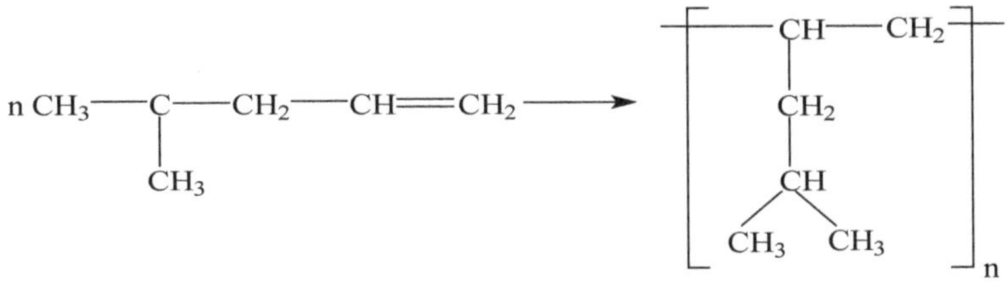

$$n \; CH_3 - \underset{\underset{CH_3}{|}}{C} - CH_2 - CH = CH_2 \longrightarrow \left[\begin{array}{c} CH - CH_2 \\ | \\ CH_2 \\ | \\ CH \\ \diagup \; \diagdown \\ CH_3 \quad CH_3 \end{array} \right]_n$$

Poly(methylpentene) is the lightest plastic material with the following characteristics:

contents of the crystallites [%]	40–65
density [g/cm^3]	0.83
tensile strength [MPa]	28
elongation at rupture [%]	15
softening temperature (Vicat) [°C]	179
glass transition temperature [°C]	−40
melting temperature of crystalites [°C]	242
dielectric constant,	2.12

Poly(methylopentene) is produced since 1968 by the English company ICI in a variety of transparent and opaque materials. It can be sterilized and utilized up to the temperature of 170 °C.

3.5   POLYSTYRENE (PS)

Polystyrene is one of the oldest known thermoplastics.

It is obtained by radical polymerization of styrene (vinylbenzene) in bulk or in suspension. The reaction proceeds according to the scheme:

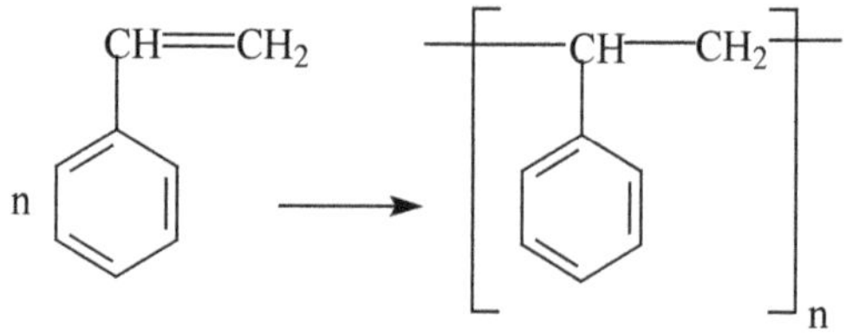

Polystyrene is characterized by a linear structure with "head to tail" arrangement of molecules and by a high molecular weight. The material is hard, brittle and transparent, soluble in aromatic hydrocarbons, chlorinated organic derivatives, esters, ketones, and carbon disufide. The mechanical properties of polystyrene depends on its molecular weight as well as on temperature and diminish when approaching its softening point.

The average values characterizing the properties of polystyrene are as follows:

density [g/cm^3]	1.05
tensile strength [MPa]	55
modulus of elasticity [MPa]	3.4×10^3
bending strength [MPa]	105
impact strength by Charpy [J/m^2]	
- without notch	25×10^3
- notched	2.5×10^3
cross-resistance [W cm]	10^{18}
dielectric breakdown [kV/mm]	60
dielectric permeability	2.5
dielectric loss factor (tan δ)	0.0003

The specific heat of polystyrene is lower than that of majority of other polymers and is 1.34 [J/gK]. The coefficient of linear thermal expansion of polystyrene is of 8×10^{-6}[K^{-1}]. It diminishes with addition of fillers and increases as a result of the softeners additition.

The visible light transmittance of polystyrene is about 90%, so that it can be used in optics. Indeed, polystyrene is widely used in the manufacturing of optical fibers. Its refractive index n_D^{20} = 1.5916–1.5927 and the critical incidence angle of light is of 38° 55². Thanks to that the phenomenon of the total internal reflection of light occurs even at a small radius of curvature. Major disadvantage of polystyrene for application in optical elements is its low durability and easy scratching of surface.

The biggest disadvantages of polystyrene are its fragility and combustibility. In order to counter these shortcomings, limiting application of polystyrene, new polystyrene plastics, with improved properties, were developed.

High-impact polystyrenes (HIPS), containing physically (type K) or chemically (type G) bound rubber by the grafting reaction, are characterized by good mechanical properties. The addition of decabromodiphenyl to polystyrene plastic makes it so much self-estinguishing that the material can be used to manufacture the rear walls of TV receivers.

The polystyrene foam with a cellular structure, known as Styrofoam, is used as insulating and packaging material.

3.6   STYRENE COPOLYMERS

Styrene is a highly reactive comonomer and forms easily series of copolymers, whose properties are often much better than those of polystyrene. Depending on the reactivity of the second comonomer Q and the double bond polarization e different reactivity ratios r_1 and r_2 in the copolymerization process, which determine the structure of the resulting copolymer, are observed.

The monomers used in the copolymerization reactions with styrene can be divided into three main groups.

Group I includes monomers for which the products of reactivity coefficients with styrene are equal to zero. These monomers copolymerize easilywith styrene and show a distinct tendency for alternatingin the growing chain. They are characterized by the presence of electronegative groups in the molecule. This type of monomers include: maleic anhydride, methyl fumarate, fumaronitrile, acrylonitrile, methacrylonitrile, methylvinyl ketone, and p-cyanostyrene.

Group II of monomers, among the others, includes methyl methacrylate, substituted in the ring styrene derivatives and butadiene These monomers copolymerize easily with styrene but show a weak tendency to alternate.

Group III includes monomers, which hardlycopolymerize with styrene. A privilegied reaction is that of styrene homopolymeryzation. Example of this type of monomers are vinylidene chloride, methyl maleate, trichlorethylene, vinyl chloride, and vinyl acetate.

3.7 POLY(VINYL CHLORIDE) (PVC)

Poly(vinyl chloride) is produced by the free-radical polymerization of the vinyl chloride:

$$\text{n } CH_2\!=\!CHCl \longrightarrow -\!\!\left[CH_2\!-\!CHCl\right]_n$$

The rate of polymerization of vinyl chloride in the presence of peroxide initiator increases gradually to about 40% of the reacted monomer and then remains stable up to about 80% of conversion. At the end of the reaction the rate decreases, what is due to the lack of solubility of the formed polymer in the monomer. Further polymerization takes place in the headed nodules of polymer. The completion of the growth of poly(vinyl chloride) macromolecules takes place mainly by the transfer of chain reaction onto monomer or polymer molecules. The latter reaction is followed by the branching of the chain.

The average molecular weight of poly(vinyl chloride) depends on the method of polymerization used and decreases with increasing initiator concentration and reaction temperature.

Poly(vinyl chloride) is a thermoplastic polymer, resistant to the action of water, acids, bases, mineral oils, hydrocarbons, oxygen, and ozone. It dissolves in esters, ketones, chlorinated organic derivatives, tetrahydrofuran, pyridine, and carbon disulfhide.

In order to characterize the properties of poly(vinyl chloride), a constant called number K, known also as the Fikentscher's number, was introduced. This constant depends on the size and the shape of macromolecules and is proportional to the molecular weight.

The relationship between the relative viscosity and the K constant is given by the equation:

$$\log \frac{\eta}{\eta_o} = \frac{75K^2}{1+1.5c} + K,$$

where
    $\eta 0$—viscosity of cyclohexanone at a temperature of 25 °C;
    h—viscosity of 1g PVC solution in 100 cm3 of cyclohexanone at 25 °C;
    K—Fikentscher's constant
    c—concentration

The K number is derived from viscosity measurements. It is read from special nomograms or from tables for the measured viscosity. Therefore, it is unnecessary to solve the above cited equation to get the value of K. Most often the

K values are in the range from 50–80. It corresponds to the average molecular weight between 40,000 and 129,000.

Average values characterizing the properties of PVC:

density [g/cm^3]	1.38
bursting strength [MPa]	45–60
tensile elongation at break [%]	8–20
glass transition temperature [°C]	80
softening temperature [°C]	145–170

The products made from poly(vinyl chloride) can be found in commerce in form of vinidur (hard poly(vinyl chloride)) or viniplast, containing 20–60% of plasticizer. Vinidur is processed at the temperature of 160 °C. A small amount of hydrogen chloride is then split off. It catalyzes the decomposition, while causing browning of the material. To avoid degradation, thermal stabilizers are added to PVC in order to bind the evolved hydrogen chloride. Among them the metal carboxylates, for example, lead, cadmium, or calcium, stearates, organotin compounds, and: organic phosphites are most frequently used.

3.8   POLYTETRAFLUOROETHYLENE (PTFE)

The discovery of polytetrafluoroethylene occurred by chance. Dr. Roy J. Plunkett, who worked for Du Pont company, intended to use the tetrafluoroethylene as a nontoxic cooling agent in the manufacturing of compressors for refrigerators. But it turned out that in one container, instead of the gaseous tetrafluoroethylene he found a white powder with a wax appearance. This powder was characterized by a very good chemical resistance and lack of solubility in all known solvents. This discovery coincided with the World War II and the studies conducted on the fission of uranium. General Leslie R. Groves, responsible on behalf of the US army for construction of the atomic bomb, was looking for material to seal the apparatus for the distillation of highly corrosive uranium hexafluoride. It was needed to obtain the uranium isotope with the mass number of 235, necessary to produce the uranium bomb. Because of the high importance of the case, he funded the research on polytetrafluoroethylene and implementation of its production.

Currently, known as Teflon, the polytetrafluoroethylene is obtained by free radical, emulsion polymerization of tetrafluoroethylene in the presence of peroxide initiators. This polymerization reaction takes place more easily than the polymerization of ethylene and the emitted heat of reaction is of 196 [kJ/mol].

The reaction proceeds according to the scheme:

$$n\ CF_2 = CF_2 \longrightarrow \left[CF_2 - CF_2 \right]_n$$

Polytetrafluoroethylene is a linear polymer with a high degree of crystallinity (80–85%). The polymer is thermoresistant and can be used up to the temperature of 250 °C, and in some circumstances even up to 327 °C. At this temperature, the crystalline structure disappears and the material becomes to be amorphous and transparent.

Physical properties of polytetrafluoroethylene:

density [g/cm^3]	2.1–2.3
glass transition temperature [°C]	−120
bursting strength [MPa]	14–20
bending strength [MPa]	11–14
impact strength [J/m^2]	10^5
cross-resistance, [Ωcm]	10^{18}
dielectric breakdown, [kV/mm]	25–27
dielectric loss tan δ	0.0002
coefficient of friction	0.06–1

Polytetrafluoroethylene is processed by sintering at temperatures of 360–380 °C. It is used in aviation, chemical, radiotechnic, radio engineering refrigeration, food and pharmaceutical industries as well as a coating and fiber-forming material.

One of the unconventional uses of polytetrafluoroethylene are its reactions with metals such as aluminum, magnesium, and titanium, which are accompanied by the secretion of large amounts of heat energy in a very short time. This allows to produce very high temperatures, allowing for cutting metal or other materials.

Another application of Teflon powders are the manufacturing methods of submicron silicon carbide or tungsten carbide during powders during the self-propagating reaction between silicon or tungsten and the powdered graphite with the addition of Teflon powder.

3.9   POLY(VINYLIDENE FLUORIDE) (PVDF)

The poly(vinylidene fluoride)is obtained by the free-radical polymerization of vinylidene fluoride

$$n\ CH_2 = CF_2 \longrightarrow \left[CH_2 - CF_2 \right]_n$$

The polymerization process is carried out by emulsion method in water as well as in aqueous solutions of acetone, alcohols (e.g., t-butyl), or ethers. The polymerization reaction proceeds with good yield at temperatures of 50–130 °C and at a pressure of 2.1–8.4 MPa. As initiators potassium persulfate, hydrogen peroxide as well as peroxides of dicarboxylic acids are used. Tetrachlorophthalic and pentachlorobenzoic acids as well as the ammonium salts of chloroorfluorocarboxylic acids are used as emulsifiers.

A method of suspensive polymerization of vinylidene fluoride in the presence of 0.01–3 wt.% of dicarboxylic perdicarbonates at a temperature of 0–50 °C at and 4 MPa pressure was also elaborated. In this method, the methylcellulose in an amount of 0.1% is used as the suspension stabilizer. Acetone, 2-butanone, or saturated linear hydrocarbons containing 3–12 carbon atoms serve as molecular weight regulators. The suspensive polymerization reaction proceeds with the yield of 98%.

The poly(vinylidene fluoride)polymer is thermally stable up to temperatures of 300–350 °C (573–623K). Its melting point is 170–180 °C. It enables its processing using the typical methods for thermoplastics within the temperature range of 200–250 °C. Heating of poly(vinylidene fluoride) to a temperature above 430 °C causes its pyrolysis combined with its degradation into short chain fragments of polymer and decomposition with emission of hydrogen fluoride. The poly(vinylidene fluoride) decomposition takes also place during its irradiation with gamma rays in the presence of oxygen. The action of these rays in the absence of oxygen (in vacuum) can cause a crosslinking of polymer.

The poly(vinylidene fluoride) resists to the action of acids, bases, strong oxidizers, halogens and a very large number of organic compounds.

It is not resistant to polar solvents, in which the poly(vinylidene fluoride) dissolves as well as to acetone, oleum and other sulfonating agents at elevated temperature.

The poly(vinylidene fluoride) dissolves in dimethylformamide, dimethylacetamide, and dimethyl sulfoxide. The polymer dissolution is carried out at 40–50 °C.

PVDF is the hardest plastic material with the highest mechanical strength of all currently known fluoropolymers and can be compared with polyamides and with Penton. It breaks under the tension of 50–60 MPa at room temperature. Its relative tensile elongation at break is of 200–300%. The deflection temperature of poly(vinylidene fluoride) is of 150 and 98 °C under the load of 0.46 and 1.86 MPa, respectively.

The poly(vinylidene fluoride)is currently used as a construction and lining material. It is utilized in the form of tubes and foils in the chemical industry, machine building, construction, precision industry, and in medicine.

The polymer is recommended for use in the building of chemical equipment, more particularly for making bearing shells, valves, pipes, cocks, gaskets, carpets for autoclaves, tanks, and pumps. It may be also used for fabrication of elements the apparatus resistant to bromine action. It is also used for the manufacture of sealing joints in the rocket devices. The poly(vinylidene fluoride) is also used to make thermoshrinkable hoses (trade name Termofit) designed for the isolation of electronic circuits and electro heating elements.

Metal tapes coated with poly(vinylidene fluoride) are used in architecture as cladding element of buildings for various destinations. Strong foils of PVDF found application for packing chemical reagents, pharmaceuticals, and medical instruments.

3.10   0POLY(VINYL ACETATE) (PVAC)

Poly(vinyl acetate) is obtained by the free-radical polymerization of vinyl acetate

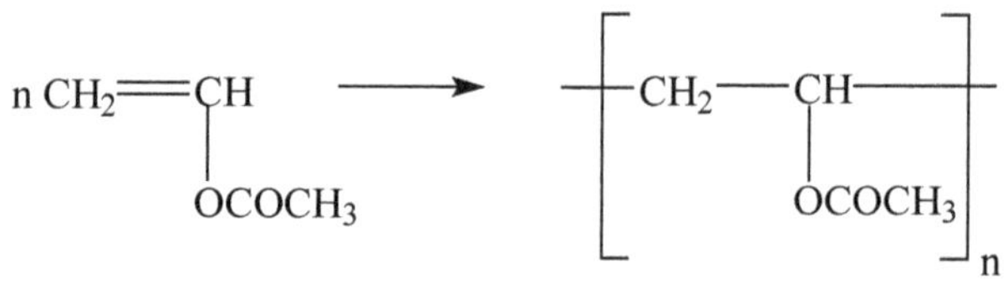

The emitted heat of reaction is of 88 kJ/mol. The period of induction and the kinetics of the process depend on the impurities contained in the vinyl acetate, which inhibits the polymerization.

The presence of a tertiary carbon atom in the molecule of vinyl acetate may cause the formation of a branched product at high temperature and high degree of conversion. Branching can occur at both carbon atom of vinyl group and the rest of acetate.

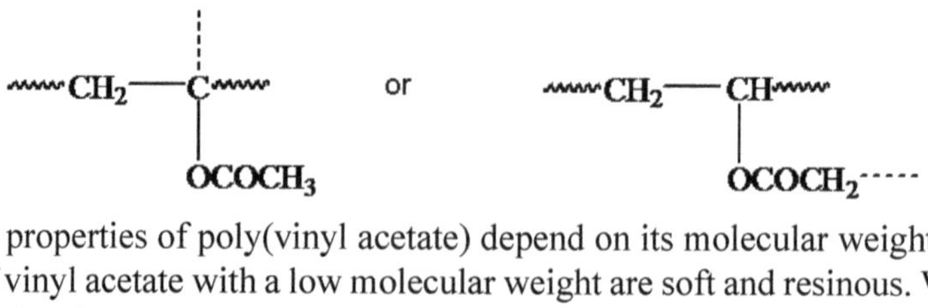

The properties of poly(vinyl acetate) depend on its molecular weight. Polymers of vinyl acetate with a low molecular weight are soft and resinous. With its growth, they become to be hard and brittle.

Typical characteristics of PVAC:

refractive index n_D^{20} 1.467

density [g/cm3] (at 20 °C)	1.191
specific heat [J/gK]	1.63
breaking strength [Pa]	0.1–0.34
softening temperature [°C]	44–86

Poly(vinyl acetate) is used as an ingredient in adhesives, plastics, coatings, binding masses, apprets, and as a raw material for the manufacture of poly(vinyl alcohol).

3.11 POLY(VINYL ALCOHOL) (PVOH)

PVOH is a vinyl polymer, which is not obtained by polymerization, but by the hydrolysis of poly(vinyl acetate). Under the term poly(vinyl alcohol), one also understands its copolymers with vinyl acetate of low content of acetate groups.

As a result of the ionic polymerization of acetaldehyde, carried out at low temperatures, one obtains poly(vinyl alcohol). However, this method has not found yet a practical use.

The hydrolysis of poly(vinyl acetate) is carried out in a methanol solution in alkaline or acidic medium at 50–70 °C. The polymer precipitates spontaneously from the solution at the time when approximately 60% of acetate groups are hydrolyzed.

The reactions proceed according to the scheme:

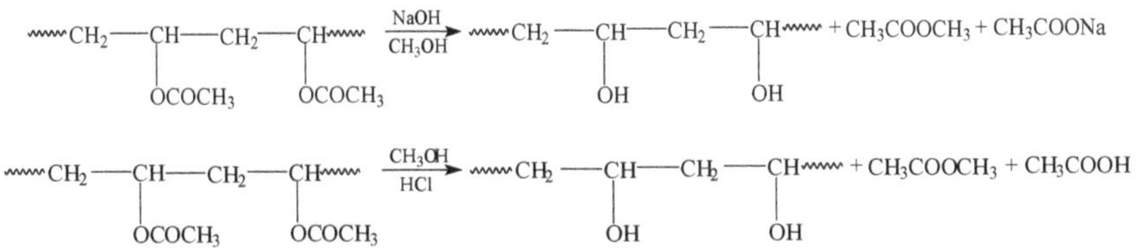

The hydrolysis process under the acidic conditions is slow. The product contains still over a dozen percent of acetate groups. For this reason, the hydrolysis reaction is carried out most frequently in an alkaline environment, in which it runs much faster and the degree of hydrolysis depends on the catalyst concentration.

Due to the presence of a large number of hydroxyl groups poly(vinyl alcohol) is soluble in water and insoluble in most of organic solvents.

The solubility of poly(vinyl alcohol) and the properties of its aqueous solutions depend on the percentage of unhydrolyzed acetate groups. In the case of atactic polymers, the best solubility exhibit the poly(vinyl alcohol) species containing 11–13% of acetate groups. A further degree of hydrolysis makes easier the formation of hydrogen bonds, which in turn cause the formation of crystalline forms and hinder the solubility.

The glass transition temperature of poly(vinyl alcohol) is of 80 °C. A further heating leads to changes in the polymer structure, which are due to intra- and intermolecular dehydration reactions leading to the formation of lactones and double as well as ether bonds.

The poly(vinyl alcohol) through the presence of hydroxy functional groups shows a high chemical reactivity. It is relatively easy for performing esterification reactions, etherification, acetalisation, oxyethylation, cyanoethylation, dehydration, chlorination, bromination, and reaction with isocyanates.

Poly(vinyl alcohol) is used in the manufacturing of adhesives, finishes, protective colloids, photographic emulsions, as a thickener and gelling ingredient in cosmetics and pharmaceuticals, pipes, tubes, plates, aprons and gloves resistant to gasoline and other solvents, surgical sutures, coatings, and membranes, which are only in a small extent permeable for gases. Large amounts of polyvinyl alcohol is processed into polyacetals. PVOH was also used to obtain synthetic fibers.

3.12 POLYACRYLAMIDE (PAM)

The polyacrylamide is obtained during the radical polymerization of acrylamide in aqueous solution.

$$n\ CH_2{=}CHCONH_2 \longrightarrow \left[CH_2{-}\underset{\underset{CONH_2}{|}}{CH} \right]_n$$

The emitted heat of reaction is of 81.7 kJ/mol. The molecular weight of the formed polymer can be adjusted by choosing the ratio of oxidant to reductant concentrations, used to initiate the reaction in the oxidation-reducing system as well as by the addition of regulators such as the isopropyl alcohol or the sulfur compounds.

A very important parameter in the acrylamide polymerization reaction is temperature. The process should be conducted at a temperature below 60 °C, since at higher temperatures the hydrolysis of amide groups to carboxylic ones takes place and as a reaction product one can obtain copolymers of acrylamide with ammonium acrylate.

The industrial process of polyacrylamide synthesis is carried out in a reactor with a mixer fitted with a heating–cooling system in an aqueous medium or in water–isopropanolic solution. As initiators, a redox system consisting of potassium persulfate and sodium hydrogen sulfate is used. It allows to carry out the reaction at an appropriate temperature during 5–7 h at pH = 7–85. The course of reaction is determined by following the amount of the emitted heat. The cessation of the temperature growth in reactor signals the end of the process.

Is also possible to carry out the polymerization of acrylamide in the presence of potassium persulfate at the temperature of 75 °C within 2h. The product separates from the aqueous solution by precipitation with acetone or methanol.

Polyacrylamide is a white powder. Because of its polar groups, it is easily soluble in water and is insoluble in acetone, alcohols, and hydrocarbons. It mixes well with surface-active substances as well as with many other water-soluble polymers. Similarto poly(vinyl alcohol), the polyacrylamide can be plastified with glycerine. Heating polyacrylamide to a temperature above 373K (100 °C) leads to changes in the structure combined with emission of nitrogen.

Polyacrylamide is a reactive compound and can be chemically modified. After a treatment with formalin or glyoxal, the polymer becomes insoluble in water. As a result of Hofmann degradation under the influence of sodium hypobromite, 94% of amide groups convert into amino groups with the formation of poly(vinyl amine).

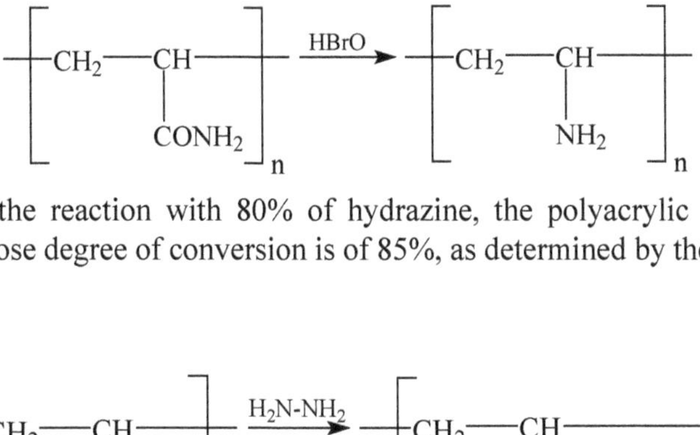

During the reaction with 80% of hydrazine, the polyacrylic hydrazid is formed, whose degree of conversion is of 85%, as determined by the iodometry method

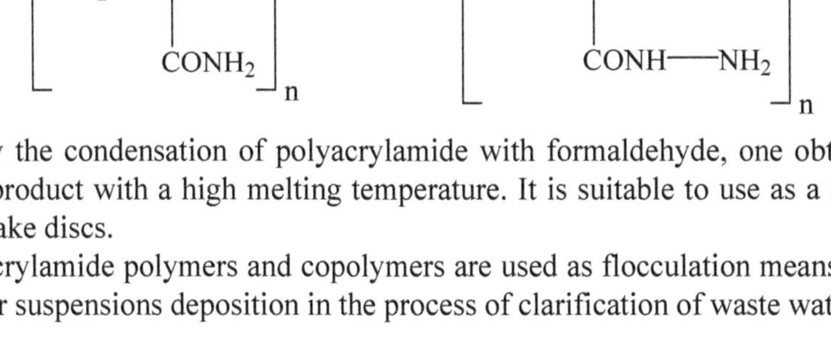

By the condensation of polyacrylamide with formaldehyde, one obtains a hard product with a high melting temperature. It is suitable to use as a binder for brake discs.

Acrylamide polymers and copolymers are used as flocculation means serving for suspensions deposition in the process of clarification of waste water and improving the quality of drinking water. They are also applied in soil condition-ing, to improve its structure, in the manufacture of adhesives, dispersants, as auxiliaries for textile, thickening agents as well as as to stabilize the natural rub-ber, latex and poly(vinyl acetate) emulsion. Polyacrylamide is a better emulsion stabilizer than poly(vinyl alcohol).

3.13   POLY(METHYL METHACRYLATE) (PMMA)

The poly(methyl methacrylate) is obtained by a free-radical polymerization of methyl methacrylate. The polymerization reaction is exothermic and proceeds according to the scheme:

$$n\ CH_2\!\!=\!\!\underset{\underset{CH_3}{|}}{C}\!\!-\!\!COOCH_3 \longrightarrow \left[\!\!-CH_2\!\!-\!\!\underset{\underset{CH_3}{|}}{\overset{\overset{COOCH_3}{|}}{C}}\!\!-\!\!\right]_n + 60\ kJ/mol$$

A characteristic property of poly(methyl methacrylate) is its good transpar-ency. It allows its use as an organic glass, known under the commercial name of Plexiglass. For this reason, its production technology is tailored to those needs.

The bulk method of polymerization of methyl methacrylate is carried out in a periodic way. In the first stage, one obtains a prepolymer, formed by the initial polymerization of methyl methacrylate at the presence of benzoyl peroxide or by dissolution of pieces of poly(methyl methacrylate) waste in monomer. The content of polymer in prepolymer is of 5–10%.

The obtained prepolymer is used to fill the forms made from plates of mir-ror glass, separated by dividers, carefully cleaned and covered by a thick paper. Forms are filled to 98–99% of their volume. Then the inlet is sealed and the prepolymer is subjected to polymerization in these cells by raising gradually the temperature from 298 to 363 K (25–90 °C). After completion of the polymeriza-tion the forms are exposed to hot water, which detach the paper and facilitates separation of glass panels from the formed sheets of poly(methyl methacrylate). Obtained in this way, polymer plates are subject to a technical control in order to eliminate the products with blisters. Then both sides of the PMMA plates are wrapped with thin paper, protecting them against scratches.

Currently, on the industrial scale, this method has been displaced by the method of suspensive polymerization of methyl methacrylate and the organic glass plates are obtained by extruding the polymer in extruding machines with a nozzle slot.

Properties of poly(methyl methacrylate) are as follows:

density [g/cm^3]	1.18
refractive indexn_D^{20*}	1.58–1.69
absorptiveness of water [%]	0.25
tensile strength [MPa]	60–70
softening temperature (Vicat) [°C]	105–110
flow rate, [g/10min]	0.8
cross-resistance [Ω m]	10^{16}
dielectric loss factor	
tanδ at 50Hz	0.07

*index of refraction at the wavelength of sodium doublet line (589 nm) and at 20 °C

3.14 POLY(N-VINYL-PYRROLIDONE) (PVP)

Poly(N-vinyl-pyrrolidone) is formed as a result of the free radical polymerization of N-vinylpyrrolidone in aqueous or alcohol solution. Its average molecular weight is in the range 10,000–360,000.

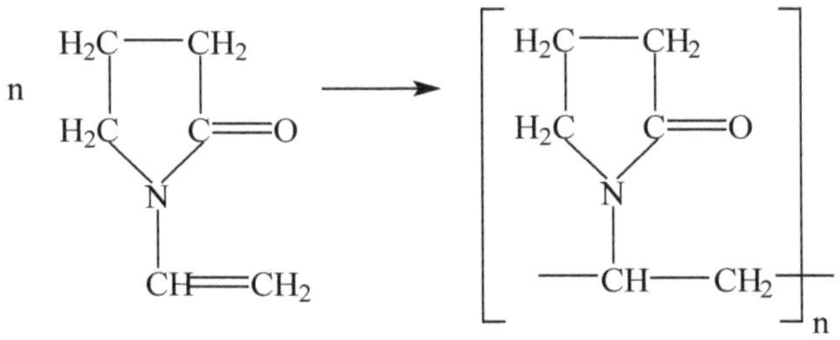

In industry, the polymerization process of N-vinylpyrrolidone is most frequently carried out in the 30%aqueous solution at a temperature of 50–70 °C, and in the presence of hydrogen peroxide, as the initiator and ammonia as the activator (regulator). One uses 0.1 wt.% of hydrogen peroxide and 0.1 wt.% of ammonia with respect to the amount of monomer. The reaction proceeds at this temperature during 2–3 h.

Poly(N-vinyl pyrrolidone) is isolated from the post reaction solution by precipitation in acetone, salted out with sodium sulfate or spray-drying in an oven.

Poly(N-vinyl pyrrolidone) is a colorless, amorphous, and hygroscopic polymer with the density of 1190 kg/m^3. It is well soluble in water, alcohols, higher ketones, organic acids, and in aromatic and chlorinated hydrocarbons. It does not dissolve in ethers as well as in aliphatic and alicyclic hydrocarbons.

The polymer softens at temperature 140–169 °C, but it becomes insoluble and decomposes at the temperature of 230 °C. The aqueous solutions of this polymer are weakly acidic (pH ~5) and are characterized by persistence in a neutral and weakly acidic environment, even when heated to the temperature of 100 °C.

In an alkaline medium poly(N-vinyl pyrrolidone) undergoes a hydrolysis associated with the formation of the poly-N-vinyl-γ-aminobutyric acid.

Poly(N-vinyl pyrrolidone) forms complexes with many inorganic and organic compounds, in particular with dyes, vitamins, medicines, and poisonous substances. It is characterized by the miscibility with many resins, polymers, and plasticizers.

One of the main applications of poly(N-vinyl pyrrolidone), produced under the trade names Subtosan and Periston is the production of synthetic plasma (blood plasma substitute) It is possible due to its nontoxicity and suitable speed of excretion from the body. This preparation was extensively tested and used on a large scale in hospitals in Germany during the World War II. The poly(N-vinyl pyrrolidone) solutions are also used to treat the poisoning occurring in children with diphtheria, diabetes insipidus, a fatty degeneration of the kidneys, and other ailments. The Periston-type solutions have particular advantages in the treatment of shock caused by heavy burning, when the blood may be concentrated due to the loss of water and of plasma by the walls of damaged capillary cells. These solutions serve also to strengthen the injections of insulin, penicillin, novocaine, hormones, salicylates, and other drugs.

Poly(N-vinyl pyrrolidone) is also used as a binding agent in the pharmaceutical industry in the tableting of drugs, the production of hair lacquers, to bleach badly dyed fabrics, and as an activator to improve the capacity of synthetic fibers for staining. It is also used as a component of adhesives.

There exists also a cross-linked form of PVP known as cross-linked polyvinyl pyrrolidone (PVPP), crospovidone, or crospolividone. PVPP is insoluble in water but still absorbing it. It swells very rapidly, generating a swelling force. Therefore, PVPP is used as a disintegrant in pharmaceutical tablets. It is also used to bind impurities to remove them from solutions. In beer processing, PVPP is used to remove polyphenols to clear them and to stabilize foam. Because of its binding properties, PVPP is also applied as a drug against diarrhea.

3.15 POLY(N-VINYLCARBAZOL) (PVK)

The poly(N-vinylcarbazol)is obtained by the polymerization of n-vinylcarbazole in the presence of ionic polymerization catalysts and radical initiators:
The polymerization proceeds according to the following scheme:

The rate of polymerization reaction depends on the purity of the monomer and on the amount of the added initiator or catalyst.

Poly(N-vinylcarbazol)is a chain polymer, in which the "head to tail" type structurepredominates. It is characterized by a high degree of crystallinity, suggesting a low branching and a more symmetrical structure. Due to the presence of large chain rings that prevent a free rotation ofchain segments, its glass transition temperature is high, of about 426 K (173 °C).

Poly(N-vinylcarbazol) ischaracterized by a high heat resistance. It is weakly polar, hydrophobic, chemically resistant, and exhibits good insulating properties. PVK is soluble in ketones, esters, aromatic, and chlorinated hydrocarbons.

PVK is used in electrical engineering as a substitute for mica and asbestos. In chemical industry, it is used for the manufacture of chemical equipment resistant to bases, acids and fluorinated compounds at a temperature of 393 K (120 °C).

Vinylcarbazole copolymers, with acrylonitrile, replace the printing metal in printing industry.

The recent applications of PVK base on the use of the electric, photoelectric, and photocunducting properties of charge-transfer complexes of poly(N-vinylcarbazol)with a photosinsitizer like trinitrofluorene (TNF). The PVK–TNF complexes are used in electrophotography, the manufacture of thermistors, phototermistors, as elements of thin film electronic components, photorefractive materials and as elements of optical memories. These are the prospective applications of poly(N-vinylcarbazol) in which the production costs and the product price are not crucial, that is, the development of manufacturing technologies is not hindered by the economic considerations. The production of this polymer shows an upward trend.

3.16   POLY(VINYL ETHER) (PVE)

In contrary to alkyl–vinyl and aryl–vinyl esters the alkyl–vinyl ethers do not polymerize in the presence of free radicals but polymerize by the cationic mechanism.

The reaction proceeds according to the scheme:

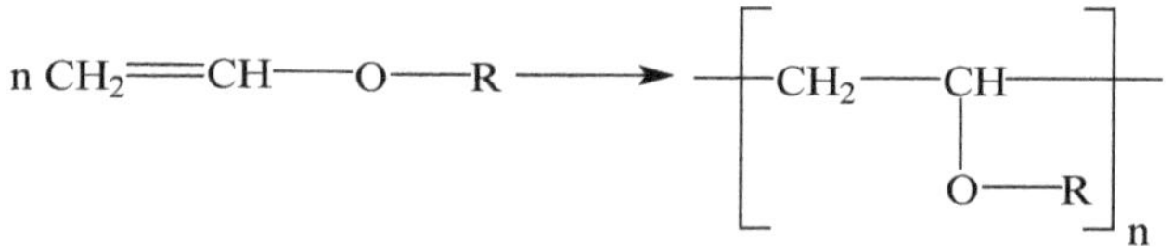

$$n \; CH_2\!\!=\!\!CH\text{---}O\text{---}R \longrightarrow \left[\begin{array}{c} CH_2\text{---}CH \\ \\ O\text{---}R \end{array} \right]_n$$

As catalysts metal fluorides and chlorides, such as BF_3, $SnCl_2$, $SnCl_4$, $FeCl_3$, and $AlCl_3$, are used.

During the room temperature polymerization of alkyl–vinyl ethers, liquid or viscous oligomers are formed. At lower temperature, below 263 K(-10 °C) one obtains solid, similar to rubber, polymers. The polymerization reaction is the most frequently carried out in solution, in liquid propane.

The molecular weight of vinyl polymers depends on the type of catalyst, its concentration and on the reaction temperature.

In industry, the polymerization of vinyl ethers is carried out by the batch method. The reaction proceeds in an apparatus fitted with stirrer, heating–cooling jacket and a reflux condenser. In the polymerization of n-butyl ether–vinyl, the ferric chloride (III) solution in butyl alcohol, in an amount of 0.03–0.15 wt.% is used as catalyst. After reaction, the formed polymer is detached from the catalyst and dried.

During the low-temperature (about 200 K) bulk polymerization of alkyl–vinyl ethers, stereoregular polymers with high mechanical strength are formed.

Polymers of vinyl ethers are produced both as concentrated solutions in hydrocarbon solvents, in aliphatic and aromatic hydrocarbons, as well as stabilized solids.

Known are also vinyl ether copolymers with various vinyl monomers, which are obtained by radical polymerization initiated with peroxides.

The copolymerization is conducted mostly by the suspensive emulsion method.

The poly(vinyl ethers)are well soluble in organic solvents except ethyl alcohol. Poly(methyl vinyl ethers)are soluble even in cold water. PVE's are resistant to the action of aqueous solutions of acids and bases. They adhere to glass, metal, wood, fabrics, and other materials. They are used in production of paints and lacquers, in treatment, and impregnation of fabrics for the manufacture of artificial leather, adhesives, plastyfying additives, and as thickeners. Poly(butyl vinyl ether), known under the trade name of Szostakowski balsam, is used in medicine for the treatment of burns.

Copolymers of vinyl ethers with maleic acid are used as additives in lacquers and impregnating agents of electrical insulation materials. Copolymers of vinyl ethers with methyl methacrylate can be used as adhesive layers in the production of safety glass.

3.17   IONOMERS

The term ionomers denote polymers, which in their macromolecule contain a small (up to 8%) number of ionic groups.

The first ionomers, which have found industrial importance are copolymers of ethylene and sodium acrylate with trade name "Surelen A."

Polymers of this type are obtained by a partial neutralization of α-olefin copolymers with acrylic acids by metal compounds of I, II, and III group of periodic table. Due to that only a partial ionization of carboxyl groups, which is a measure of the degree of ionization of the system, takes place.

In ionomer molecules the non-neutralized carboxyl groups form hydrogen bonds and the metal cations ionic bonds with carboxyl anions, respectively. It causes, in particular, the strengthening of intermolecular bonds. The degree of ionization of such a polymer, that is, the fraction of ionized carboxyl groups, contributes substantially to the properties of the product. Introduction to the chain of the ionic poly(α-olefin) groups have a particular impact on the morphology of crystal structures of these polymers. With the increasing content of carboxyl groups the structure changes from the sferulite type to the amorphous one.

Selected properties of poly(α-olefin) ionomers:

density [g/cm^3]	0.93–0.96
hardness (after Shore) [MPa]	60–65
tensile strength, [kg/cm^3]	245–390
relative elongation at break [%]	200–600
modulus [kg/cm^3]	1800–5200
shrinkage under pressure [%]	0.3
friction coefficient for the foil	0.6–0.9

softening temperature (after Vicat) [°C]	71–96
[K]	344–369
maximum processing temperature [°C]	330
[K]	603
dielectric constant	2.5
dielectric loss factor tan δ	0.0015
specificresistance [Ωcm3]	0.5×1017
dielectric breakdown [kV/cm]	2250

Ionomers have a higher tensile strength of the starting copolymers and also have good chemical resistance. At room temperature, they are insoluble in organic solvents. At elevated temperatures, they dissolve partially only. In some cases to obtain a ionomer solution, one can use a mixture of such solvents as decalin and dimethylacetamide. The extremely high resistance of these polymers to the action of oils and greases make them a good material for packaging. Ionomers are characterized by absorption of water.The moisture deteriorates their mechanical properties.

Ionomeric materials are suitable for the manufacture of numerous products by extrusion, injection, blow molding, and vacuum forming.

The olefin ionomers have found the largest application in production of packaging for food and for medicines. The packaging materials made from ionomers are transparent and highly resistant at low temperatures. They are resistant also to abrasion and they are more resistant to puncture with a sharp object than films made from other polymers.

The large adhesion of ionomers to metals and nonmetals allows to produce the laminates, which are characterized by durability and elasticity.

The ionomers of this type can be used to fabricate the protective glass, fancy goods, shoe soles, gaskets, helmets, syringes, insulating materials, etc.

Ionomers with a-olefins are characterized by easily connecting to other polymers, pigments, and fillers. A color effect of ionomer with a simultaneous transparency can be obtained by a proper selection of ions or of mixtures of metal ions used to neutralize the carboxyl groups. The miscibility of ionomer with fillers is so good that the fivefold increase in material stiffness does not decrease its toughness. However, when selecting the fillers, antioxidants, stabilizers, and other additives, a special attention has to be paid to avoid that they react with the cross linking ionic bonds.

3.18   TERPOLYMERS ETHYLENE/PROPYLENE/CARBON OXIDE

Terpolymer are obtained as a result of copolymerization of ethylene and propylene with carbon monoxide. Its structure is that of an aliphatic polyketone in which the carbon monoxide mers in chain are arranged alternately with the olefin mers as presented in the following scheme:

$$\left(\!\!-CH_2\!-\!CH_2\!-\!\underset{\displaystyle \overset{O}{\|}}{C}\!\!-\!\right)_{\!\!n}\ \left(\!\!-CH_2\!-\!\underset{\displaystyle \underset{CH_3}{|}}{\overset{\displaystyle \overset{O}{\|}}{CH}}\!-\!C\!\!-\!\right)_{\!\!m}$$

The obtained terpolymer is partially crystalline with a density of 1.235 g/cm^3 (density of the amorphous phase is 1.206 g/cm^3), the melting temperature of 493K (220 °C) and the glass transition temperature of 288 K (15 °C). The polymer contains 30–40% of crystalline areas. The packing density of atoms in the amorphous phase is larger than in other polymers of this type. Thus, the terpolymer exhibits good barrier properties. This material is resistant to most organic solvents and to not highly concentrated acids as well as bases.

The poly(ethylene/propylene/carbon monoxide) is a new plastic that can be used to build the plant for fuel in cars, for the manufacture of parts of mechanisms (including gears), and in fabrication of electrical and electronic equipment. It is produced by the company "Shell Chemical Company" under the trade name "Carilon."

REFERENCES

Abusleme, J.; Giannetti, E. Emulsion Polymeryzation. *Macromolecules* 1991, 24.

Adesamya, I. Classification of Ionic Polymers. *Polym. Eng. Sci.* 1988, 28, 1473.

Akbulut, U.; Toppare, L.; Usanmaz, A.; Onal, A. Electroinitiated Cationic Polymeryzation of Some Epoksides. *Makromol. Chem. Rapid. Commun.* 1983, 4, 259.

Andreas, F.; Grobe, K. *Chemia propylenu*; WNT:Warszawa, 1974.

Barg, E. I. *Technologia tworzyw sztucznyc*; PWT:Warszawa, 1957.

Billmeyer, F. W.*Textbook of Polymer Science*; John Wiley and Sons: New York, 1984.

Borsig, E.; Lazar, M. Chemical Modyfication of Polyalkanes. *Chem. Listy* 1991, 85, 30.

Braun, D.; Czerwiński, W.; Disselhoff, G.; Tudos, F. Analysis of the Linear Methods for Determining Copolymeryzation Reactivity Ratios. *Angew. Makromol. Chem.* 1984, 125, 161.

Burnett, G. M.; Cameron, G. G.; Gordon, G.; Joiner, S. N. Solvent Effects on the Free Radical Polymerization of Styrene. *J. Chem. Soc. Faraday Trans. I* 1973, 69, 322.

Challa, R. R.; Drew, J. H.; Stannet, V. T.; Stahel, E. P. Radiationindueed Emulsion Polymeryzation of Vinyl Acetate. *J. Appl. Polym. Sci.* 1985, 30, 4261.

Candau, F. Mechamism and Kinetics of the Radical Polymeryzation of Acrylamide In Inverse Micelles. *Makromol. Chem., Makromol Symp.* 1990, 31, 27.

Ceresa, R. J. *Kopolimery blokowe i szczepione*; WNT: Warszawa, 1965.

Charlesby, A. *Chemia radiacyjna polimerów*; WNT: Warszawa, 1962.

Chern, C. S; Poehlein, G. W. Kinetics of Cross-Linking Vinyl Polymerization. *Polym. Plast. Technol. Eng.*1990, *29*, 577.

Chiang, W. Y.; Hu, C. M. Studies of Reactions with Polymers. *J. Appl. Polym. Sci.*1985, *30*, 3895–4045.

Chien, J. C.; Salajka, Z. Syndiospecific Polymeryzation of Styrene. *J. Polym. Sci. A*1991, *29*, 1243.

Chojnowski, J.; Stańczyk, W. E. Procesy chlorowania polietylenu. *Polimery*1972, *17*, 597.

Collomb, J.; Morin, B.; Gandini, A.; Cheradame, H. Cationic Polymeryzation Induced by Metal Salts. *Europ. Polym. J.* 1980, *16*, 1135.

Czerwiński, W. K. Solvent Effects on Free-Radical Polymerization. *Makromol. Chem.* 1991, *192*, 1285–1297.

Dainton, F. S. *Reakcje łańcuchowe*; PWN:Warszawa, 1963.

Ding, J.; Price, C.; Booth, C. Use of Crown Ether in the Anionic Polymeryzation of Propylene Oxide. *Eur. Polym. J.* 1991, *27*, 895–901.

Dogatkin, B. A. *Chemia elastomerów*; WNT:Warszawa, 1976.

Duda, A.; Penczek, S. Polimeryzacja jonowa i pseudojonowa. *Polimery*1989, *34*, 429.

Dumitrin, S.; Shaikk, A. S.; Comanita, E.; Simionescu, C. Bifunetional Initiators. *Eur. Polym. J.* 1983, *19*, 263.

Dunn, A. S. Problems of Emulsion Polymeryzation. *Surf. Coat. Int.* 1991, *74*, 50.

Eliseeva, W. I.; Asłamazova, T. R. Emulsionnaja polimeryzacja v otsustivie emulgatora. *Uspiechi Chimii*1991, *60*, 398.

Erusalimski, B. L. Mechanism dezaktivaci rastuszczich ciepiej v procesach ionnoj polimeryzacji vinylovych polimerov. *Vysokomol. Soed. A.* 1985, *27*, 1571.

Fedtke, M. *Chimiczieskijereakcjipolimerrov*; Khimia:Moskwa, 1990.

Fernandez-Moreno, D.; Fernandez-Sanchez, C.; Rodriguez, J. G.; Alberola, A. Polymeryzation of Styrene and 2-vinylpyridyne Using AlEt-VCl Heteroaromatic Bases. *Eur. Polym. J.* 1984, *20*, 109.

Fieser, L. F.; Fieser, M. *Chemia organiczna*; PWN: Warszawa, 1962.

Flory, P. J. Nauka o makrocząsteczkach. *Polimery* 1987, *32*, 346.

Frejdlina, R. Ch.; Karapetian, A. S. *Telomeryzacja*; WNT: Warszawa, 1962.

Funt, B. L.; Tan, S. R. The Fotoelectrochemical Initiation of Polymeryzation of Styrene. *J. Polym. Sci. A*1984, *22*, 605.

Gerner, F. J.; Hocker, H.; Muller, A. H.; Shulz, G. V. On the Termination Reaction In The Anionic Polymeryzation of Methyl Methacrylate in Polar Solvents. *Eur. Polym. J.* 1984, *20*, 349.

Ghose, A.; Bhadani, S. N. Elektrochemical Polymeryzation of Trioxane in Chlorinated Solvents. *Indian J. Technol.* 1974, *12*, 443.

Giannetti, E.; Nicoletti, G. M.; Mazzocchi, R. Homogenous Ziegler-Natta Catalysis. *J. Polym. Sci. A*1985, *23*, 2117.

Goplan, A.; Paulrajan, S.; Venkatarao, K.; Subbaratnam, N. R. Polymeryzation of N,N'- Methylene Bisacrylamide Initiated by Two New Redox Systems Involving Acidic Permanganate. *Eur. Polym. J.* 1983, *19*, 817.

Gózdz, A. S.; Rapak, A. Phase Transfer Catalyzed Reactions on Polymers. *Makromol. Chem. Rapid Commun.* 1981, *2*, 359.

Greenley, R. Z. An Expanded Listing of Revized Q and e Values. *J. Macromol. Sci. Chem. A*1980, *14*, 427.

Greenley, R. Z. Recalculation of Some Reactivity Ratios. *J. Macromol. Sci. Chem. A*1980, *14*, 445.

Guhaniyogi, S. C.; Sharma, Y. N. Free-Radical Modyfication of Polipropylene. *J. Macromol. Sci. Chem. A*1985, *22*, 1601.

Guo, J. S.; El Aasser, M. S.; Vanderhoff, J. W. Microemulsion Polymerization of Styrene. *J. Polym. Sci. A*1989, *27*, 691.

Gurruchaga, M.; Goni, I.; Vasguez, M. B.; Valero, M.; Guzman, G. M. An Approach to the Knowledge of The Graft Polymerization of Acrylic Monomers Onto Polysaccharides Using Ce (IV) as Initiator. *J. Polym. Sci. Part C*1989, *27*, 149.

Hasenbein, N.; Bandermann, F. Polymeryzation of Isobutene With $VOCl_3$ in Heptane. *Makromol. Chem.* 1987, *188*, 83.

Hatakeyama, H. et al. Biodegradable Polyurethanes from Plant Components. *J. Macromol. Sci. A.* 1995, *32*, 743.

Higashimura, T.; Law, Y. M.; Sawamoto, M. Living Cationic Polymerization. *Polym. J.*1984, *16*, 401.

Hirota, M.; Fukuda, H. Polymerization of Tetrahydrofuran. *Nagoya-ski Kogyo Kenkynsho Kentyn Hokoku*,1984, 51.

Hodge, P.; Sherrington, D. C.*Polymer-Supported Reactions in Organic Syntheses*; J. Wiley and Sons: New York,1980, tłum. na rosyjski Mir: Moskwa, 1983.

Inoue, S. Novel Zine Carboxylates as Catalysts for the Copolymeryzation of CO_2 With Epoxides. *Makromol. Chem. Rapid Commun.*1980, *1*, 775.

Janović, Z. *Polimerizacije I polimery*; Izdanja Kemije Ind.: Zagrzeb 1997.

Jedliński, Z. J. Współczesne kierunki w dziedzinie syntezy polimerów. *Wiadomości Chem.* 1977, *31*, 607.

Jenkins, A. D., Ledwith, A., Ed.;*Reactivity, Mechanism and Structure in Polymer Chemistry*; John Wiley and Sons: London, 1974;tłum. na rosyjski, Mir: Moskwa 1977.

Joshi, S. G.; Natu, A. A. Chlorocarboxylation of Polyethylene. *Angew. Makromol. Chem.* 1986, *140*, 99–115.

Kaczmarek, H. polimery, a środowisko. *Polimery* 1997, *42*, 521.

Kang, E. T.; Neoh, K. G. Halogen Induced Polymerization of Furan. *Eur. Polym. J.* 1987, *23*, 719.

Kang, E. T.; Neoh, K. G.; Tan, T. C.; Ti, H. C. Iodine Induced Polymerization and Oxidation of Pryridazine. *Mol. Cryst. Lig. Cryst.* 1987, *147*, 199.

Karnojitzki, V. Organiczieskije pierekisy, I. I. L.:Moskwa, 1961.

Karpiński, K.; Prot, T. Inicjatory fotopolimeryzacji i fotosieciowania. *Polimery*1987, *32*, 129.

Keii, T. Propene Polymerization with $MgCl_2$-Supported Ziegler Catalysts. *Makromol. Chem.* 1984, *185*, 1537.

Khanna, S. N.; Levy, M.; Szwarc, M. Complexes formed by Anthracene with 'Living' Polystyrene. *Dormant Polymers. Trans Faraday Soc.*1962,*58*, 747–761.

Koinuma, H.; Naito, K.; Hirai, H. Anionic Polymerization of Oxiranes. *Makromol. Chem.* 1982, *183*, 1383.

Korniejev, N. N.; Popov, A. F.; Krencel, B. A. *Komplieksnyje metalloorganiczeskije katalizatory*; Khimia: Leningrad, 1969.

Korszak, W. W. Niekotoryje problemy polikondensacji. *Vysokomol. Soed. A*1979, *21*, 3.

Korszak, W. W. *Technologia tworzyw sztucznych*; WNT: Warszawa, 1981.

Kowalska, E.; Pełka, J. Modyfikacja tworzyw termoplastycznych włóknami celulozowymi. *Polimery*2001, *46*, 268.

Kozakiewicz, J.; Penczek, P. Polimery jonowe. *Wiadomości Chem.* 1976, *30*, 477.

Kubisa, P. Polimeryzacja żyjąca. *Polimery*1990, *35*, 61.

Kubisa, P. Neeld, K.; Starr, J.; Vogl, O. Polymerization of Higher Aldehydes. *Polymer*1980, *21*, 1433.

Kucharski, M. Nitrowe i aminowe pochodne polistyrenu. *Polimery*1966, *11*, 253.

Kunicka, M.; Choć, H. J. Hydrolitic Degradation and Mechanical Properties of Hydrogels. *Polym. Degrad. Stab.* 1998, *59*, 33.

Lebduska, J.; Dlaskova, M. Roztokova kopolymerace. *Chemicky Prumysl*1990, *40*, 419–480.

Leclerc, M.; Guay, J.; Dao, L. W. Synthesis and Characterization of Poly(alkylanilines). *Macromolecules*1989, *22*, 649.

Lee, D. P.; Dreyfuss, P. Triphenylphesphine Termination of Tetrahydrofuran Polymerization. *J. Polym. Sci. ,Polym. Chem. Ed.* 1980, *18*, 1627.

Leza, M. L.; Casinos, I.; Guzman, G. M.; Bello, A. Graft Polymerization of 4-vinylpyridineOnto Cellulosic Fibres by the Ceric on Method. *Angew. Makromol. Chem.* 1990, *178*, 109–119.

Lopez, F.; Calgano, M. P.; Contrieras, J. M.; Torrellas, Z.; Felisola, K. Polymerization of Some Oxiranes. *Polym. Int.* 1991, *24*, 105.

Lopyrev, W. A.; Miaczina, G. F.; Szevaleevskii, O. I.; Hidekel, M. L. Poliacetylen, *Vysokomol. Soed.* 1988, *30*, 2019.

Mano, E. B.; Calafate, B. A. L. Electrolytically Initiated Polymerization of N-vinylcarbazole. *J. Polym. Sci. ,Polym. Chem. Ed.* 1983, *21*, 829.

Mark, H. F. *Encyclopedia of Polymer Science and Technology*; Concise 3rd ed.; Wiley-Interscience: Hoboken, New York, 2007.

Mark, H.; Tobolsky, A. V. *Chemia fizyczna polimerów*; PWN: Warszawa, 1957.

Matyjaszewski, K.; Davis, T. P. *Handbook of Radical Polymerization*; Wiley-Interscience: New York, 2002.

Matyjaszewski,K.; Mülle, A. H. E. 50 years of Living Polymerization. *Prog. Polym. Sci.* , **2006**, *31*, 1039–1040

Mehrotra, R.; Ranby, B. Graft Copolymerization Onto Starch. *J. Appl. Polym. Sci.* 1977, *21*, 1647, 3407 i 1978, 22, 2991.

Morrison, R. T.; Boyd, N. *Chemia organiczna*; PWN:Warszawa,1985.

Morton, M. *Anionic Polymerization*; Academic Press: New York, 1983.

Munk, P. *Introduction to Macromolecular Science*; John Wiley and Sons: New York, 1989.

Naraniecki, B. Ciecze mikroemulsyjne. *Przem. Chem.* 1999, *78(4)*, 127.

Neoh, K. G.; Kang, E. T.; Tan, K. L. Chemical Copolymerization of Aniline with Halogen-Substituted Anilines. *Eur. Polym. J.* 1990, *26*, 403.

Nowakowska, M.; Pacha, J. Polimeryzacja olefin wobec kompleksów metaloorganicznych. *Polimery*1978, *23*, 290.

Ogata, N.; Sanui, K.; Tan, S. Synthesis of Alifatic Polyamides by Direct Polycondensation with Triphenylphosphine. *Polymer J.* 1984, *16*, 569.

Onen, A.; Yagci, Y. Bifunctional Initiators. *J. Macromol. Sci-Chem.* A1990, *27*, 743, *Angew. Makromol. Chem.* 1990, *181*, 191.

Osada, Y. Plazmiennaja polimerizacia i plazmiennaja obrabotka polimerov. *Vysokomol. soed.* A1988, *30*, 1815.

Pasika, W. M. Cationic Copolymerization of Epichlorohydrin With Styrene Oxide And Cyclohexene Oxide. *J. Macromol. Sci-Chem.* A1991, *28*, 545.

Pasika, W. M. Copolymerization of Styrene Oxide and Cyklohexene Oxide. *J. Polym. Sci. Part* A1991, *29*, 1475.

Pielichowski, J.; Puszyński, A. Preparatyka monomerów, W. P. Kr. Kraków.

Pielichowski, J.; Puszyński, A. Preparatyka polimerów, TEZA WNT 2005 Kraków.

Pielichowski, J.; Puszyński, A. Technologia tworzyw sztucznych, WNT 2003, Warszawa.

Pielichowski, J.; Puszyński, A. Wybrane działy z technologii chemicznej organicznej. W. P. Kr. Kraków.

Pistola, G.; Bagnarelli, O. Electrochemical Bulk Polymeryzation of Methyl Methacrylate in the Presence of Nitric Acid. *J. Polym. Sci. , Polym. Chem. Ed.* 1979, *17*, 1002.

Pistola, G.; Bagnarelli, O.; Maiocco, M. Evaluation of Factors Affecting The Radical Electropolymerization of Methylmethacrylate in the Presence of HNO_3. *J. Appl. Electrochem.* 1979, *9*, 343.

Połowoński, S. Techniki pomiarowo-badawcze w chemii fizycznej polimerów; W. P. L. : Łódź, 1975.

Porejko, S.; Fejgin, J.; Zakrzewski, L. *Chemia związków wielkocząsteczkowych*; WNT: Warszawa 1974.

Praca zbiorowa: Chemia plazmy niskotemperaturowej; WNT: Warszawa, 1983.

Puszyński, A. Chlorination of Polyethylene in Suspension. *Pol. J. Chem.* 1981, *55*, 2143.

Puszyński, A.; Godniak, E. Chlorination of Polyethylene in Suspension in the Presence of Heavy Metal Salts. *Macromol. Chem. Rapid Commun.* 1980, *1*, 617.

Puszyński, A.; Dwornicka, J. Chlorination of Polypropylene in Suspension. *Angew. Macromol. Chem.* 1986, *139*, 123.

Puszyński, J. A.; Miao, S. Kinetic Study of Synthesis of SiC Powders and Whiskers In the Presence $KClO_3$ and Teflon. *Int. J. SHS*1999, *8(8)*, 265.

Rabagliati, F. M.; Bradley, C. A. Epoxide Polymerization. *Eur. Polym. J.* 1984, *20*, 571.

Rabek J. F. : Podstawy fizykochemii polimerów, W. P. Wr. , Wrocław 1977

Rabek T. I. : *Teoretyczne podstawy syntezy polielektrolitów i wymieniaczy jonowych*; PWN: Warszawa 1962.

Regas, F. P.; Papadoyannis, C. J. Suspension Cross-Linking of Polystyrene With Friedel-Crafts Catalysts. *Polym. Bull.* 1980, *3*, 279.

Rodriguez, M.; Figueruelo, J. E. On the Anionic Polymeryzation Mechanism of Epoxides. *Makromol. Chem.* 1975, *176*, 3107.

Roudet, J.; Gandini, A. Cationic Polymerization Induced by Arydiazonium Salts. *Makromol. Chem. Rapid Commun.* 1989, *10*, 277.

Sahu, U. S.; Bhadam, S. N. Triphenylphosphine Catalyzed Polimerization of Maleic Anhydride. *Makromol. Chem.* 1982, *183*, 1653.

Schidknecht, C. E. *Polimery winylowe*; PWT: Warszawa 1956.

Szwarc, M.; Levy, M.; Milkovich, R. Polymerization Initiated by Electron Transfer to Monomer. A New Method of Formation of Block Copolymers. *J. Am. Chem. Soc.* **1956**,*78*, 2656–2657.

Szwarc, M. 'Living' Polymers. *Nature*1956, *176*, 1168–1169.

Sen, S.; Kisakurek, D.; Turker, L.; Toppare, L.; Akbulut, U. Elektroinitiated Polymerization of 4-Chloro-2,6-Dibromofenol. *New Polym. Mater.* 1989, *1*, 177.

Sikorski, R. T. Chemiczna modyfikacja polimerów, Prace Nauk. *Inst. Technol. Org. Tw. Szt. P. Wr.* 1974, *16*, 33.

Sikorski, R. T. *Podstawy chemii i technologii polimerów*; PWN: Warszawa, 1984.

Sikorski, R. T.; Rykowski, Z.; Puszyński, A. Elektronenempfindliche Derivate von Poly-Methylmethacrylat. *Plaste und Kautschuk*1984, *31*, 250.

Simionescu, C. I.; Geta, D.; Grigoras M. : Ring-Opening Isomerization Polymeryzation of 2-Methyl-2-Oxazoline Initiated by Charge Transfer Complexes, *Eur. Polym. J.* 1987, *23*, 689.

Sheinina, L. S.; Vengerovskaya, Sh. G.; Khramova, T. S.; Filipowich, A. Y. Reakcji piridinov i ich czietvierticznych soliej s epoksidnymi soedineniami. *Ukr. Khim. Zh.* 1990, *56*, 642.

Soga K. , Toshida Y. , Hosoda S. , Ikeda S. : A Convenient Synthesis of a Polycarbonate, Makromol. Chem. 1977, 178, 2747

Soga K. , Hosoda S. , Ikeda S. : A New Synthetic Route to Polycarbonate, J. Polym. Sci. ,Polym. Lett. Ed. 1977, 15, 611

Soga, K.; Uozumi, T.; Yanagihara, H.; Siono, T. Polymeryzation of Styrene with Heterogeneous Ziegler-Natta Catalysts. *Makromol. Chem. Rapid. Commun.* 1990, *11*, 229.

Soga, K.; Shiono, T.; Wu, Y.; Ishii, K.; Nogami, A.; Doi, Y. Preparation of a Stable Supported Catalyst for Propene Polymeryzation. *Makromol. Chem. Rapid Commun.* 1985, *6*, 537.

Sokołov, L. B.; Logunova, W. I. Otnositielnaja reakcionnosposobnost monomerov i prognozirovanic polikondensacji v emulsionnych sistemach. *Vysokomol. Soed. A.* 1979, *21*, 1075.

Soler, H.; Cadiz, V.; Serra, A. Oxirane Ring Opening with Imide Acids. *Angew. Makromol. Chem.* 1987, *152*, 55.

Spasówka, E.; Rudnik, E. Możliwości wykorzystania węglowodanów w produkcji biodegradowalnych tworzyw sztucznych. *Przemysł Chem.* 1999, *78*, 243.

Stevens, M. P. *Wprowadzenie do chemii polimerów*; PWN: Warszawa, 1983.

Stępniak, I.; Lewandowski, A. Elektrolity polimerowe. *Przemysł Chem.* 2001, *80(9)*, 395.

Strohriegl, P.; Heitz, W.; Weber, G. Polycondensation Using Silicon Tetrachloride. *Makromol. Chem. Rapid Commun.* 1985, *6*, 111.

Szur, A. M. *Vysokomolekularnyje soedinenia*; Vysszaja Szkoła: Moskwa, 1966.

Tagle, L. H.; Diaz, F. R.; Riveros, P. E. Polymeryzation by Phase Transfer Catalysis. *Polym. J.* 1986, *18*, 501.

Takagi, A.; Ikada, E.; Watanabe, T. Dielectric Properties of Chlorinated Polyethylene and Reduced Polyvinylchloride. *Memoirs Faculty Eng. ,Kobe Univ.* 1979, *25*, 239.

Tani, H. Stereospecific Polymeryzation of Aldehydes and Epoxides. *Adv. Polym. Sci.* 1973, *11*, 57.

Vandenberg, E. J. Coordynation Copolymeryzation of Tetrahydrofuran and Oxepane with Oxetanes and Epoxides. *J. Polym. Sci. Part A*1991, *29*, 1421.

Veruovic, B. *Navody pro laboratorni cviceni z chemie polymer*; SNTL: Praha, 1977.

Vogdanis, L.; Heitz, W. Carbon Dioxide as a Mmonomer. *Makromol. Chem. Rapid Commun.* 1986, *7*, 543.

Vollmert, B. *Grundriss der Makromolekularen Chemic*; Springer-Verlag: Berlin, 1962.

Wei, Y.; Tang, X.; Sun, Y. A study of the Mechanism of Aniline Polymeryzation. *J. Polym. Sci. Part A*1989, *27*, 2385.

Wei, Y.; Sun, Y.; Jang, G. W.; Tang, X. Effects P-Aminodiphenylamine on Electrochemical Polymeization of Aniline. *J. Polym. Sci. ,Part C*1990, *28*, 81.

Wei, Y.; Jang, G. W. Polymerization of Aniline and Alkyl Ringsubstituted Anilines in the Presence of Aromatic Additives. *J. Phys. Chem.* 1990, *94*, 7716.

Wiles, D. M. Photooxidative Reactions of Polymers. *J. Appl. Polym. Sci. , Appl. Polym. Symp.* 1979, *35*, 235.

Wilk, K. A.; Burczyk, B.; Wilk, T. Oil in Water Microemulsions as Reaction Media. 7th International Conference Surface and Colloid Science, Compiegne, 1991.

Winogradova, C. B. Novoe v obłasti polikondensacji. *Vysokomol. Soed. A*1985, *27*, 2243.

Wirpsza, Z. *Poliuretany*; WNT: Warszawa, 1991.

Wojtala, A. Wpływ właściwości oraz otoczenia poliolefin na przebieg ich fotodegradacji. *Polimery* 2001, *46*, 120.

Xue, G. Polymerization of Styrene Oxide with Pyridine. *Makromol. Chem. Rapid Commun.* 1986, *7*, 37.

Yamazaki, N.; Imai, Y. Phase Transfer Catalyzed Polycondensation of Bishalomethyl Aromatic compounds. *Polym. J.* 1983, *15*, 905.

Yasuda, H. *Plasma Polymerization*; Academic Press, Inc. : Orlando, 1985.

Yuan, H. G.; Kalfas, G.; Ray, W. H. Suspension polymerization. *J. Macromol. Sci. Rev. Macromol. Chem. -Phys.* 1991, C 31, 215.

Zubakowa, L. B.; Tevlina, A. C.; Davankov, A. B. *Sinteticzeskije ionoobmiennyje materiały*;Khimia: Moskwa, 1978.

Zubanov, B. A. *Viedienie v chimiu polikondensacionnych procesov*; Nauka: Ałma Ata, 1974.

Żuchowska, D. Polimery konstrukcyjne. WNT: Warszawa, 2000.

Żuchowska, D. *Struktura i właściwości polimerów jako materiałów konstrukcyjnych*; W. P. Wr. : Wrocław, 1986.

Żuchowska, D.; Steller, R.; Meisser, W. Structure and Properties of Degradable Polyolefin—Starch Blends. *Polym. Degrad. Stab.* 1998, *60*, 471.

CHAPTER 4

PRACTICAL HINTS ON POLYMERIZATION METHODS

G. E. ZAIKOV

Russian Academy of Sciences, Moscow, Russia

CONTENTS

4.1 MIGRATION POLYADDITION

In contrary to the previously discussed polymerization methods of vinyl compounds, the migration polyaddition is not based on the polarization of the unsaturated bonds. It is due to the displacement of the mobile hydrogen atom originating from another reactive chemical compound to this bond.

A typical example of such a reaction is the synthesis of polyurethane from diisocyanate and glycol. The reaction of the isocyanate group with the hydroxyl group is as follows:

$$\text{{\sim}\!\!\sim\!CH_2\!-\!OH} + \text{O}\!=\!\text{C}\!=\!\text{N}\!-\!\text{CH}_2\!\!\sim\!\!\sim \longrightarrow \text{{\sim}\!\!\sim\!CH_2\!-\!O\!-\!\overset{\displaystyle O}{\overset{\|}{C}}\!-\!\overset{\displaystyle H}{\overset{|}{N}}\!-\!CH_2\!\!\sim\!\!\sim}$$

The reaction is a gradual process. It means that the molecular weight of polymers obtained by this method increases steadily during the subsequent reaction steps and the growing macromolecule after each act of addition is completely stable.

However, it is not a reversible reaction. The addition of successive monomer molecules to the growing chain is very fast. Therefore, the separation of the simple addition products such as dimers and trimers is very difficult. For this reason, the reaction is very similar to the previously discussed reaction of polymerization. Migration polyaddition catalysts are salts, acids, bases (III-row amines), and water. Linear polymers of high molecular weight can be obtained only in the case of equimolecular ratio of reactants, namely diisocyanate and glycol. In the case of excess of one of them, the macromolecules with the same end groups are formed. It limits or even prevents the further growth of the chain.

The use in the reaction of monomers containing at least three functional groups results in formation of polymers with cross-linked spatial structure.

Other types of migration polymerization include:

- Synthesis of polyurethanes from diisocyanates and diamines

$$n\,\text{NCO}\!-\!\text{R}\!-\!\text{NCO} + n\,\text{H}_2\text{N}\!-\!\text{R'}\!-\!\text{NH}_2 \longrightarrow \text{NCO}\!\!\left[\!-\!\text{R}\!-\!\overset{\displaystyle H}{\overset{|}{N}}\!-\!\overset{\displaystyle O}{\overset{\|}{C}}\!-\!\overset{\displaystyle H}{\overset{|}{N}}\!-\!\text{R'}\!-\!\right]_n\!\!-\!\text{NH}_2$$

- Synthesis of polyamides by polymerization of N-substituted acrylamide derivatives

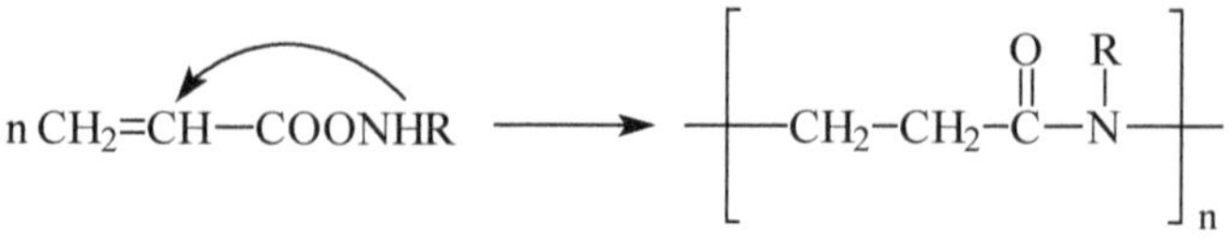

$$n\ CH_2=CH-COONHR \longrightarrow \left[-CH_2-CH_2-\overset{\overset{O}{\|}}{C}-\overset{\overset{R}{|}}{N}-\right]_n$$

- Synthesis of polyamides by a reaction of carbon suboxide with diamines:

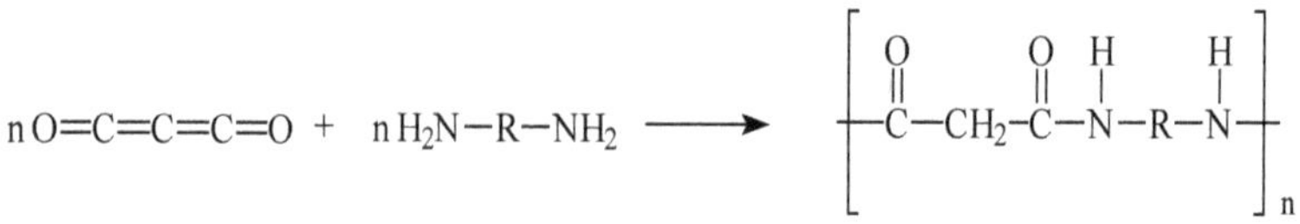

$$n\ O=C=C=C=O\ +\ n\ H_2N-R-NH_2 \longrightarrow \left[-\overset{\overset{O}{\|}}{C}-CH_2-\overset{\overset{O}{\|}}{C}-\overset{\overset{H}{|}}{N}-R-\overset{\overset{H}{|}}{N}-\right]_n$$

- Polyquaternization of pyridine derivatives

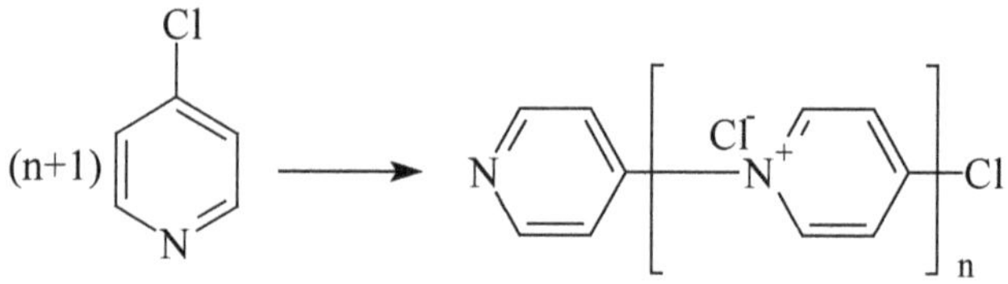

4.2 CYCLOPOLYMERIZATION

The cyclopolymerization is a special case of tetrafunctional polymerization of unsaturated compounds containing two double bonds in which there is no cross-linking reaction and a linear polymer containing rings in the main chain is formed.

A classic example is that the cyclopolymerization reaction of methyl is the polymerization of N-diallyl diethyloammonium bromide conducted in the presence of persulfates and iodine as initiator:

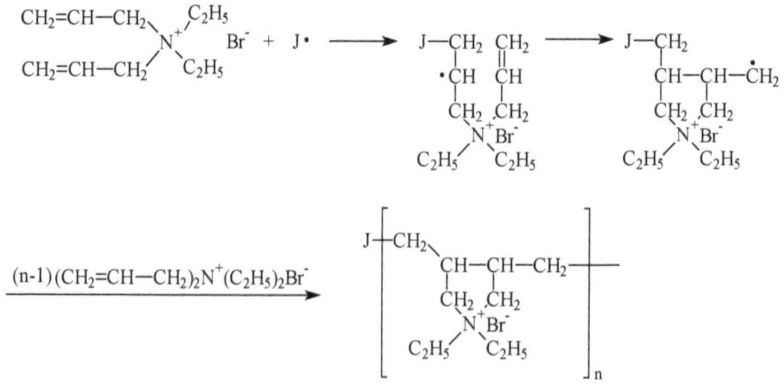

In a similar way runs the cyclopolymerization of anhydrous methacrylic acid:

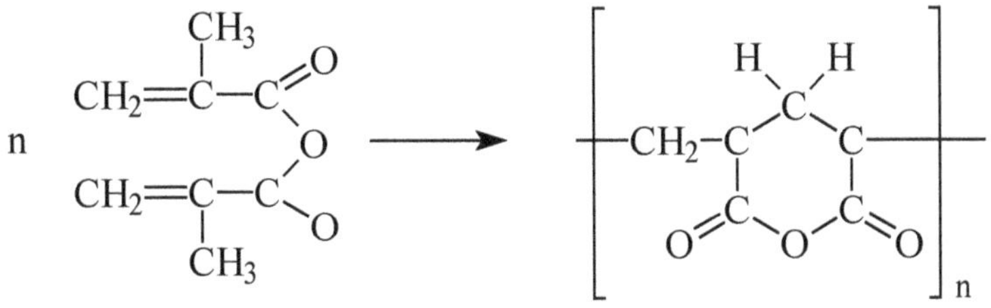

Some of dienes containing isolated double bonds may also polymerize with a formation of cyclic polymers. This reaction takes place when there is a possibility of formation of rings with five or six members.

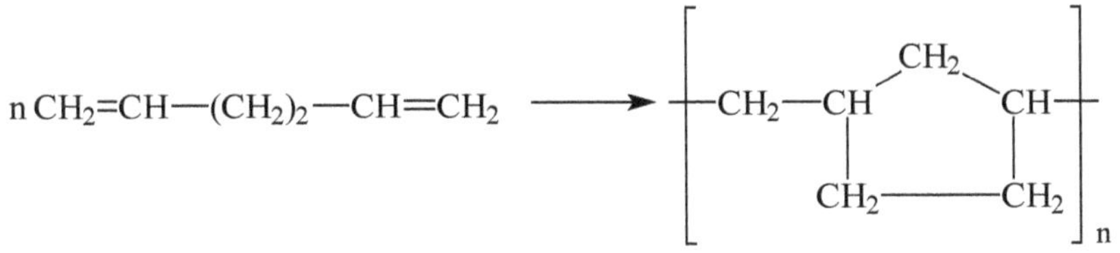

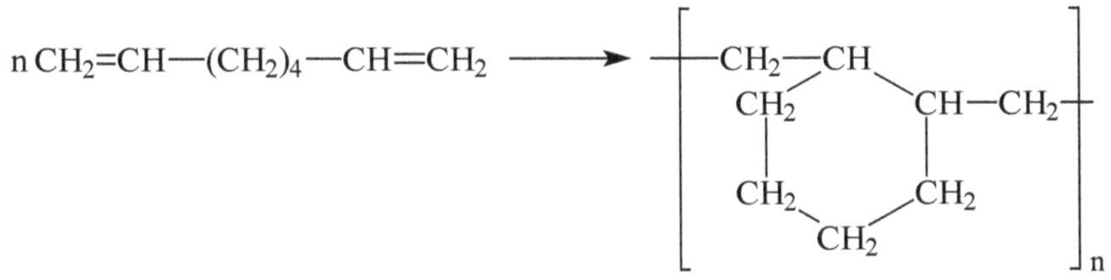

Dienes with isolated double bonds containing less than five methylene groups are sporadically and incompletely cyclizated. The formed polymers contain a number of multiple member rings and linear segments in the chain with the unsaturated bonds in side chains. Such polymers can undergo a crosslinking reaction.

An interesting example of cyclopolymerization is the reaction of divinyl ether with the maleic anhydride:

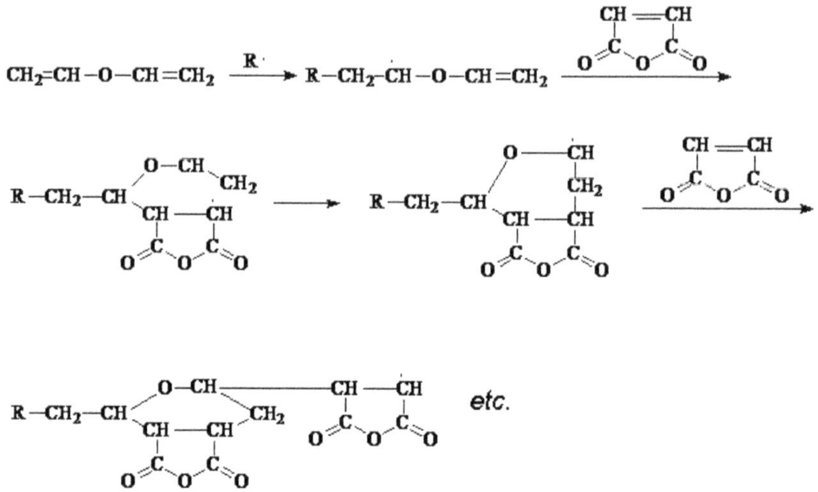

$$CH_2=CH-O-CH=CH_2 \xrightarrow{R\cdot} R-CH_2-\overset{\cdot}{C}H-O-CH=CH_2 \longrightarrow$$

and of divinyl ether with divinyl sulfone that passes spontaneously through the stage of a charge-transfer complex:

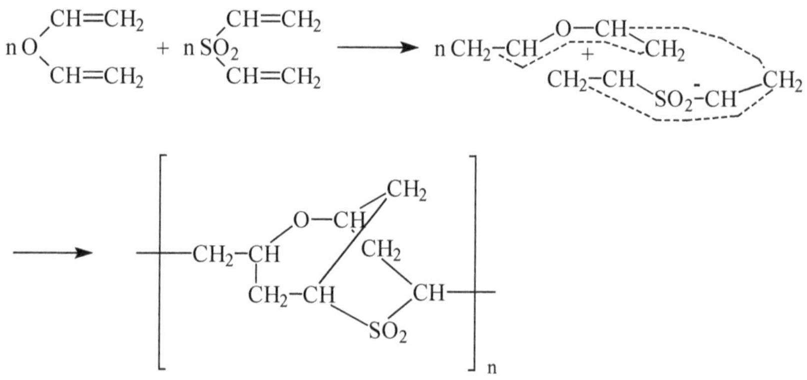

An important method of cyclopolymerization is the cycloaddition polymerization of Diesel–Alder.

The classical Diesel–Alder reaction is the reaction of a nucleophilicdiene with electrophilic alkene, called dienophile, with the formation of the cyclohexane ring.

In practice, the most spectacular are dienophiles with double electron attracting substituents to the atoms forming double bonds. The diene reactivity increases if the monomer is in a more reactive cisoidal configuration.

Of course the dienophile used for the synthesis of polymers cannot be a simple alkene, but must be a ternary functional monomer, that is, it has to contain two unsaturated bonds.

The Diesel–Alder polymerization allows for the formation of the ladder polymers. An example of such reaction is the cycloaddition of 2-vinyl-buta-1,3-diene to p-benzoquinone:

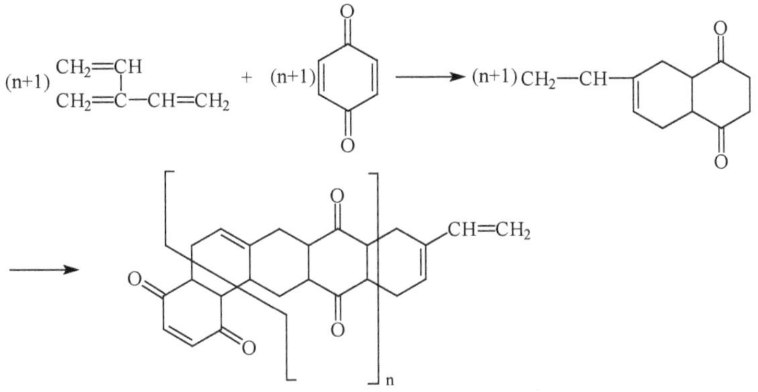

In a similar way reacts N, N'-alkylene-bis (cyclopentadiene) with p-benzoquinone or a suitable bismaleimide.

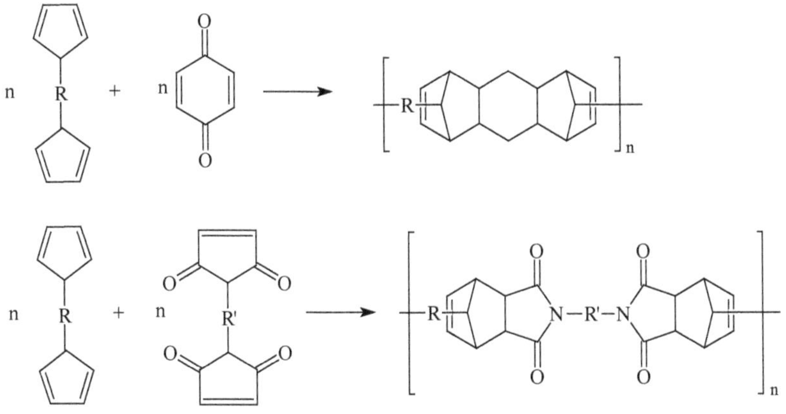

In the cyclopolymerization reaction of Diesel–Alder, one can also use monomers being both a diene and a dienophileat the same time. An example of such a reaction is the spontaneous cyclopolimerization of biscyclopentadiene:

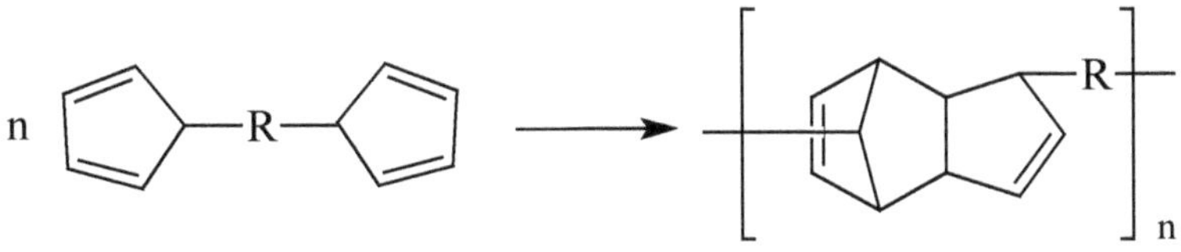

The above-mentioned structure of the emerging polymer is a probable one. In fact the structure of poly(dicyclopentadiene) is more complex because in the Diesel–Alder reaction, different ways of addition or of crosslinking are possible.

4.3   OXIDATIVE POLYMERIZATION

The oxidative polymerization is a new method of synthesis of polymers from phenols or from aromatic amines. As it turned out, the oxidation of phenols leads to the formation of free radicals, which react together to form a poly(phenylene oxide) (PPO) or its derivatives. This reaction is carried out especially easy when using phenol derivatives having substituents at positions 2 and 6. As oxidizing agents, mild oxidants such as potassium ferrycyanide (hexacyanoferrate (III)), lead (II) oxide, and silver oxide may be used. However,the oxidation is performed most frequently with oxygen from the air in the presence of catalysts derived from copper (II) chloride and pyridine or other tertiary amines. Probably the alkaline salt of copper, obtained by oxidation of copper (I) chloride and coordinatively bound with two amino groups can be used as an active catalyst too.

The resulting catalytic complex reacts with the dialkylphenol. The reaction is associated with the formation of an appropriate phenolate:

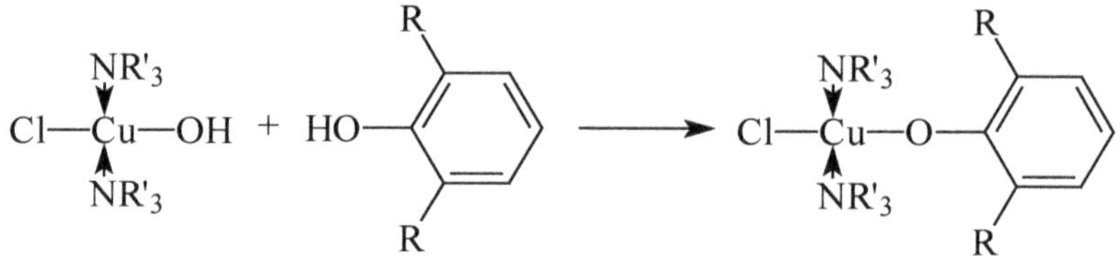

The formed phenoxyl anion is oxidized in a reaction to two mesomeric radicals, which recombine together with the formation of a dimer:

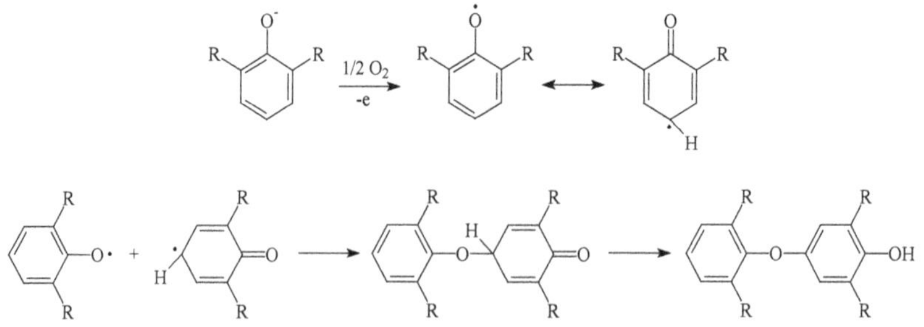

The resulting dimer reacts again with the catalytic complex and the reaction is repeated successively, until the formation of polymer. So the chain is created in a progressive manner and a large distribution of molecular weight is result of the transfer of hydrogen atom reaction.

Known are also the ways to carry out the oxidative copolymerization of different phenols. For example, it was shown that during the oxidative copolymerization of 2,6-diphenylphenol with 2,6-dimethylphenol, one can obtain block copolymers as a result of the initial oxidation of 2,6-diphenylphenol and after its exhaustion, the next introduced portions of 2,6-dimethylphenol. Changing the order of addition of these monomers or use of their mixtures in the process leads to the formation of a statistical copolymer.

In recent years, a number of research works were devoted to the oxidative polymerization process of aromatic amines. In fact the resulting polymers react readily with acids, forming conducting ionic polymers.

Aniline and its derivatives, substituted in the phenyl ring, can polymerize by oxidizing both chemically and electrochemically.Most frequently the oxidation is performed with the solution of ammonium persulfate in an acidic environment.

The reaction yield and some properties of ionic polymers obtained by the oxidation of aniline and its derivatives are given in Table 4.1.

TABLE 4.1   Reaction Yield and Some Properties of Ionic Polymers Formed during the Oxidation of Aniline and its Derivatives by Ammonium Persulfate.**No**

S. No.	Monomer	Yield [%]	Content of ionic systems Cl/N	Electrical conductivity $[\Omega^{-1} \, cm^{-1}]$
1	Aniline	82	0.44	5
2	2-methylaniline	80	0.65	0.3
3	3-methylaniline	29	0.7	0.3
4	2-ethylaniline	16	0.68	1
5	3-ethylaniline	1	–	0
6	2-n-propylaniline	2	–	0

Source: Adapted from Leclerc, M.; Guay, J.; Dao, L. H.*Macromolecules*1989, *22*, 649.

Molecular mass of synthesized polymers depends on the reaction conditions and on the nature of monomer used.

TABLE 4.1   Reaction Yield and Some Properties of Ionic Polymers Formed during the Oxidation of Aniline and its Derivatives by Ammonium Persulfate.**No**

S. No.	Monomer	Yield [%]	Content of ionic systems Cl/N	Electrical conductivity $[\Omega^{-1} cm^{-1}]$
1	Aniline	82	0.44	5
2	2-methylaniline	80	0.65	0.3
3	3-methylaniline	29	0.7	0.3
4	2-ethylaniline	16	0.68	1
5	3-ethylaniline	1	–	0
6	2-n-propylaniline	2	–	0

Source: Adapted from Leclerc, M.; Guay, J.; Dao, L. H.*Macromolecules*1989, *22*, 649.

Table 4.2compares the molecular mass of polyanilines obtained by the chemical and electrochemical oxidation method. Polymers formed in the electrochemical process are characterized by a lower molecular mass. Similarly, the increase in volume of the substituent and its distance from the amino group lowers the efficiency of polymerization products with a simultaneous decrease of the molecular mass of the formed polymer.

The study of the reaction mechanism of oxidative polymerization of aniline has shown that, in the first stage, a split of proton takes place, associated with the formation of the nitrene radical cation ($C_6H_5N_2^{\cdot+}$). As a result of the electrophilic substitution of the formed the nitrene radical cation to the next molecule of aniline, the substrates dimerization to p-aminodiphenylamine takes place. Then, similarly as in the case of oxidation of phenols, the process is repeated several times until the formation of the polymer. The course of the synthesis of polyaniline may be shown schematically as follows:

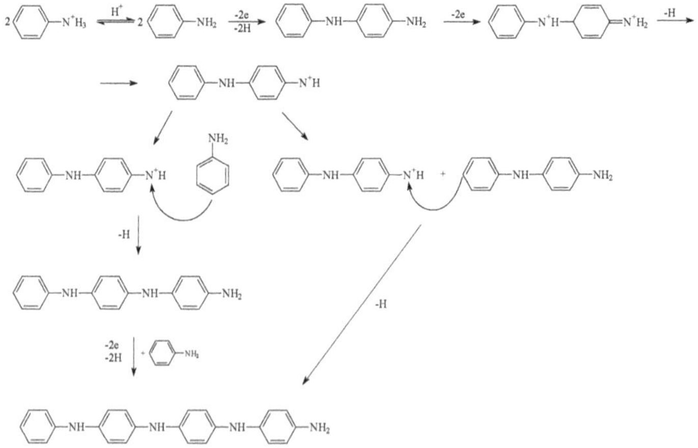

The oxidative polymerization method was also applied to the synthesis of conductive poly(heterocycles) like polypyrrole, polythiophene, polyfuran, poly-selenophene, and polypiridazine.

For example, during the oxidative polymerization of piridazine, a good oxi-dizing agent may be iodine. The reaction is conducted in a polar solvent such as acetonitrile at room temperature or at 0 °C.

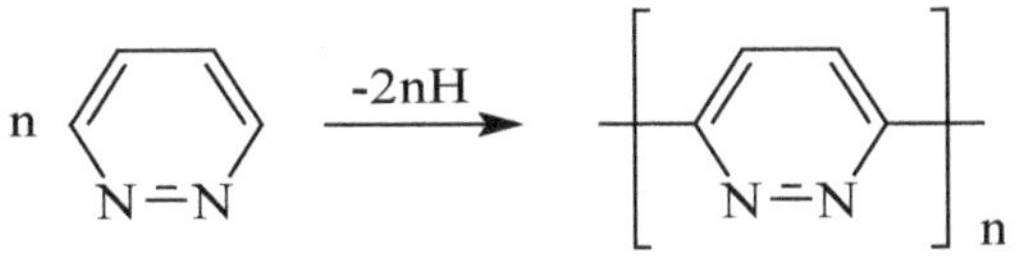

The resulting polymer forms a charge-transfer type complex with iodine. Complexes of poly (polypiridazine-J) are black with electric conductivity of 2–3 S/cm.

4.4 POLYMERIZATION OF ALDEHYDES

Aldehydes are highly reactive compounds, undergoing easily the oxidation and reduction reactions. In particular, equally in acidic and in alkaline environment, the aldehyde molecules react with each other and, depending on the process conditions, are subject to aldol condensation, cyklization, oligomerization, or polymerization.

An example of the aldol condensation is the synthesis of crotonaldehyde, run-ning through the stage of dimerization of the acetaldehyde as shown below:

$$5\ CH_3\text{-}CHO \longrightarrow CH_3\text{-}CHOH\text{-}CH_5\text{-}CHO \longrightarrow CH_3\text{-}CH\text{-}CH\text{-}CHO + H_5O$$

In an acidic environment, the aldehydes containing up to 10 carbon atoms in the molecule undergo easily the cyclization process and most frequently the cyclotrimerization:

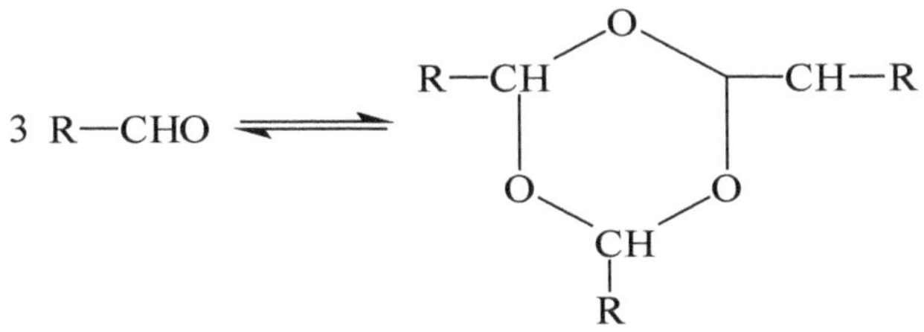

A pure trimer of this type doesnot undergo either depolymerization during the distillation nor during the storage.

However, after adding an acid catalyst, it can depolymerize to the starting aldehyde state.

At temperatures below 273 K in acidic environment, the aldehydes polymerize with the formation of crystalline oligomers (most frequently the tetramers) called meta aldehydes. Aldehyde oligomers heated to the temperature above 100 °C depolymerize to the starting compound.

By using appropriate catalysts and the conduction process conditions, one can obtain from many aldehydes polymers with high molecular weight. The resulting polymers can be made resistant to temperature by blocking the terminal hydroxyl group by esterification, most frequently with the acetic anhydride.

Polymerization of aldehydes takes place by opening the double carbon–oxygen bond, whose energy is almost four times smaller than that of the double carbon–carbon bond and is about 20.9 kJ/ mol.

Polymerization of aldehydes always takes place below the limit temperature T_c, at which the free energy of polymerization is equal to zero.

The characteristic thermodynamic values of the enthalpy and the entropy changes at the limit temperature measured during the polymerization of some aldehydes are listed in Table 4.3.

TABLE 4.3 Thermodynamic Data for Polymerization of Aldehydes.

S. No	Aldehyde	Chemical formula	Solvent	$-\phi H$ [J/mol$\cdot 10^{-3}$]	$-DS$ [J/mol$\cdot$K]	$T_c = \phi H_{pol}/DS_{pol}$[K]
1	Acetalaldehyde	CH_3CHO	–	27.6	118	234
2	Butyraldehyde	C_3H_7CHO	Hexane	35.5	122	–
3	Isobutyr aldehyde	C_3H_7CHO	–	46	–	–
4	Isobutyr aldehyde	C_3H_7CHO	Tetrahydrofuran	15.5	74	–
5	Valeralaldehyde	C_4H_9CHO	–	22.6	97	231
6	Methoxypropionic aldehyde	CH_3OCH_2- $-CH_2CHO$	–	19.7	82	238

TABLE 4.3 *(Continued)*

7	Trichloroacetic aldehyde	CCl_3CHO	Tetrahydrofuran	14.6	52	284
8	Trichloroacetic aldehyde	CCl_3CHO	Toluene	37.8	134	282
9	Trifluoroacetic aldehyde	CF_3CHO	Toluene	54.9	155	354
10	Tribromoacetaldehyde	CBr_3CHO	Toluene	19.1	100	196

*Tc—limit temperature** at which the free energy of polymerization is equal to zero.
Source: Adapted from Kubis, P.;Neeld, K.; Starr, J.; Vogl, O.Polymer1980, 21, 1433.

The aldehyde polymerization reaction process is always exothermic. The heat of the most frequently performed formaldehyde polymerization is of 71.23 kJ/mol.

The mechanism of this process in an acidic environment consists on addition of a proton to the oxygen atom of carbonyl group with the formation of an active carbocation. This carbocation reacts with other aldehyde molecules leading to the formation of polymer molecule:

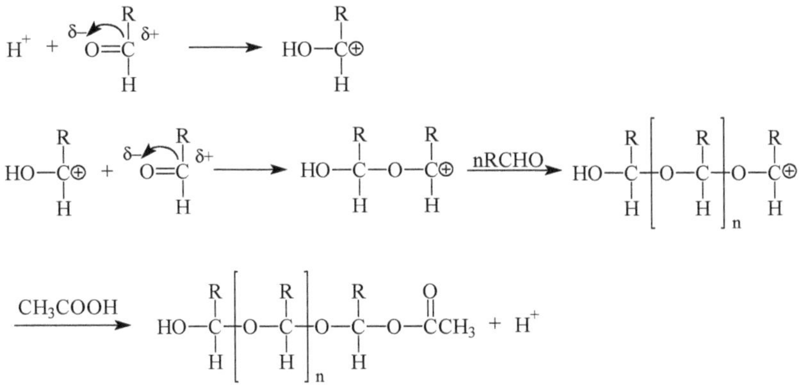

$$H^+ + \; O=C\overset{R}{\underset{H}{|}} \; \longrightarrow \; HO-\overset{R}{\underset{H}{\overset{|}{C}}}\oplus$$

$$HO-\overset{R}{\underset{H}{\overset{|}{C}}}\oplus + \; O=C\overset{R}{\underset{H}{|}} \; \longrightarrow \; HO-\overset{R}{\underset{H}{\overset{|}{C}}}-O-\overset{R}{\underset{H}{\overset{|}{C}}}\oplus \; \xrightarrow{nRCHO} \; HO-\left[\overset{R}{\underset{H}{\overset{|}{C}}}-O-\overset{R}{\underset{H}{\overset{|}{C}}}\right]_n-O-\overset{R}{\underset{H}{\overset{|}{C}}}\oplus$$

$$\xrightarrow{CH_3COOH} \; HO-\overset{R}{\underset{H}{\overset{|}{C}}}\left[-O-\overset{R}{\underset{H}{\overset{|}{C}}}\right]_n-O-\overset{R}{\underset{H}{\overset{|}{C}}}-O-\overset{O}{\overset{||}{C}}CH_3 + H^+$$

The mechanism of polymerization of aldehyde in an alkaline medium is based on the attachment of the Nu⁻ nucleofile to carbon atom and in this way creating a negative charge on the oxygen atom. The resulting anion reacts with the next aldehyde molecules with formation of a macroanion, which can be stabilized in reaction with the acetic anhydride.

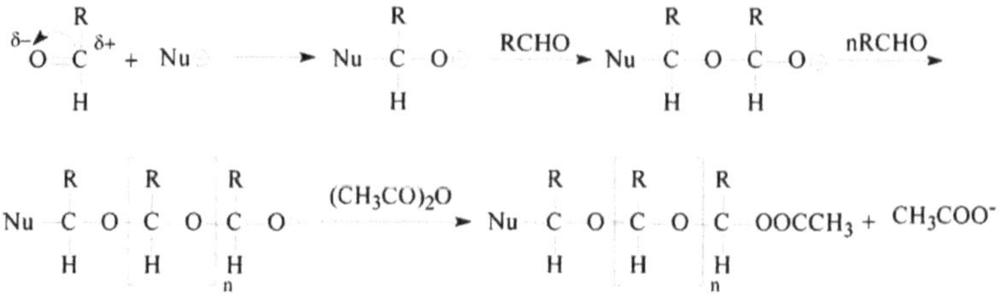

where Nu ⊖ – nukleofile factor

Similarly, as in the case of anionic polymerization of vinyl compounds, there is no proper termination reaction here. "Living" polymers, suitable for the block copolymerization, can be formed too. In the reaction with acetic anhydride, one obtains the hydrophobic chain termination and, in this way, a stabilized polymer with increased thermal resistance.

The low limit temperature in aldehydes in the polymerization is utilized in the synthesis of poly (trichloroacetic aldehyde) (polichloral), characterized by noncombustibility and good mechanical properties.

In order to obtain this product, chloral is mixed with a catalyst at a temperature above the limit temperature, which amounts to 58 °C. Then the mixture is poured into the mold and cooled to the temperature of about 10 °C to obtain polymers with high molecular weights:

In the discussed group of polymers, a great importance is also given to the stereospecific polymerization of aldehydes on coordination catalysts.

As it was shown in previous studies, only theisotactic polymers are obtained. The reaction is assumed to run as follows:

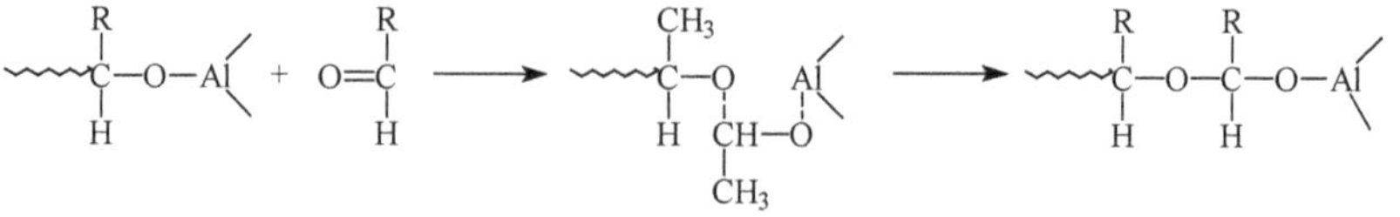

The results obtained during the polymerization of acetaldehyde in the presence of the triethylaluminum complex with different complexing compounds are given in Table 4.4.

TABLE 4.4   Polymerization of Acetaldehyde in the Presence of TriethylaluminumComplex.

S. No	Complexing factor	Chemical formula	Yield* [%]	Stereospeçifity indicator** [%]
1	Cycloheksanon	$(CH_2)_5CO$	70	96
2	2-pyrrolidone	$OC(CH_2)_3NH$	49	85
3	Acetamide	CH_3CONH_2	87	95
4	Methylacetamidee	$CH_3CONHCH_3$	69	88
5	Phenylacetamide	$CH_3CONHC_6H_5$	87	95
6	Phenylbenzamide	$C_6H_5CONHC_6H_5$	85	92

* **Fraction** insoluble in hexane.
** **Ratio** of the fraction insoluble in chloroform to the fraction insoluble inhexane.
Source: Adapted from Tani, H.Polym. Adv. Sci. 1973, 11, 1957.

Thus,polymers obtained in this way are characterized by a high degree of isotacticity.

During the polymerization of dialdehydes, the formed polymers contain cyclic ether groups. For example, during the polymerization, glyoxal in the presence of sodium naphthalene, performed in tetrahydrofuran at −78 °C, a polyglyoxal with the following structure is formed:

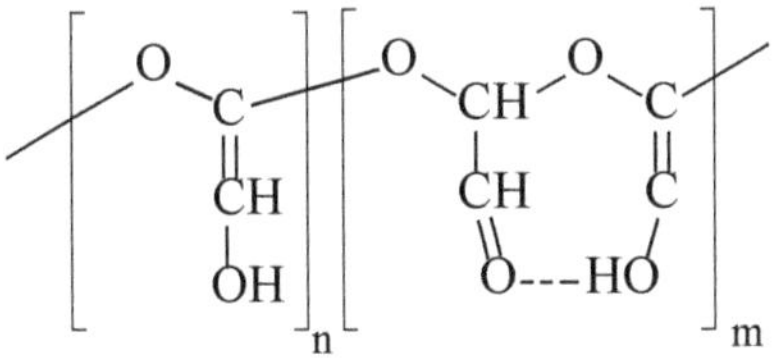

Aldehydes such as glutaric, succinic, and phthalic, polymerized in the presence of Lewis acids, give amorphous polymers with low molecular weights.The chemical structure of the polymer formed from glutaraldehyde is as following:

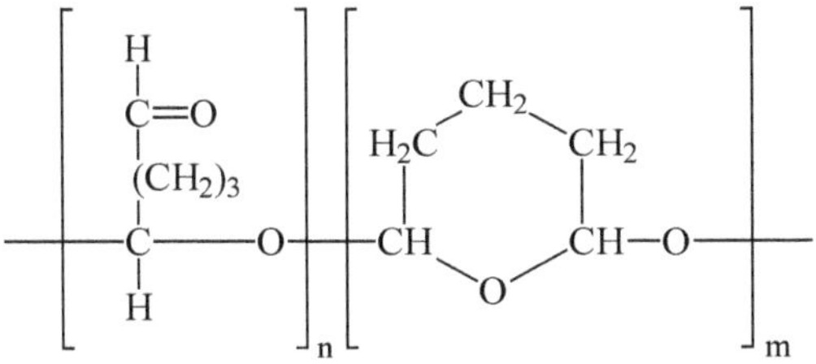

Terephthalic and isophthalic aldehydes engage only one aldehyde groupin the process of polymerization, whereas the o-phthalic aldehyde undergo a partial cyclopolymerization at 0 °C. It cyclopolimerizes almost exclusively at the temperature of −78 °Caccording to the following scheme:

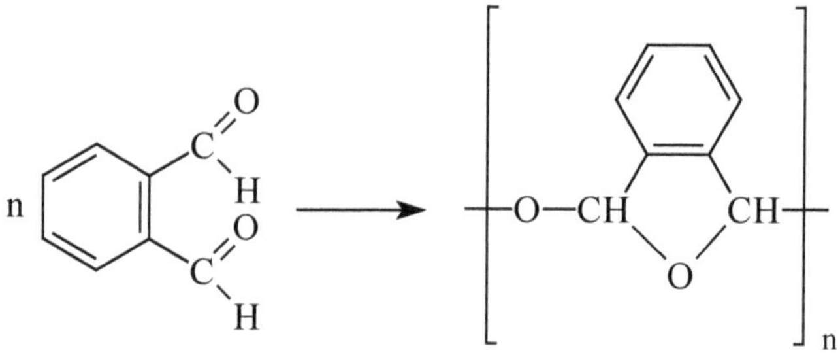

4.5   POLYMERIZATION OF KETONES

Ketones are less reactive compounds than aldehydes and harder to polymerize. The polymers formed from them have no practical significance.

Acetone polymerize at low temperatures at the presence of Ziegler–Natta catalystsor under the γ irradiation. The polyacetone obtained in this way is unstable. More stable is a block copolymer of acetone with propylene.

Significantly better monomers are ketenes. Depending on the reaction environment, various polymers are formed during the polymerization of dimethylketenes.

In polar solvents, polyethers are formed. In nonpolar solvents and in the presence of lithium, magnesium, and aluminum ions,polyketones are formed. In nonpolar solvents and in the presence of sodium and potassium ions, polyesters are formed:

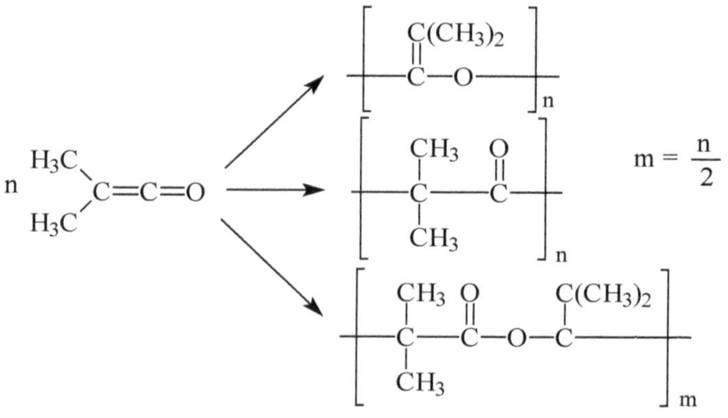

4.6 POLYMERIZATION OF HETEROCYCLIC COMPOUNDS WITH RING OPENING

Some heterocyclic compounds undergo the ring-opening polymerization relatively easy resulting in the formation of linear polymers.

As catalysts for these reactions, acidic (usually Lewis acids), alkaline, or coordination compounds may be used.

The method of the ring-opening polymerization was widely used for the synthesis of polyethers from oxacyclic connections, polyamides from lactams, polyesters from lactones,and polyorganosiloxanes.

4.7 POLYMERIZATION OF CYCLIC ETHERS

The most commonly used oxacyclic monomers include the following compounds or their derivatives:

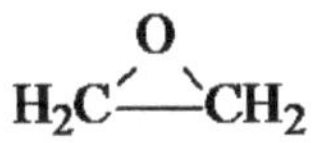

oxirane, epoxyethane
(ethylene oxide)

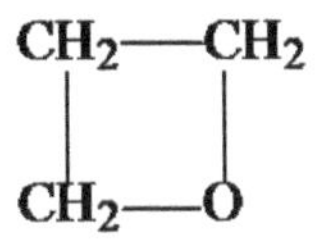

oxetane
(oxacyclobutane)

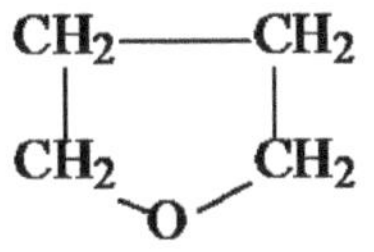

tetrahydrofuran

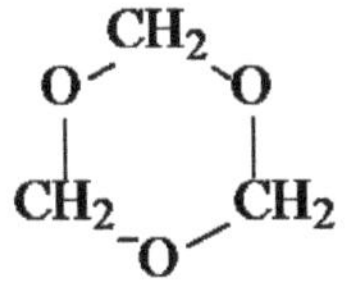

trioxymethylene

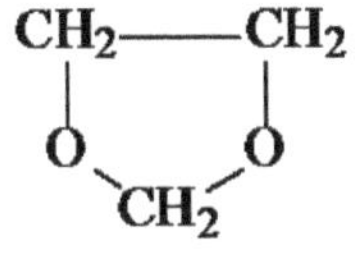

1,3 - dioxalon

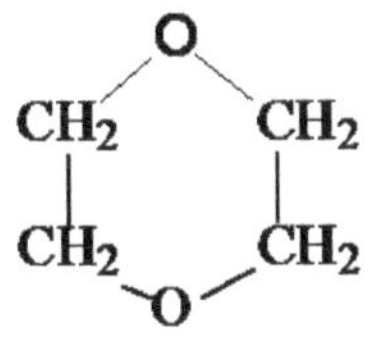

1,4 dioxane

4.7.1 CATIONIC POLYMERIZATION OF CYCLIC ETHERS

Polymerization of cyclic ethers by cationic mechanism is triggered by various protic catalysts and Lewis acids. In the event of proton donors, the polymerization reaction proceeds through the stage of an oxoion, but as a result of this reaction, mostly oligomers or polymers with small molecular weights are formed. The course of polymerization of ethylene oxide under the influence of acid or boron trifluoride in the presence of water is given by the following scheme:

$$H_2C\overset{O}{\underset{}{\diagup\!\!\diagdown}}CH_2 \;+\; HX \;\rightleftharpoons\; H_2C\overset{\overset{H}{|}}{\underset{}{\overset{+}{O}}}\!\!\!\overset{X^-}{\underset{}{\diagdown}}CH_2$$

$$H_2C\overset{\overset{H}{|}}{\underset{}{\overset{+}{O}}}\!\!\!\overset{X^-}{\underset{}{\diagdown}}CH_2 \;+\; H_2C\overset{O}{\underset{}{\diagup\!\!\diagdown}}CH_2 \;\rightleftharpoons\; HO\!-\!CH_2\!-\!CH_2\!-\!O\overset{X^-\,CH_2}{\underset{CH_2}{\diagup\!\!\!\overset{+}{}}}$$

The completion of the chain growth takes place by the reaction of macrocation with a water molecule:

$$H\!\!\left[\!-\!O\!-\!CH_2\!-\!CH_2\!-\!\right]_n\!\!\overset{X^-\,CH_2}{\underset{CH_2}{O^+\!\!\diagup}} \;+\; H_2O \;\longrightarrow\; H\!\!\left[\!-\!O\!-\!CH_2\!-\!CH_2\!-\!\right]_n\!\!O\!-\!CH_2\!-\!CH_2\!-\!OH \;+\; HX$$

In the case of using other cationic catalysts, the polymerization mechanisms are often complex and not completely understood. Some of these catalysts most likely react with monomer, forming metal alkoxides. It means that the polymerization proceeds according to the coordination anionic mechanism.

During the polymerization of ethylene oxide in the presence of tin tetrachloride, the reaction mechanism differs from the mechanism initiated by boron(III) fluoride and leads to the high molecular weight polymer. The kinetic studies indicate that each particle of tin chloride initiates two chains of polymer.

The reaction may be represented schematically as follows:

$$SnCl_4 \;+\; 4\,H_2C\overset{O}{\underset{}{\diagup\!\!\diagdown}}CH_2 \;\longrightarrow\; \overset{H_2C}{\underset{H_2C}{\diagdown\!\!\!\diagup}}\!\!O^+CH_2CH_2O\!-\!SnCl_4^{-2}\!-\!OCH_2CH_2\!-\!O\overset{CH_2}{\underset{CH_2}{\diagup\!\!\!\overset{+}{}}}$$

The "living" polytetrahydrofuran is formed by the reaction of alkyl halide with the silver hexafluoro antimony:

$$RX \;+\; AgSbF_6 \;+\; \underset{CH_2\underset{O}{\diagdown}CH_2}{\overset{CH_2\!-\!\!-\!CH_2}{|\qquad|}} \;\longrightarrow\; R\!-\!\overset{+}{O}\overset{CH_2\!-\!CH_2}{\underset{SbF_6^-\;CH_2\!-\!CH_2}{\diagup\!\!\!\diagdown}} \;+\; AgX$$

The chain growth completion can take place in reaction with the triphenylphosphine:

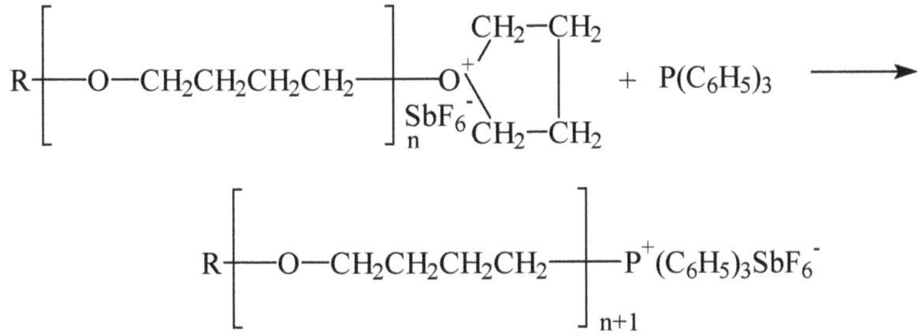

The above-mentioned reaction has also found application in the synthesis of graft copolymers.

The hexachloro antymony of triphenylmethyl has proved to be a good catalyst, enabling the cationic copolymerization of cyclohexene oxide with styrene oxide. The initiation reaction in this case consists on formation of the carbonion cation:

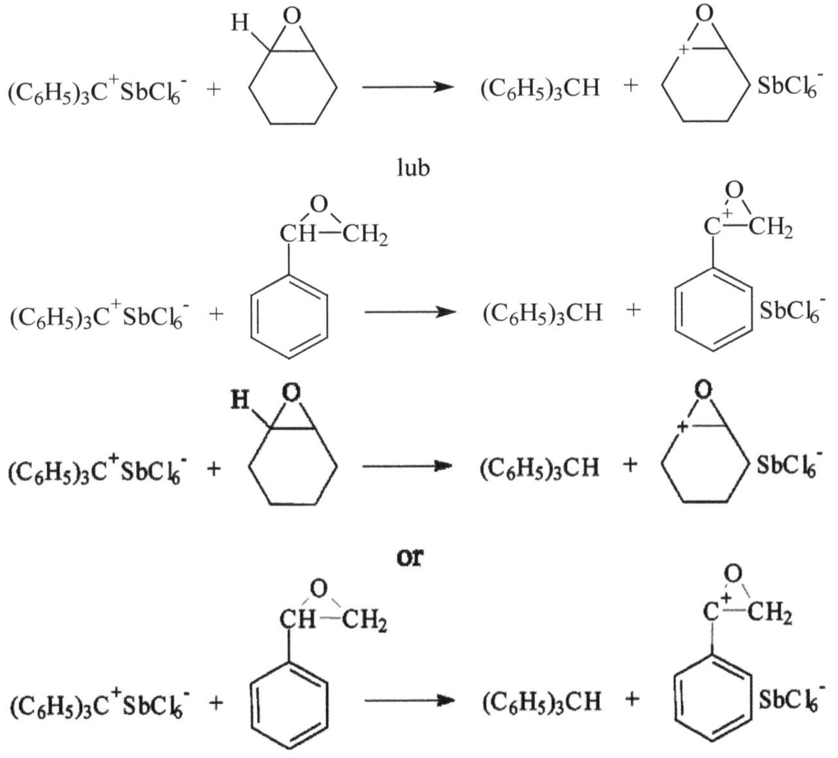

lub

or

The further increase of the copolymer chain proceeds through the stage of the oxonium ion:

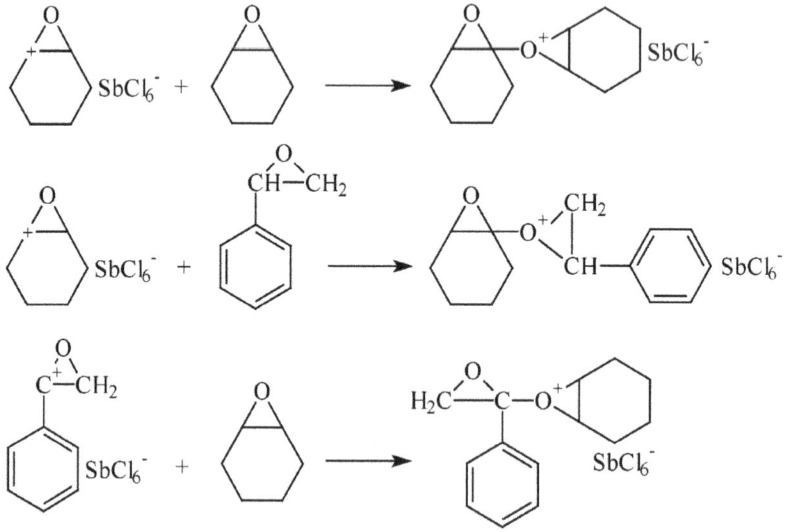

The resulting oxonium cations react with next monomer molecules, with ring opening. As a consequence, linear copolyethers are formed.

The results obtained during the course of cationic copolymerization of cyclohexene oxide with styrene oxide and the measured reactivity ratios are given in Tables 4.5 and 4.6.

TABLE 4.5 Course of Cationic Copolymerization of Cyclohexene Oxide (M1) with Styrene Oxide (M2) within 1 h and at the Given Initiator Concentration.

S. No	Composition		Reaction temp. [°C]	Yield [weight %]	Copolymer composition		Molecular mass of copolymer $M_n \times 10^2$
	M_1 [% mol]	M_2 [% mol]			M_1 [% mol]	M_2 [% mol]	
1	11.1	88.9	0	1	30.7	69.3	901
2	22.5	77.5	0	1.3	51.7	48.3	1048
3	38.7	61.3	0	1.1	68.6	31.4	1296
4	53.5	46.5	0	1.48	80.4	19.6	1188
5	66.6	33.4	0	2.2	90.5	9.5	1550

TABLE 4.5 (*Continued*)

6	82.1	17.9	0	3.8	95.1	4.9	1278
7	11.1	88.9	30	16.4	27	73	1858
8	22.5	77.5	30	18	46.9	53.1	1298
9	38.7	61.3	30	16.3	67	33	1137
10	53.5	46.5	30	17.8	78.9	21.1	889
11	66.6	33.4	30	18.1	88.7	11.3	778
12	82.1	17.9	30	17.6	94.9	5.1	668
13	11.1	88.9	46	6.3	25.3	74.7	1655
14	22.5	77.5	46	9.1	45.1	54.9	1672
15	38.7	61.3	46	13.4	66.8	33.2	1490
16	53.5	46.5	46	13.7	78.9	21.1	1256
.17	66.6	33.4	46	15.6	86.3	13.7	1159
18	82.1	17.9	46	14.6	95	5	1481

Source: Adapted from Pasika, W. M.; Chen, D. J.J. Polym. Sci.A1991, 29, 1457.

TABLE 4.6 Reactivity Ratios of Cyclohexane Oxide (r1) with Styrene Oxide (r2).

S. No.	Temperature [°C]	Kelen–Tudos method		Fineman–Ross method	
		r_1	r_2	r_1	r_2
1	0	4.24	0.4	4.4	0.35
2	30	4.19	0.42	4.15	0.4
3	46	3.94	0.46	3.9	0.45

Source: Adaptedfrom Pasika,W. M.;Chen, D. J.J. Polym. Sci.A1991, 29, 1457.

One of the cationic polymerization of cyclic ethers method is the polymerization initiated by a constant electric current. It can be applied to both oxirane derivatives and tetrahydrofuran. Polymerization oxirane derivatives is performed in the dichloromethane solution using a platinum electrode. As electrolyte, the tetrabutylammonum hexafluorophosphate with concentration of 0.1 mol/dm^{3}is used.

The results obtained during the electroinitiated cationic polymerization of some epoxy compounds are presented in Table 4.7.

TABLE 4.7 Electroinitiated Cationic Polymerization of Some Epoxides.

S. No.	Monomer	Anode poten. [V]	Poten. polym. [V]	Max. current inten. [mA]	Temp. [°C]	Reaction time [min]	Monomer conversion degree. [%]	Polymer melting temp. [°C]	Polymer solution viscosity in benzene 0.1 dm³/g at 35 °C
1	1.2-epoxy-4-epoxy--ethyl cyclohexane	2.7	2.3	2	–37	5	5	>340	–
2	Epoxycyclohexane (cyclohexene oxide)	3.01	2.3	4.5	–37	90	68.8	93	0.071
3	Epoxycyclopentane (cyclopentane oxide)	2.92	2.2	5	–35	90	69	–10	0.13
4	Epoxyethylbenzene (styrene oxide)	2.39	2.2	5	–20	75	>95	15	0.092

Source: Adapted fromAkbult, U. et alMacromol. Chem. Rapid Commun.1983, 4, 259.

4.7.2   ANIONIC POLYMERIZATIONOF OXIRANE AND ITS DERIVATIVES

Oxiranes polymerize by the anionic mechanism too. Catalysts for these reactions are usually hydroxides or alkoxides of sodium and potassium. The oxirane ring-opening reaction proceeds by S_N2 substitution with the formation of an alkoxylane ion. As consequence, the chain growth takes place either by the nucleophilic rearrangement of the newly formed alkoxylane ion or in a gradual process, consisting on transfer or termination following the addition of alcohol, as illustrated by the following reaction scheme:

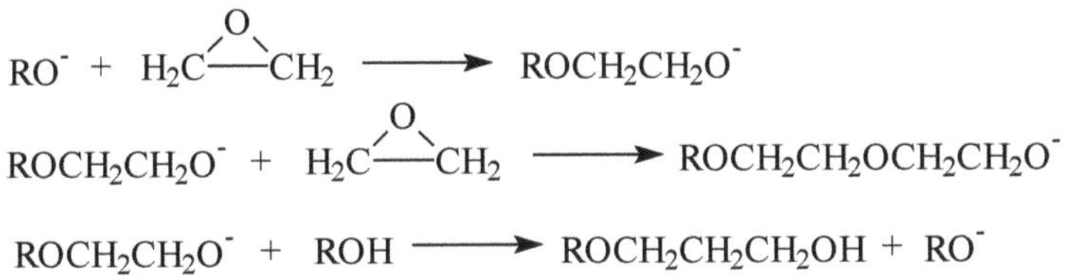

$$RO^- + H_2C\overset{O}{-}CH_2 \longrightarrow ROCH_2CH_2O^-$$

$$ROCH_2CH_2O^- + H_2C\overset{O}{-}CH_2 \longrightarrow ROCH_2CH_2OCH_2CH_2O^-$$

$$ROCH_2CH_2O^- + ROH \longrightarrow ROCH_2CH_2CH_2OH + RO^-$$

The initiation reaction takes place faster in this case than the growth chain reaction. It causes the formation of polyethers of low molecular weight. The curse of this reaction is largely influenced by the type of alcohol used during for its initiation. When using a tertiary alcohol, a more reactive primary alcohol is formed in the stage of initiating. It causes growth rate advantage over the rate of chain reaction initiation. Thus, it allows to obtain polymers with higher molecular weight. The course of the anionic polymerization of epoxides is also affected by the type of solvent used and the temperature at which the reaction is conducted.

It was shown that during the reaction carried out in dimethylsulfoxide a methylsulfinyl the methanide anion is formed in the reaction environment,

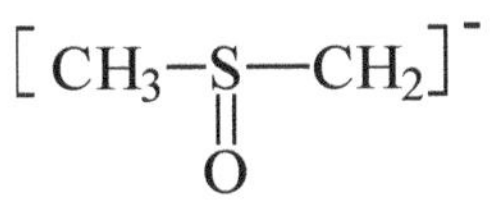

$$\left[CH_3-\underset{\underset{O}{\|}}{S}-CH_2 \right]^-$$

which participates in the polymerization process and affects the kinetics of the reaction.

This anion is in equilibrium with the alkoxyl anion:

$$\left[CH_3\text{--}\underset{\underset{O}{\|}}{S}\text{--}CH_2 \right]^- + ROH \rightleftharpoons RO^- + CH_3\text{--}\underset{\underset{O}{\|}}{S}\text{--}CH_3$$

The results of kinetic studies of anionic polymerization of oxirane and its homologues in the dimethyl sulfoxide environment are given in Table 4.8.

TABLE 4.8 Kinetic Data of Oxirane and its Homologues Polymerization in Dimethyl Sulfoxide.

No	Monomer		Temp. °C	Order with respect to		Literature
				mono-mer	initi-ator	
1	Oxirane (ethylene oxide)	Tert-butylan of potassium	25	1	1	Polym. 1969, *10*, 653.
2	Oxirane	Tert-butylan of potassium	20	1	1	J. Am. Chem. Soc.**1966**, *88*, 4039
3	Methyloxirane (propylene oxide)	1-ethoxy-3-oxapentylan of cesium	50	1	1.9	Makromol. Chem. **1975**, *176*, 3107
4	Methyloxirane (propylene oxide)	Tert-butylan of potassium	30	1	1	J. Am. Chem. Soc.**1966**, *88*, 4039
5	Methyloxirane (propylene oxide)	Tert-butylan of potassium	40	1	1.7	J. Polym. Sci. A-1**1972**,*10*, 1353
6	Ethyloxirane (1,2-butylene oxide)	Tert-butylan of sodium	30–60	1	1.8	J. Polym. Sci. Polym. Chem. Ed. **1972**, *10*, 3089

In a completely different way, runs the anionic polymerization of oxirane derivatives in the presence of 4-ethylopyridine. In that case, no polyethers are formed, but aromatic type polymers containing pyridine groups in the main chain.

The reaction of 4-ethylopyridine with styrene oxide can be presented as follows:

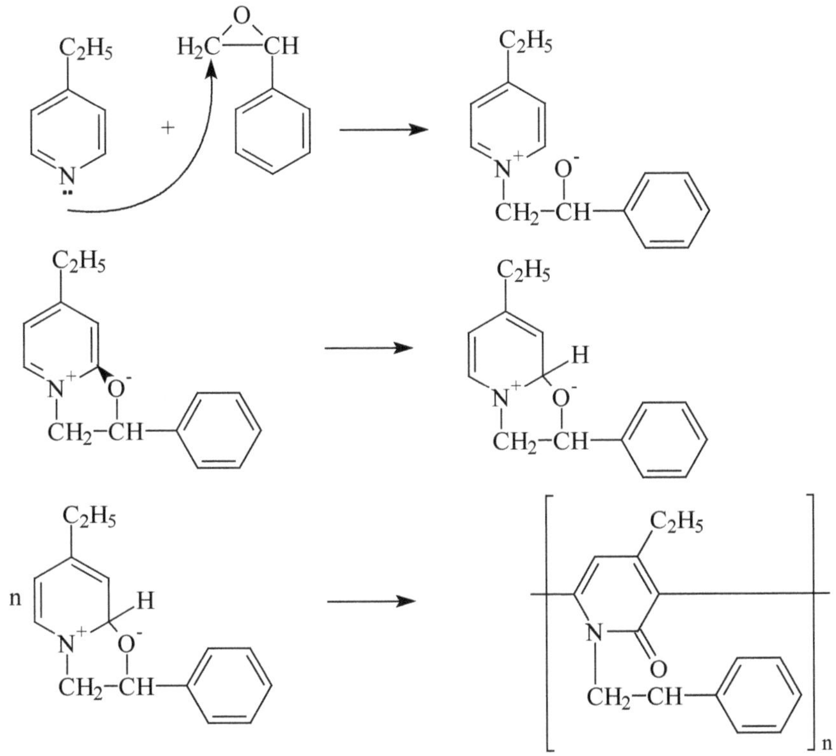

Known are also methods of anionic polymerization of propylene oxide in the presence of crown ethers and copolymerization of epoxides with carbon dioxide in the presence of zinc salts of carboxylic acids.

4.7.3   STEREOSPECIFIC POLYMERIZATION OF CYCLIC ETHERS

During the polymerization of cyclic ethers in the presence of coordination catalysts, polyethers with high molecular weights and a stereoregularspatial structure are formed. The mechanism of formation of such polymers is similar to the already described mechanism for obtaining stereoregular vinyl polymers with Ziegler–Natta catalysts.

The stereoregularity effect depends on the type of catalyst and the reaction conditions.

The coordination catalysts used for polymerization of cyclic ethers are chelates of β-hydroxyesters or β-diketones, calcium ethoxy amide, Ziegler–Natta catalysts, trialkylaluminum, dialkylzinc, and a complex of ferric chloride with 1,2-epoxypropane.

Some coordination catalysts (mainly alkylmetals) increase their activity in the presence of cocatalysts such as water or ketones.

The addition of a cocatalyst changes the structure of the basic catalyst and renders it more active. For example, the addition of water to diethylzinc causes likely its partial hydrolysis with the formation of a compound of the following chemical structure:

$$C_2H_5-Zn-O-Zn-C_2H_5$$

The formed catalyst may react with the propylene oxide in the following way:

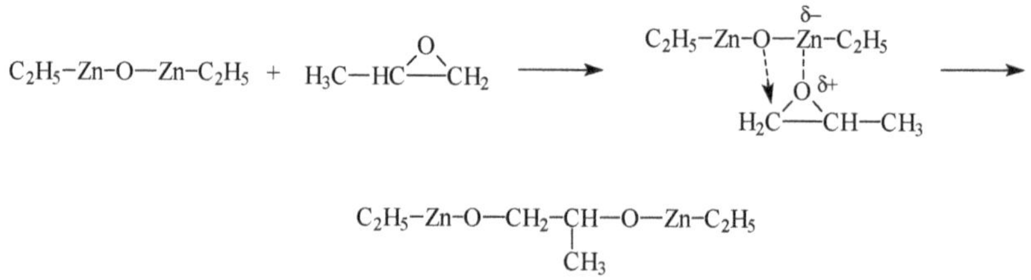

A further chain growth takes place as a result of an intramolecular rearrangement. The mechanism of this reaction has not yet been clearly determined and requires a further research. The polymerization reaction of oxirane and its derivatives in the presence of dialkylzinc is conducted mostly in a benzene solution.

The results obtained during the polymerization of ethylene oxide in benzene at the temperature of 60 °C and in the presence of water activated by zinc diphenyl are presented in Table 4.9. The data presented there show that the increase in the amount of ethylene oxide with respect to the catalyst and the extension of the reaction time increases the efficiency of the reaction and favors the formation of a polymer with a higher molecular weight. The optimal molar ratio of water to zinc diphenyl is 1:1.

TABLE 4.9 Polymerization of Oxirane in the Presence of $(C6H5)2Zn-H2O$ in Benzene at 60 °C.

No	Zinc diphenyl [mmol]	Oxirane $(C_6H_5)_2Zn$ [mol/mol]	Reaction time [H]	Yield [%]	Intrinsic viscosity [0.1 dm^3/g]	Molecular mass M_{V_5} ×10	Polymer flow temperature [°C]
1	0.285	199.7	2	1	–	–	–
2	0.285	199.7	4	6	3.04	3	62–64
3	0.285	199.7	6	24.7	1.44	1.5	61–63

TABLE 4.9   *(Continued)*

4	0.285	199.7	12	30.3	3.7	6.1	63–65
5	0.285	214	24	38.2	3.28	5.1	59–63
6	0.278	200	36	56.8	3.07	4.7	54–60
7	0.285	199.7	48	63.4	5.44	10.7	54–59
8	0.285	199.7	72	83.9	4.93	9.3	52–63
9	0.143	214.6	120	93.3	5.98	12.3	59–63
10	0.278	200	144	97.1	6.96	15.4	57–61

Source: **Adapted** from Rabagliati, F. M.; Bradley, C.Eur.Polym. J.1984,20, 571.

The molecular weight of formed polyethers was determined by the viscosimetry technique in water at 298 K using the following formulas:

$$[\eta] = 3.97 \times 10^{-4} (\bar{M}_v)^{0.686} \qquad \text{for ethylene oxide}$$

$$[\eta] = 1.12 \times 10^{-4} (\bar{M}_v)^{0.77} \qquad \text{for propylene oxide}$$

Tetrahydrofuran and 2-methyl-2-oxazoline can polymerize through the formation of complexes such as "charge-transfer" complex (CTC).

An example of such a reaction is the polymerization of tetrahydrofuran in the presence of tetracyanoethylene (TCNE):

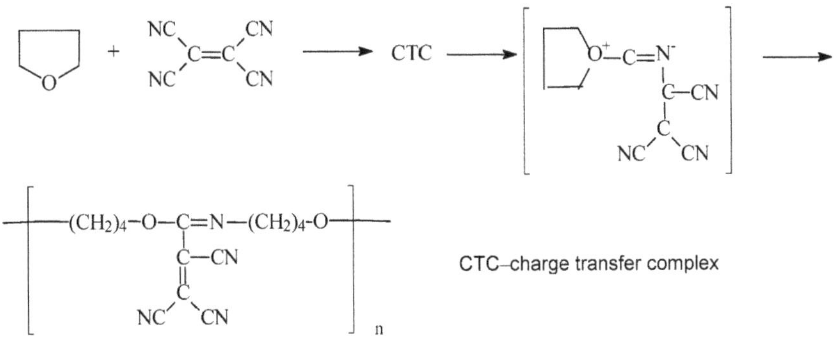

CTC–charge transfer complex

This method can be also used to obtain polymers from 2-methyl-2-oxazoline with molecular mass between 2200 and 18,800.

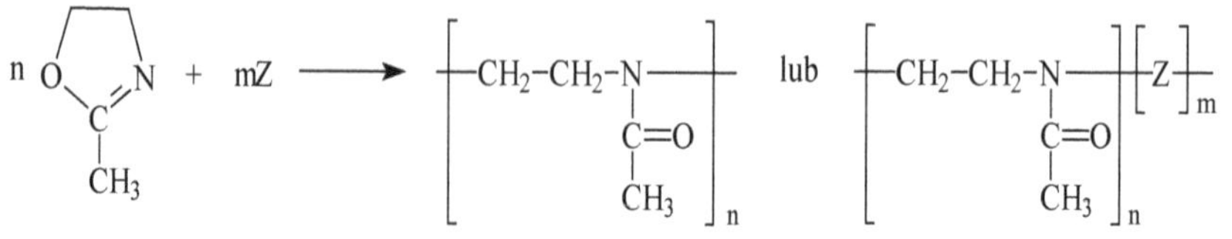

Z—acceptor CTC

4.8   POLYMERIZATION OF CYCLIC IMINES

Cyclic imines polymerize similarly as oxirane and its derivatives. Propyleneimine copolymerize with carbon dioxide without the use of a catalyst. As a result of this reaction, polyurethanes containing 10–35 mol% of urethane groups are formed. The amount depends on the temperature of conducting the process. The structure of the resulting copolymer can be represented by the following formula:

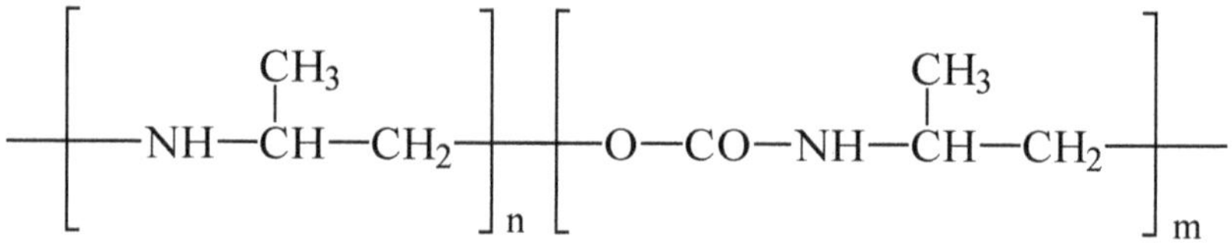

Table 4.10 presentsthe results showing the influence of temperature and the reaction time on the course of copolymerization of propylenimine with carbon dioxide at the molar ratio of reagents 1:5.

From the obtained results, one can conclude that increasing the temperature and prolonging the reaction time results in the increase of the degree of substrates conversion and of the molecular weight of formed polymer, as it follows from the observed increase of the limit viscosity value.

TABLE 4.10   Influence of Temperature and Reaction Time on the Course of Copolymerization of Propylenimine with Carbon Dioxide.

No	Temperature [°C]	Reaction time [H]	Conversion rate	Intrinsic viscosity[g/0.1dm^3]
1	−78	3	0	–
2	−20	3	27.4	–
3	0	3	32.9	0.114
4	0	4.5	90.9	0.238
5	0	16	91.7	0.282

TABLE 4.10   (*Continued*)

6	20	3	58.1	0.285
7	50	0.5	46.7	0.232
8	50	1	52.1	0.208
9	50	3	78.5	0.434
10	50	5	70.1	0.312
11	80	3	75.9	0.528
12	100	3	93.9	0.515

4.9 POLYMERIZATION OF LACTAMS

$$n(CH_2) \left\langle {\overset{\overset{\displaystyle O}{\|}}{\underset{NH}{C}}} \right.$$

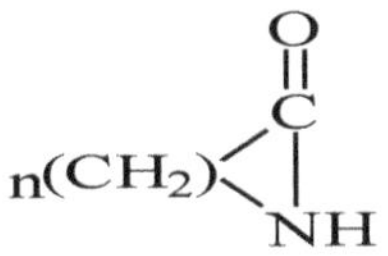

Lactams, similarly as cyclic ethers, can polymerize with ring opening using acidic or alkaline catalysts. The reaction is carried out in an anhydrous medium to prevent the hydrolysis of the ε-aminocarboxylic acid.

4.9.1 CATIONIC POLYMERIZATION OF LACTAMS

During the polymerization reaction of caprolactam in the presence of hydrogen chloride, a tautomeric equilibrium between two forms of cationic caprolactam is fixed: the mesomeric stabilized imide form (A) on one side and the form containing the amide cation form (B) on the other one.

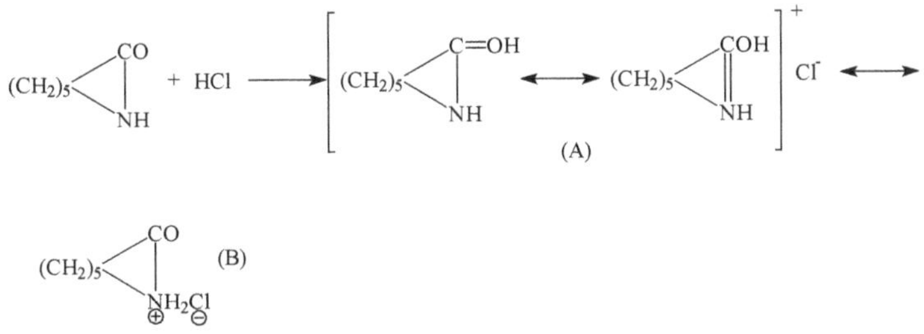

Form (B) is more active because of the absence of mesomeric stabilization. It reacts with the next caprolactam molecule with formation of hydrochloride of amino caprolactam,which initiates a further polymerization:

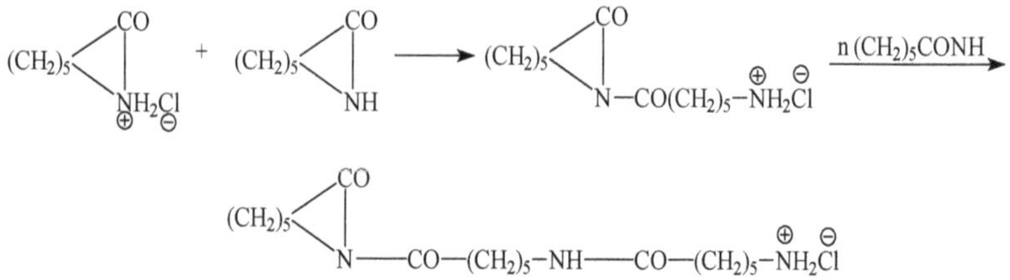

The cationic polymerization of lactams has no practical significance. Industrial application has found the anionic polymerization of lactams.

This method gives from caprolactam, the polyamide with a molecular weight many times higher than that obtained during the polycondensation of e-aminocaproic acid.

4.9.2 ANIONIC POLYMERIZATION OF LACTAMS

Typical catalysts for anionic polymerization of lactams are alkali metals and their hydroxides or alkoxides.

The mechanism of anionic polymerization of lactams has been thoroughly tested on the example of caprolactam. Lactams are very weak acids. Acidity of the hydrogen atom of the lactam amide group is a little greater than the acidity of the hydrogen atom of water or alcohol. For this reason, in the reaction of caprolactam with the metal hydroxide or alkoxide, an equilibrium state is reached:

The reaction of anionic polymerization of lactams is a complex process and can be carried out only in an anhydrous and alcohol-free medium in order to prevent the hydrolysis of lactam.

The mechanism of chain growth during the anionic polymerization of caprolactam is as follows:

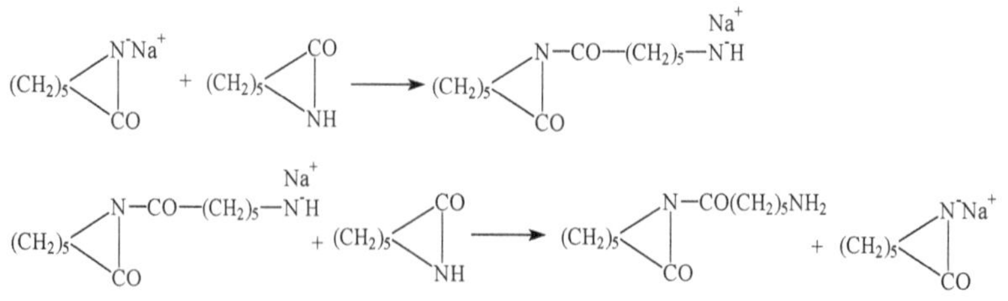

The unstable in alkaline medium acyl caprolactam reacts with caprolactam, acylating it at the expense of ripping its own ring:

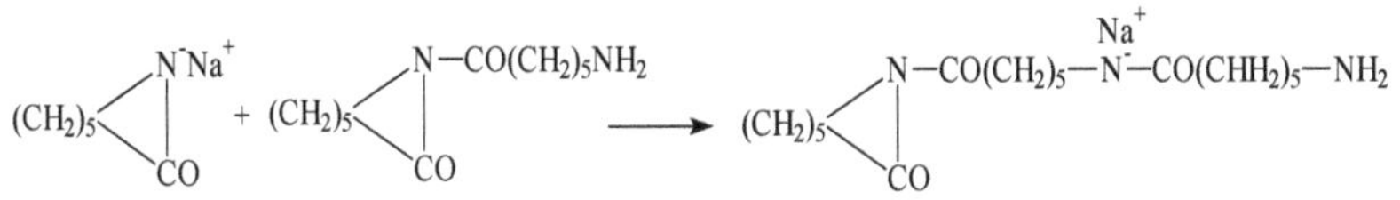

The cycle of sodium cation exchange with the next molecule of caprolactam and its acylation with the rising macroanion is repeated several times until the polymer is obtained.

From the above considerations, it follows that the above-mentioned reaction proceeds through the stages of acyl derivatives. This is confirmed by experimental data, which show that the added to the system chlorides and acid anhydrides are activators of the process, because they shorten the induction time and allow the reaction to run at lower temperature. The anionic polymerization of caprolactam without using an activator is carried out at temperatures above 190 °C.

Polymerization of five- and six-member lactam rings is difficult. However, under the influence of catalysts, one can obtain polymers at temperatures below 60 °C. For example, polyamide-4 is obtained by anionic polymerization of 2-pyrrolidone (butyrlactam), catalyzed by silicon tetrachloride.

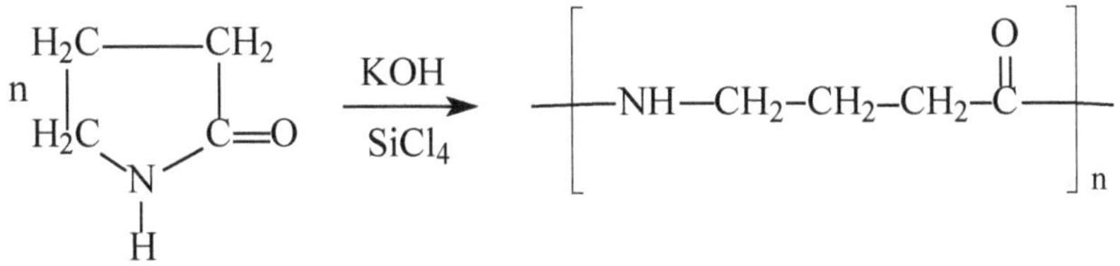

The substituents in the lactam ring, particularly those connected to the nitrogen atom, contribute in reducing the activity of monomer.

4.10 POLYMERIZATION OF CYCLIC SILOXANES

During the hydrolysis of organic derivatives of dichlorosilane, beside the linear polyorganosiloxanes, cyclic siloxanes with six- or more members are also formed. The cyclic siloxanes polymerize easily by ionic mechanism, both by cationic and anionic, with formation of polyorganosiloxanes of large molecular weights.

4.10.1 CATIONIC POLYMERIZATION OF CYCLIC SILOXANES

The cationic polymerization of cyclic siloxanes is usually performed in the presence of acids—sulfuric, chlorosulfon, phosphoric, or boric—and also in systems containing the Lewis acids such as tin tetrachloride, aluminum chloride, boron trifluoride, and zinc chloride.

The relatively well-studied reaction of cationic polymerization of hexamethylcyclotrisiloxane proceeds along the following scheme:

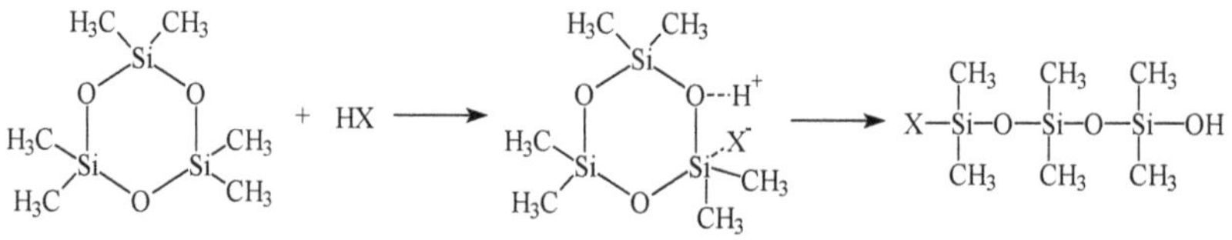

The resulting linear compound reacts with the next molecules of hexamethylcyclotrisiloxane forming a linear polymer.

4.10.2 ANIONIC POLYMERIZATION OF CYCLIC SILOXANES

The cyclic siloxanes polymerize by an anionic mechanism in the presence of bases. The catalytic activity of metal hydroxides in this process decreases in the following order:

$$Ca(OH)_2 > KOH > NaOH > LiOH$$

An example of such a reaction can be the anionic polymerization of hexamethylcyclotrisiloxane in the presence of potassium hydroxide. At the first stage of this process, as a result of the nucleophilic attack of hydroxyl anion on the silicon atom in cyclosiloxane, a transition compound is formed, in which the weakened silicon–oxygen bond of siloxane are broken:

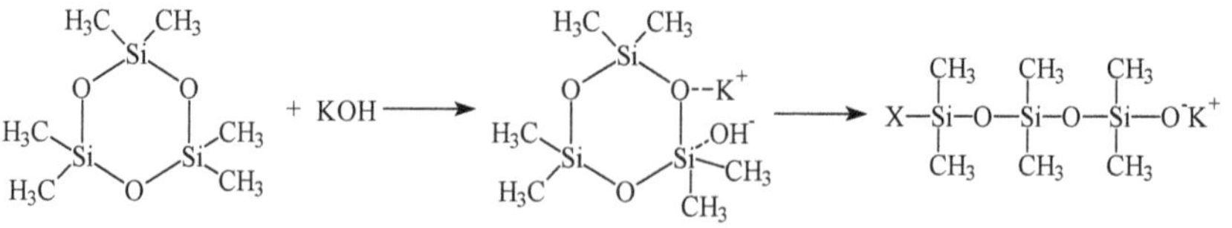

The formed active anion reacts sequentially with the next hexamethylcyclotrisiloxane molecules, causing a rupture of the ring and chain increase with the creation of a new active center at the end of the chain. The reaction proceeds until obtaining a polymer with high molecular weight, which is a "living polymer."

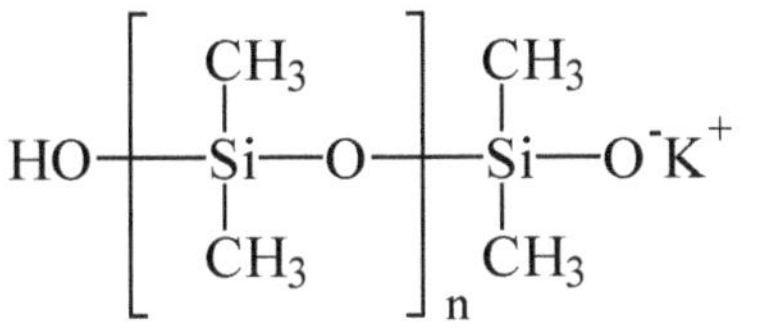

$$\text{HO} - \left[\begin{array}{c} CH_3 \\ | \\ Si - O \\ | \\ CH_3 \end{array} \right]_n \begin{array}{c} CH_3 \\ | \\ Si - O^-K^+ \\ | \\ CH_3 \end{array}$$

4.11 POLYMERIZATION OF ALKYNES (POLYACETYLENE)

Acetylene is the simplest representative of the homologous series of alkynes. The first attempts to carry out the polymerization of acetylene resulted in obtaining low molecular weight compounds. During the transmission of acetylene through the glass tubes, heated to the red heat, acetylene undergo a trimerization process, which results in the formation of benzene:

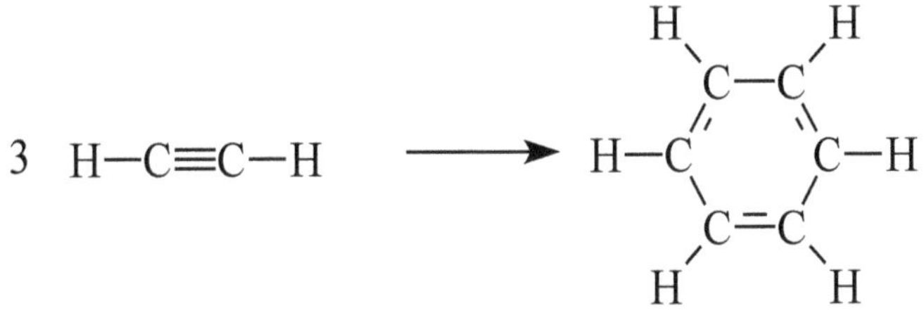

$$3 \; H-C{\equiv}C-H \quad \longrightarrow$$

In the presence of Cu_2Cl_2 and NH_4Cl catalysts at 50 °C, two molecules of acetylene dimerize with the formation of vinylacetylene:

$$2 \; H-C{\equiv}C-H \quad \longrightarrow \quad CH_2{=}CH-C{\equiv}CH$$

It was also found that acetylene passed through a copper sponge polymerize to a mass similar to that of cork. This form didnot find practical significance. Research over the course of the polymerization of acetylene and other alkynes have gained importance after finding in 1977 by H. Shirakawa and C. K. Chiang that the electrical conductivity of polyacetylene, after an appropriate doping, increases by 11–12 orders of magnitude and is comparable to that of metals.

The use of Ziegler–Natta catalysts, modified by Shirakawa, allows synthesis of different species of polyacetylene with an average molecular weight ranging from several to tens of thousands. For this purpose, one most commonly uses the soluble coordination catalysts prepared withtetrabutoxytitanum and triethylaluminum with the chemical formula:

$Ti(OC_4H_9)_4$—$Al(C_2H_5)_3$, **where** the ratio Al:Ti = 4.

Thin films of polyacetylene are obtained on catalysts, which are complexes of titanium triacetylacetonate withtriethylaluminum or chromium triacetylacetonate with triethylaluminum.

During the study of the catalytic activity of metal complexes of metal acetylacetonate complexes with triethylaluminum, the following order was found:

$$Ti, V >> Cr > Fe > Co$$

The temperature at which the acetylene polymerization is performed has a large impact on its course, and more particularly, on the structure of the resulting polyacetylene. Table 4.11 shows that at low temperatures in the overwhelming amount, the "cis" polyacetylene is formed. With increasing polymerization temperature, the amount of "trans" form rises too. At the temperature of 150 °C, only trans-polyacetylene is formed.

TABLE 4.11   Effect of Temperature on the Structure of Polyacetylene Obtained in the Presence of Shirakawa*Catalyst with Concentration of 10 mol/dm3.

Polymerization temperature	Isomer content	
[°C]	cis [%]	trans [%]
−78	98.1	1.9
−18	95.4	4.6
0	78.6	21.4
18	59.3	40.7
50	32.4	67.6
100	7.5	92.5
150	0	100

*Catalyst: Ti(OC4H9)4–Al(C2H5)3, Al/Ti = 4.
Source: Adapted from Ito, T.; Shirakawa, H.; Ikeda, S. J. Polym. Sci. Polym. Chem. Ed. 1974, 12, 11.

The mechanism of acetylene polymerization on coordination catalysts can be illustrated by the following scheme:

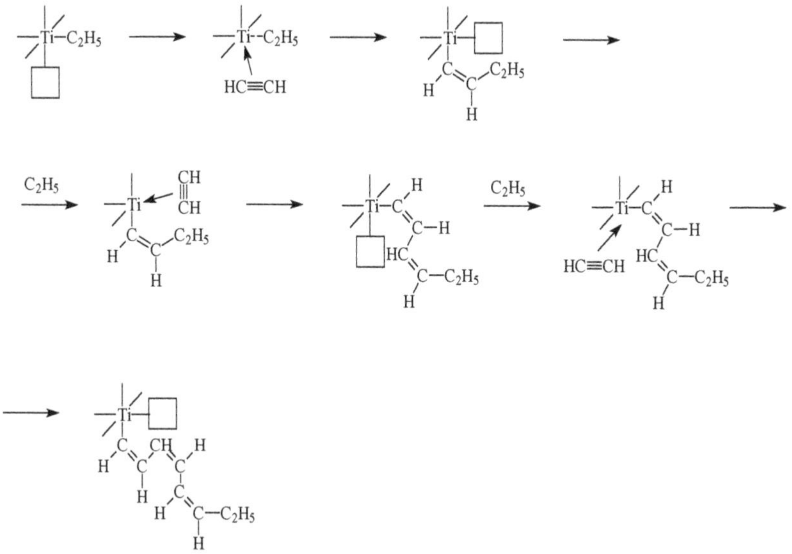

According to this scheme, polyacetylene with cis-transoidal structure only is formed. It is obtained exclusively from the polymerization of acetylene at low temperatures.

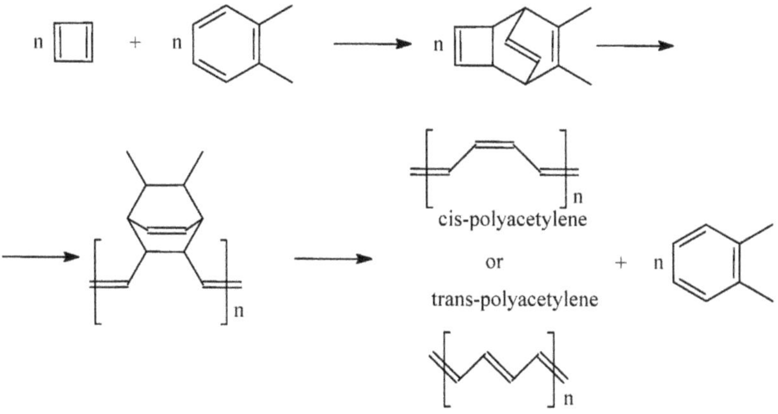

Polyacetylene can be also obtained by applying the Diels–Alder synthesis reaction. One example of the polyacetylene synthesis using this method is the addition of cyclobutadiene to aromatic compounds. The polymerization of the obtained in this way adduct and the thermal elimination of aromatic compounds from the resulting polymer gives polyacetylene. The reaction can be presented as follows:

REFERENCES

Abusleme J.; Giannetti, E. Emulsion Polymeryzation. *Macromolecules* 1991,*24.*

Adesamya I. Classification of Ionic Polymers. *Polym. Eng. Sci.* 1988, *28*, 1473.

Akbulut, U.; Toppare, L.; Usanmaz, A.; Onal, A. Electroinitiated Cationic Polymerization of Some Epoksides. *Macromol. Chem. Rapid. Commun.* 1983, *4*, 259.

Andreas, F.; Grobe, K. *Chemia propylenu;* WNT:Warszawa, 1974.

Barg, E. I. Technologia tworzyw sztucznych; PWT:Warszawa, 1957.

Billmeyer, F. W. *Textbook of Polymer Science*; J. Wiley and Sons: New York, 1984.

Borsig, E.; Lazar, M. Chemical Modification of Polyalkanes. *Chem. Listy* 1991, *85*, 30.

Braun, D.; Czerwiński, W.; Disselhoff, G.; Tudos, F. Analysis of the Linear Methods for Determining Copolymerization Reactivity Ratios. *Angew. Makromol. Chem.* 1984, *125*, 161.

Burnett, G. M.; Cameron, G. G.;Gordon, G.;Joiner, S. N. Solvent Effects on the Free-Radical Polymeryzation of Styrene. *J. Chem. Soc. Faraday Trans.* 1 1973, *69*, 322.

Challa, R. R.;;Drew, J. H.; Stannet, V. T.; Stahel, E. P. Radiation-Induced Emulsion Polymerization of Vinyl Acetate. *J. Appl. Polym. Sci.* 1985, *30*, 4261.

Candau, F. Mechamism and Kinetics of the Radical Polymerization of Acrylamide in Inverse Micelles. *Makromol. Chem., Makromol Symp.* 1990, *31*, 27.

Ceresa, R. J. :*Kopolimery blokowe i szczepione*; WNT:Warszawa, 1965.

Charlesby, A. *Chemia radiacyjna polimerów*; WNT:Warszawa, 1962.

Chern, C. S.; Poehlein, G. W. Kinetics of Cross-Linking Vinyl Polymerization. *Polym. Plast. Technol. Eng.* 1990, *29*, 577.

Chiang, W. Y.; Hu, C. M. Studies of Reactions with Polymers. *J. Appl. Polym. Sci.* 1985, *30*, 3895–4045.

Chien, J. C.; Salajka, Z. Syndiospecific Polymeryzation of Styrene. *J. Polym. Sci. A*1991, *29*, 1243.

Chojnowski, J.; Stańczyk, W. E. Procesy Chlorowania Polietylenu. *Polimery*1972, *17*, 597.

Collomb, J.; Morin, B.; Gandini, A.; Cheradame, H. Cationic Polymeryzation Induced by Metal Salts. *Europ. Polym. J.* 1980, *16*, 1135.

Czerwiński, W. K. Solvent Effects on Free-Radical Polymerization. *Makromol. Chem.* 1991, *192*, 1285–1297.

Dainton, F. S. *Reakcje Lańcuchowe*; PWN: Warszawa, 1963.

Ding, J.; Price, C.; Booth, C. Use of Crown Ether in the Anionic Polymeryzation of Propylene Oxide. *Eur. Polym. J.* 1991, *27*, 895–901.

Dogatkin, B. A. *Chemia elastomerów*; WNT:Warszawa, 1976.

Duda, A.; Penczek, S. Polimeryzacja jonowa i pseudojonowa. *Polimery*1989, *34*, 429.

Dumitrin, S.; Shaikk, A. S.; Comanita, E.; Simionescu, C. Bifunetional Initiators. *Eur. Polym. J.* 1983, *19*, 263.

Dunn, A. S. Problems of Emulsion Polymeryzation. *Surf. Coat. Int.* 1991, *74*, 50.

Eliseeva, W. I.; Asłamazova, T. R. Emulsionnaja polimeryzacja v otsustivie emulgatora. *Uspiechi. Chimii.* 1991, *60*, 398.

Erusalimski, B. L. *Mechanism dezaktivaci rastuszczich ciepiej v procesach ionnoj polimeryzacji vinylovych polimerov. Vysokomol. Soed. A.* 1985, *27*, 1571.

Fedtke, M. *Chimiczieskije reakcji polimerrov*; Khimiya: Moskwa, 1990.

Fernandez-Moreno, D.; Fernandez-Sanchez, C.; Rodriguez, J. G.; Alberola, A. Polymeryzation of Styrene and 2-vinylpyridyne Using AlEt-VCl Heteroaromatic Bases. *Eur. Polym. J.* 1984, *20*, 109.

Fieser, L. F.; Fieser, M. *Chemia organiczna;* PWN:Warszawa, 1962.

Flory, P. J. Nauka o makrocząsteczkach. *Polimery* 1987, *32*, 346.

Frejdlina, R. Ch.; Karapetian, A. S. *Telomeryzacja*; WNT: Warszawa, 1962.

Funt, B. L.; Tan, S. R. The Photoelectrochemical Initiation of Polymeryzation of Styrene. *J. Polym. Sci. A* 1984, *22*, 605.

Gerner, F. J.; Hocker, H.; Muller, A. H.; Shulz, G. V. On the Termination Reaction in the Anionic Polymeryzation of Methyl Methacrylate in Polar Solvents. *Eur. Polym. J.* 1984, *20*, 349.

Ghose, A.; Bhadani, S. N. Electrochemical Polymeryzation of Trioxane in Chlorinated Solvents. *Indian J. Technol.* 1974, *12*, 443.

Giannetti, E.; Nicoletti, G. M.; Mazzocchi, R. Homogenous Ziegler-Natta Catalysis. *J. Polym. Sci. A*1985, *23*, 2117.

Goplan, A.; Paulrajan, S.; Venkatarao, K.; Subbaratnam, N. R. Polymeryzation of N,N'- methylene Bisacrylamide Initiated by Two New Redox Systems Involving Acidic Permanganate. *Eur. Polym. J.* 1983, *19*, 817.

Gózdz, A. S.; Rapak, A. Phase Transfer Catalyzed Reactions on Polymers. *Makromol. Chem. Rapid Commun.* 1981, *2*, 359.

Greenley, R. Z. An Expanded Listing of Revized Q and e Values. *J. Macromol. Sci. Chem. A*1980, *14*, 427.

Greenley, R. Z. Recalculation of Some Reactivity Ratios. *J. Macromol. Sci. Chem. A*1980, *14*, 445.

Guhaniyogi, S. C.; Sharma, Y. N. Free-radical Modyfication of Polipropylene. *J. Macromol. Sci. Chem. A*1985, *22*, 1601.

Guo, J. S.; El Aasser, M. S.; Vanderhoff, J. W. Microemulsion Polymerization of Styrene. *J. Polym. Sci. A*1989, *27*, 691.

Gurruchaga, M.; Goni, I.; Vasguez, M. B.; Valero, M.; Guzman, G. M. An Approach to the Knowledge of the Graft Polymerization of Acrylic Monomers onto Polysaccharides using Ce (IV) as Initiator. *J. Polym. Sci. Part C*1989, *27*, 149.

Hasenbein, N.; Bandermann, F. Polymeryzation of Isobutene With VOCl$_3$ in Heptane. *Makromol. Chem.* 1987, *188*, 83.

Hatakeyama, H. et al. Biodegradable Polyurethanes from Plant Components. *J. Macromol. Sci. A.* 1995, *32*, 743.

Higashimura, T.; Law, Y. M.; Sawamoto, M. Living Cationic Polymerization. *Polym. J.* 1984, *16*, 401.

Hirota, M.; Fukuda, H. Polymerization of Tetrahydrofuran. *Nagoya-ski Kogyo Kenkynsho Kentyn Hokoku*1984, *51*.

Hodge, P.; Sherrington, D. C. *Polymer-Supported Reactions in Organic Syntheses;*J. Wiley and Sons: New York, 1980; tłum. na rosyjski Mir: Moskwa, 1983.

Inoue, S. Novel Zine Carboxylates as Catalysts for the Copolymeryzation of CO$_2$with Epoxides. *Makromol. Chem. Rapid Commun.* 1980, *1*, 775.

Janović, Z. *Polimerizacije I Polimery*; Izdanja Kemije Ind. : Zagrzeb, 1997.

Jedliński, Z. J. *Współczesne kierunki w dziedzinie syntezy polimerów. Wiadomości Chem.* 1977, *31*, 607.

*Reactivity, Mechanism and Structure in Polymer Chemistry;*Jenkins, A. D., Ledwith, A., Ed.; J. Wiley and sons: London, 1974; tłum. na rosyjski, Mir: Moskwa, 1977.

Joshi, S. G.; Natu, A. A. Chlorocarboxylation of Polyethylene. *Angew. Makromol. Chem.* 1986, *140*, 99–115.

Kaczmarek, H. Polimery, a Środowisko. *Polimery*1997, *42*, 521.

Kang, E. T.; Neoh, K. G. Halogen Induced Polymerization of Furan. *Eur. Polym. J.* 1987, *23*, 719.

Kang, E. T.; Neoh, K. G.; Tan, T. C.; Ti, H. C. Iodine Induced Polymerization and Oxidation of Pryridazine. *Mol. Cryst. Lig. Cryst.* 1987, *147*, 199.

Karnojitzki, V. *Organiczieskije pierekisy*; I. I. L. : Moskwa, 1961.

Karpiński, K.; Prot, T. Inicjatory fotopolimeryzacji i fotosieciowania. *Polimery*1987, *32*, 129.

Keii, T. Propene Polymerization with $MgCl_2$-Supported Ziegler Catalysts. *Makromol. Chem.* 1984, *185*, 1537.

Khanna, S. N.; Levy, M.; Szwarc, M. Complexes Formed by Anthracene with 'Living' Polystyrene. *Dormant Polym. Trans. Faraday Soc.* 1962,*58*, 747–761.

Koinuma, H.; Naito, K.; Hirai, H. Anionic Polymerization of Oxiranes. *Makromol. Chem.* 1982, *183*, 1383.

Korniejev, N. N; Popov, A. F; Krencel, B. A. *Komplieksnyje metalloorganiczeskije katalizatory*; Khimia: Leningrad, 1969.

Korszak, W. W. Niekotoryje Problemy Polikondensacji. *Vysokomol. Soed. A*1979, *21*, 3.

Korszak, W. W. Technologia tworzyw sztucznych; WNT: Warszawa, 1981.

Kowalska, E.; Pełka, J. Modyfikacja tworzyw termoplastycznych włóknami celulozowymi. *Polimery*2001, *46*, 268.

Kozakiewicz, J.; Penczek, P. Polimery jonowe. *Wiadomości Chem.* 1976, *30*, 477.

Kubisa, P. Polimeryzacja Żyjąca. *Polimery*1990, *35*, 61.

Kubisa, P.; Neeld, K.; Starr, J.; Vogl, O. Polymerization of Higher Aldehydes. *Polymer*1980, *21*, 1433.

Kucharski, M. Nitrowe i aminowe pochodne polistyrenu. *Polimery*1966, *11*, 253.

Kunicka, M.; Choć, H. J. Hydrolitic Degradation and Mechanical Properties of Hydrogels. *Polym. Degrad. Stab.* 1998, *59*, 33.

Lebduska, J.; Dlaskova, M. Roztokova kopolymerace. *Chemicky Prumysl*1990, *40*, 419–480.

Leclerc, M.; Guay, J.; Dao, L. W. Synthesis and Characterization of Poly(alkylanilines). *Macromolecules*1989, *22*, 649.

Lee, D. P.; Dreyfuss, P. Triphenylphesphine Termination of Tetrahydrofuran Polymerization. *J. Polym. Sci.,Polym. Chem. Ed.* 1980, *18*, 1627.

Leza, M. L.; Casinos, I.; Guzman, G. M.; Bello, A. Graft Polymerization of 4-vinylpyridineOnto Cellulosic Fibres by the Ceric on Method. *Angew. Makromol. Chem.* 1990, *178*, 109–119.

Lopez, F.; Calgano, M. P.; Contrieras, J. M.; Torrellas, Z.; Felisola, K. Polymerization of Some Oxiranes. *Polym. Int.* 1991, *24*, 105.

Lopyrev, W. A.; Miaczina, G. F.; Szevaleevskii, O. I.; Hidekel, M. L. Poliacetylen. *Vysokomol. Soed.* 1988, *30*, 2019.

Mano, E. B.; Calafate, B. A. L. Electrolytically Initiated Polymerization of N-vinylcarbazole. *J. Polym. Sci., Polym. Chem. Ed.* 1983, *21*, 829.

Mark, H. F. *Encyclopedia of Polymer Science and Technology*; Concise 3rd ed., WileyInterscience: Hoboken, New York, 2007.

Mark, H.;Tobolsky, A. V. *Chemia fizyczna polimerów*; PWN: Warszawa, 1957.

Matyjaszewski, K.; Davis, T. P. *Handbook of Radical Polymerization*; WileyInterscience: New York, 2002.

Matyjaszewski,K.; Mülle, A. H. E. 50 years of Living Polymerization. *Prog. Polym. Sci.* **2006**, *31*, 1039–1040.

Mehrotra, R.; Ranby, B. Graft Copolymerization onto Starch. *J. Appl. Polym. Sci.* 1977, *21*, 1647.

Morrison R. T., Boyd N. *Chemia organiczna*; PWN:Warszawa, 1985.

Morton, M. *Anionic Polymerization*; Academic Press: New York, 1983.

Munk, P. *Introduction to Macromolecular Science*; JohnWiley and Sons: New York, 1989.

Naraniecki, B. Ciecze mikroemulsyjne. *Przem. Chem.* 1999, *78(4)*, 127.

Neoh, K. G.; Kang, E. T.; Tan, K. L. Chemical Copolymerization of Aniline with Halogen-Substituted Anilines. *Eur. Polym. J.* 1990, *26*, 403.

Nowakowska, M.; Pacha, J. Polimeryzacja olefin wobec kompleksów metaloorganicznych, *Polimery* 1978, *23*, 290.

Ogata, N.; Sanui, K.; Tan, S. Synthesis of Alifatic Polyamides by Direct Polycondensation with Triphenylphosphine. *Polym. J.* 1984, *16*, 569.

Onen, A.;Yagci, Y. Bifunctional Initiators. *J. Macromol. Sci. Chem. A*1990, *27*, 743, *Angew. Makromol. Chem.* 1990, *181*, 191.

Osada, Y. Plazmiennaja polimerizacia i plazmiennaja obrabotka polimerov. *Vysokomol. Soed. A*1988, *30*, 1815.

Pasika, W. M. Cationic Copolymerization of Epichlorohydrin with Styrene Oxide and Cyclohexene Oxide. *J. Macromol. Sci. Chem. A*1991, *28*, 545.

Pasika, W. M. Copolymerization of Styrene Oxide and Cyklohexene Oxide. *J. Polym. Sci. Part A*1991, *29*, 1475.

Pielichowski, J.; Puszyński, A. *Preparatyka monomerów*; W. P. Kr. : Kraków, 2001.

Pielichowski, J.; Puszyński, A. *Preparatyka polimerów*; TEZA WNT:Kraków, 2005.

Pielichowski, J.; Puszyński, A. *Technologia tworzyw sztucznych*; WNT:Warszawa, 2003.

Pielichowski, J.; Puszyński, A. *Wybrane działy z technologii chemicznej organicznej;* W. P. Kr. : Kraków.

Pistola, G.; Bagnarelli, O. Electrochemical Bulk Polymeryzation of Methyl Methacrylate in the Presence of Nitric Acid. *J. Polym. Sci., Polym. Chem. Ed.* 1979, *17*, 1002.

Pistola, G.; Bagnarelli, O.; Maiocco, M. Evaluation of Factors Affecting the Radical Electropolymerization of Methylmethacrylate in the Presence of HNO_3. *J. Appl. Electro. Chem.* 1979, *9*, 343.

Połowoński, S. Techniki pomiarowo-badawcze w chemii fizycznej polimerów; W. P. L. : Łódź, 1975.

Porejko, S.; Fejgin, J.; Zakrzewski, L. *Chemia związków wielkocząsteczkowych*; WNT: Warszawa, 1974.

Praca zbiorowa: Chemia plazmy niskotemperaturowej; WNT: Warszawa, 1983.

Puszyński, A. Chlorination of Polyethylene in Suspension. *Pol. J. Chem.* 1981, *55*, 2143.

Puszyński, A.; Godniak, E. Chlorination of Polyethylene in Suspension in the Presence of Heavy Metal Salts. *Macromol. Chem. Rapid Commun.* 1980, *1*, 617.

Puszyński, A.; Dwornicka, J. Chlorination of Polypropylene in Suspension. *Angew. Macromol. Chem.* 1986, *139*, 123.

Puszyński, J. A.; Miao, S. Kinetic Study of Synthesis of SiC Powders and Whiskers in the Presence $KClO_3$ and Teflon. *Int. J. SHS*1999, *8(8)*, 265.

Rabagliati, F. M.; Bradley, C. A. Epoxide Polymerization. *Eur. Polym. J.* 1984, *20*, 571.

Rabek, J. F. Podstawy fizykochemii polimerów; W. P. Wr. : Wrocław, 1977.

Rabek, T. I. *Teoretyczne podstawy syntezy polielektrolitów i wymieniaczy jonowych;*PWN: Warszawa, 1962.

Regas, F. P.; Papadoyannis, C. J. Suspension Cross-linking of Polystyrene with Friedel-Crafts Catalysts. *Polym. Bull.* 1980, *3*, 279.

Rodriguez, M.; Figueruelo, J. E. : On the Anionic Polymeryzation Mechanism of Epoxides. *Makromol. Chem.* 1975, 176, 3107.

Roudet, J.;Gandini, A. Cationic Polymerization Induced by Arydiazonium Salts. *Makromol. Chem. Rapid Commun.* 1989, *10*, 277.

Sahu, U. S.; Bhadam, S. N. Triphenylphosphine Catalyzed Polimerization of Maleic Anhydride. *Makromol. Chem.* 1982, *183*, 1653.

Schidknecht, C. E. *Polimery Winylowe;* PWT: Warszawa, 1956.

Szwarc, M.; Levy, M.; Milkovich, R. Polymerization Initiated by Electron Transfer to Monomer. A new Method of Formation of Block Copolymers. *J. Am. Chem. Soc.* **1956,** 78, 2656–2657.

Szwarc, M., 'Living' Polymers. *Nature*1956, *176*, 1168–1169.

Sen, S.; Kisakurek, D.;Turker, L.;Toppare, L.; Akbulut, U. Elektroinitiated Polymerization of 4-Chloro-2,6-Dibromofenol. *New Polym. Mater.* 1989, *1*, 177.

Sikorski, R. T. : Chemiczna modyfikacja polimerów. *Prace Nauk. Inst. Technol. Org. Tw. Szt. P. Wr.* 1974, *16*, 33.

Sikorski, R. T. Podstawy chemii i technologii polimerów. PWN: Warszawa, 1984.

Sikorski, R. T.; Rykowski, Z.; Puszyński, A. Elektronenempfindliche Derivate von Poly-methyl-methacrylat. *Plaste und Kautschuk*1984, *31*, 250.

Simionescu, C. I.; Geta, D.; Grigoras, M. Ring-Opening Isomerization Polymeryzation of 2-Meth-yl-2-Oxazoline Initiated by Charge Transfer Complexes. *Eur. Polym. J.* 1987, *23*, 689.

Sheinina, L. S.; Vengerovskaya, Sh. G.; Khramova, T. S.; Filipowich, A. Y. Reakcji piridinov i ich czietvierticznych soliej s epoksidnymi soedineniami. *Ukr. Khim. Zh.* 1990, *56*, 642.

Soga, K.; Toshida, Y.; Hosoda, S.; Ikeda, S. A Convenient Synthesis of a Polycarbonate. *Makromol. Chem.* 1977, *178*, 2747.

Soga, K.; Hosoda, S.; Ikeda, S. A New Synthetic Route to Polycarbonate. *J. Polym. Sci.,Polym. Lett. Ed.* 1977, *15*, 611.

Soga, K.; Uozumi, T.; Yanagihara, H.; Siono, T. Polymeryzation of Styrene with Heterogeneous Ziegler-Natta Catalysts. *Makromol. Chem. Rapid. Commun.* 1990, *11*, 229.

Soga, K.; Shiono, T.; Wu, Y.;Ishii, K.; Nogami, A.; Doi, Y. Preparation of a Stable Supported Cata-lyst for Propene Polymeryzation. *Makromol. Chem. Rapid Commun.* 1985, *6*, 537.

Sokołov, L. B.; Logunova, W. I. Otnositielnaja reakcionnosposobnost monomerov i prognozirovan-ic polikondensacji v emulsionnych sistemach. *Vysokomol. Soed. A.* 1979, *21*, 1075.

Soler, H.; Cadiz, V.; Serra, A. Oxirane Ring Opening with Imide Acids. *Angew. Makromol. Chem.* 1987, *152*, 55.

Spasówka, E.; Rudnik, E. Możliwości wykorzystania węglowodanów w produkcji biodegradowal-nychtworzyw sztucznych. *Przemysł Chem.* 1999, *78*, 243.

Stevens, M. P. Wprowadzenie do chemii polimerów; PWN: Warszawa 1983

Stępniak, I; Lewandowski, A. Elektrolity polimerowe. *Przemysł. Chem.* 2001, *80(9)*, 395.

Strohriegl, P.; Heitz, W.; Weber, G. Polycondensation Using Silicon Tetrachloride. *Makromol. Chem. Rapid Commun.* 1985, *6*, 111.

Szur, A. M. *Vysokomolekularnyje soedinenia;* Vysszaja Szkoła: Moskwa, 1966.

Tagle, L. H.; Diaz, F. R.; Riveros, P. E. Polymeryzation by Phase Transfer Catalysis. *Polym. J.* 1986, *18*, 501.

Takagi, A.; Ikada, E.; Watanabe, T. Dielectric Properties of Chlorinated Polyethylene and Reduced Polyvinylchloride, Memoirs of the Faculty of Eng. *Kobe Univ.* 1979, *25*, 239.

Tani, H. Stereospecific Polymeryzation of Aldehydes and Epoxides. *Adv. Polym. Sci.* 1973, *11*, 57.

Vandenberg, E. J. Coordynation Copolymeryzation of Tetrahydrofuran and Oxepane with Ox-etanes and Epoxides. *J. Polym. Sci. Part A*1991, *29*, 1421.

Veruovic, B. *Navody pro laboratorni cviceni z chemie polymeru;* SNTL: Praha, 1977.

Vogdanis, L.; Heitz, W. Carbon Dioxide as a Mmonomer. *Makromol. Chem. Rapid Commun.* 1986, *7*, 543.

Vollmert, B. *Grundriss der Makromolekularen Chemic*; Springer-Verlag: Berlin, 1962.

Wei, Y.; Tang, X.; Sun, Y. A study of the Mechanism of Aniline Polymeryzation. *J. Polym. Sci. Part A*1989, *27*, 2385.

Wei, Y.; Sun, Y.; Jang, G. W.; Tang, X. Effects P-Aminodiphenylamine on Electrochemical Polymeization of Aniline. *J. Polym. Sci., Part C*1990, *28*, 81.

Wei, Y.; Jang, G. W. Polymerization of Aniline and Alkyl Ringsubstituted Anilines in the Presence of Aromatic Additives. *J. Phys. Chem.* 1990, *94*, 7716.

Wiles, D. M. Photooxidative Reactions of Polymers. *J. Appl. Polym. Sci., Appl. Polym. Symp.* 1979, *35*, 235.

Wilk, K. A.; Burczyk, B.; Wilk, T. Oil in Water Microemulsions as Reaction Media. In *7th International Conference Surface and Colloid Science;* Compiegne, 1991.

Winogradova, C. B. Novoe v obłasti polikondensacji. *Vysokomol. Soed. A*1985, *27*, 2243.

Wirpsza, Z. *Poliuretany*; WNT: Warszawa, 1991.

Wojtala, A. Wpływ właściwości oraz otoczenia poliolefin naprzebieg ich fotodegradacji. *Polimery* 2001, *46*, 120.

Xue, G. Polymerization of Styrene Oxide with Pyridine. *Makromol. Chem. Rapid Commun.* 1986, *7*, 37.

Yamazaki, N.; Imai, Y. Phase Transfer Catalyzed Polycondensation of Bishalomethyl Aromatic compounds. *Polym. J.* 1983, *15*, 905.

Yasuda, H. *Plasma Polymerization*; Academic Press, Inc. : Orlando, 1985.

Yuan, H. G.; Kalfas, G.; Ray, W. H. Suspension Polymerization. *J. Macromol. Sci. Rev. Macromol. Chem. Phys.* 1991, *C(31)*, 215.

Zubakowa, L. B.; Tevlina, A. C.; Davankov, A. B. *Sinteticzeskije ionoobmiennyje materiały*; Chimia: Moskwa, 1978.

Zubanov, B. A. Viedienie v chimiu polikondensacionnych procesov, Nauka: Ałma Ata, 1974.

Żuchowska, D. *Polimery konstrukcyjne*; WNT: Warszawa, 2000.

Żuchowska, D. Struktura i właściwości polimerów jako materiałów konstrukcyjnych, W. P. Wr. : Wrocław, 1986.

Żuchowska, D.; Steller, R.; Meisser, W. Structure and Properties of Degradable Polyolefin–Starch Blends. *Polym. Degrad. Stab.* 1998, *60*, 471.

CHAPTER 5

TECHNICAL HINTS ONPOLYCONDENSATION

G. E. ZAIKOV

Russian Academy of Sciences, Moscow, Russia

CONTENTS

5.1 PROCESS OVERVIEW

The condensation polymerization, often called as polycondensation, is a condensation reaction of a large number of monomer molecules or comonomers to the polycondensate macromolecules, during which water, hydrogen chloride, ammonia, and other simple compounds are released as byproducts.

In the polycondensate macromolecules,the main chain is formed not only of carbon atoms but it also includes atoms of other elements such as oxygen, nitrogen, phosphorus, boron, or silicon. Polycondensation is a special case of substitution reaction, in which multifunctional molecules react with each other and the equilibrium conditions do not interfere with the formation of long-chain molecules.

In contrast to the addition polymerization, the condensation polymerization is a gradual reaction. At all stages of this process,the transitional and the permanent products are formed, which can be well distinguished. If in the reaction, two types of molecules having at least two functional groups are involved, the reaction is called heteropolycondensation, whereas when reactionis among molecules of the same type (e.g.,hydroxyl acid molecules of type HO–(CH2) x–COOH), the process is called the homopolycondensation.

The increase of the macromolecule chain is slow. The kinetics of this process depends on the temperature, the rate of removal of low molecular weight byproducts and the amount and the nature of the catalyst (usually on the hydrogen ion concentration). Similarly, as in the case of normal condensation, during the polycondensation, an equilibrium is reached for each stage of the reaction. As it is generally known, the equilibrium constant is expressed as the ratio of the product of polymerization products to the product of substrates concentrations. Since the reaction of polycondensation is fixed in the equilibrium state,one takes the averaged equilibrium constant.

Assuming that the value of the equilibrium constant (K) does not change during the process duration, the equilibrium constant is given by the following equation:

$$K = \frac{(N_0 - N)N_a}{N^2},$$

where

N_a—number ofsmall molecular mass molecules of byproduct at the equilibrium state,

N_0—initial number of difunctional monomer molecules particles equal to the initial number of functional groups X and Y,

N—number of formed macromolecules, corresponding to the number of functional groups X or Y in equilibrium state,
(N_0-N)—number of new bonds created and simultaneously the number of bonds extinct,

$$\frac{N_0 - N}{N_0} = n_z$$

-is the mole fraction of bonds per the basic polymer segment, which, at the same time, is also the degree of conversion

$$n_a = \frac{N_a}{N_0}$$

- mole fraction of small molecule compound per the polymerin the equilibrium state

$$\frac{N_0}{N} = \overline{P}$$

- average degree of polymerization
Dividing the numerator and the denominator in the above equation for the equilibrium constant by N02 one gets

$$K = \frac{\left(\dfrac{N_0 - N}{N_0}\right)\dfrac{N_a}{N_0}}{\left(\dfrac{N}{N_0}\right)^2} = \frac{n_z n_a}{\left(\dfrac{1}{\overline{P}}\right)^2}$$

and after its transformation

$$\overline{P} = \sqrt{\frac{K}{n_a}\frac{1}{n_z}}\ .$$

With the high degree of polycondensation, the ratio (N/N0) is so small that it can be neglected, and then

$$\overline{P} = \sqrt{\frac{K}{n_a}}.$$

This means that the average degree of condensation polymerization is directly proportional to the square root of the equilibrium constant and inversely proportional to the square root of the molar fraction of the secreting small molecule compound. Therefore, the removal of byproduct during the reaction always increases the average degree of polycondensation.

An important parameter in the polycondensation process is the degree of conversion, given by the following formula:

$$\alpha = \frac{2(N_0 - N)}{N_0 f} = \frac{2}{f} - \frac{2}{\overline{P} f},$$

where f—functionality of monomer.

When the value of P is large then $\alpha = 2/f$.

In the case when one of the molecules of substances involved in the polycondensation reactions contain more than two functional groups, the resulting products may undergo crosslinking under the influence of temperature. It means they become thermosetting.

The value of the equilibrium constant K allows to determine the course of the polycondensation reaction. The increase in the value of equilibrium constant K results in a higher polymer yield. This issue can be illustrated by the synthesis of polyesters and urea–formaldehyde resins. The polyesterification reaction is a reaction that is typically reversible and the establishing state of equilibrium is unfavorable for the emerging polyester.

As the application in excess of one of the substrates leads to obtaining a compound of small molecular weight, the only way to shift the balance in favor of the emerging polyester, according to the Le Chatelier rule, is the removal of water as a small molecule by product. In that case, it is possible to obtain polyester of high molecular weight.

Other examples of polycondensation are the reactions of hydroxymethyl derivatives of urea or phenol. These reactions are characterized by a relatively high value of equilibrium constant also allowing the synthesis of polymer in the aquatic environment. In this case, the water removal from the reaction medium serves the synthesis of a product with higher molecular weight.

Overall, it was assumed that the polycondensation reactions, in which the equilibrium constant is less than 103, belong to the typical equilibrium reactions. If the value of the equilibrium constant is larger than 103, then such reactions are considered as unidirectional and the process is called the nonequilibrium polycondensation. An example of such a reaction is the synthesis of polyesters from glycols and dicarboxylic acid chlorides carried out in an alkaline environment:

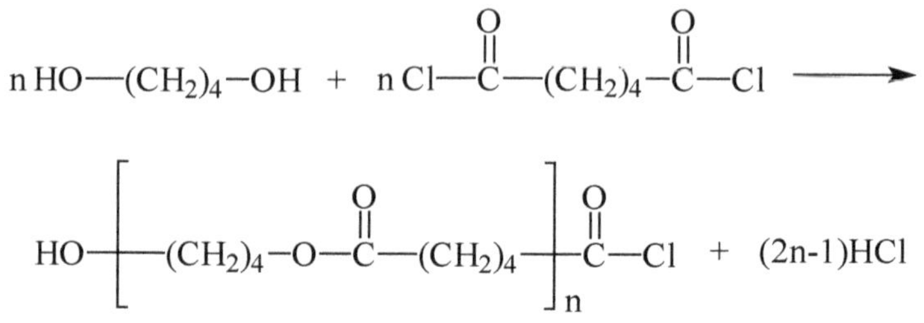

$$n\ HO-(CH_2)_4-OH\ +\ n\ Cl-\overset{\overset{O}{\|}}{C}-(CH_2)_4-\overset{\overset{O}{\|}}{C}-Cl \longrightarrow$$

$$HO\left[-(CH_2)_4-O-\overset{\overset{O}{\|}}{C}-(CH_2)_4-\right]_n \overset{\overset{O}{\|}}{C}-Cl\ +\ (2n\text{-}1)HCl$$

5.2 KINETICS OF POLYCONDENSATION

The polycondensation kinetics calculations are based on the assumption that the reactivity of the functional groups does not depend on the molecular weight of molecules in which such groups exist. This assumption was experimentally verified and confirmed.

In a classical polycondensation process, the monomers are bifunctional, used in 1:1 molar ratio of reactants.

The measure of determining the reaction rate may be the disappearance of functional groups of one of the reactants.

The general scheme of the polycondensation reaction is as follows:

$$n\ X-A-X\ +\ Y-B-Y \rightleftharpoons X\left[-A-B-\right]_n Y\ +\ (2n\text{-}1)XY$$

When using a catalyst, the process is a second-order reaction, and its rate V can be described by the following formula:

$$v = -\frac{d[X]}{d} = k[X][Y]$$

assuming that

$$[X]=[Y]=c$$

$$v = -\frac{d}{d} = k^2$$

After integration, the equation becomes

$$\kappa t = \frac{1}{c} + const$$

Because the degree of conversion s at time t is given by

$$s = \frac{c_0 - c}{c_0} \text{ then } c = c_0(1 - s),$$

where
c0—initial concentration of functional groups,
c—concentration of functional groups at the time t.

Therefore, the equation of the polycondensation reaction rate takes the following form:

$$\kappa t = \frac{1}{c_0(1 - s)} + const$$

$$ktc_0 = \frac{1}{1 - s} + const$$

If the polycondensation reaction proceeds without a catalyst and the functional groups X have catalytic properties, as it is the case in polyesterification reaction, the process may run as a second order with one respect to X.
Then

$$\frac{d[X]}{dt} = k[X]^2[Y]$$

namely

$$\frac{dc}{dt} = kc^3$$

after integration,

$$2kt = \frac{1}{c^2} + const$$

$$2c_0^2 kt = \frac{1}{(1 - s)^2} + const$$

The value of the constant k is read from the graph in the Cartesian (y, x) coordinate system. Depending on the order of reaction the y coordinate is

$$\frac{1}{1-s} \quad \text{or} \quad \frac{1}{(1-s)^2},$$

while the xaxis is the reaction time t in minutes. The reaction rate constant k is tan of the angle between the y axis and the reaction line.

The x-axis response time is in minute.

The average degree of polycondensationP is determined by the formula derived from its definition:

$$\overline{P} = \frac{N_0}{N} = \frac{c_0}{c} = \frac{1}{1-s}$$

5.2.1 EFFECT OF TEMPERATURE ON THE COURSE OF POLYCONDENSATION

Temperature is an important parameter used in the polycondensation processes. Its impact on the course of a particular reaction is various depending on the type of monomers used and the method of conducting the process. There are examples of polymer synthesis by the condensation method at low, around −40 °C, as well as at high temperatures, of the order of 350 °C.

An example of the polycondensation reaction at low temperatures is the synthesis of polyamides from dichlorides of carboxylic acids and diamines.

Other polycondensation reactions require higher temperature, whose value depends on the reactivity of certain functional groups and on an eventual use of catalysts, which reduce the activation energy. A typical example of such a reaction is the synthesis of polyesters. For example, during the reaction of glycerol with phthalic acid anhydride, it was found that the primary hydroxyl groups are more reactive than the secondary. For this reason, this reaction, conducted at the temperature of 180 °C, proceeds with the formation of a linear polyester, as if glycerol was a bifunctional monomer:

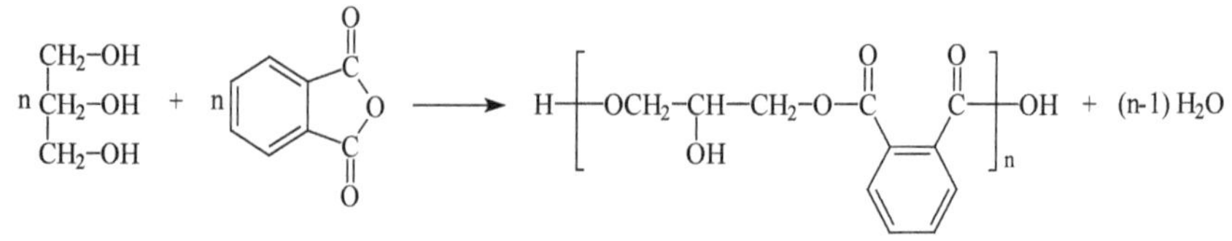

The increase of temperature up to 220 °C activates the secondary hydroxyl groups, which may, under these conditions, react with molecules of phthalic anhydride with the formation of a cross-linked polymer with spatial structure. Under these conditions, glycerol acts as a trifunctional monomer.

The polycondensation reactions are often accompanied by the cyclization reactions, which can run as secondary or side reactions.

An example of the influence of temperature on the course of synthesis of polymers with heterocyclic structure in the main chain is the synthesis of poly-imides. In the first stage of the synthesis, carried out at 20–70 °C, polyamide acids are formed. These compounds heated under a reduced pressure at temperatures above 200 °C cyclizine to the corresponding polyimides:

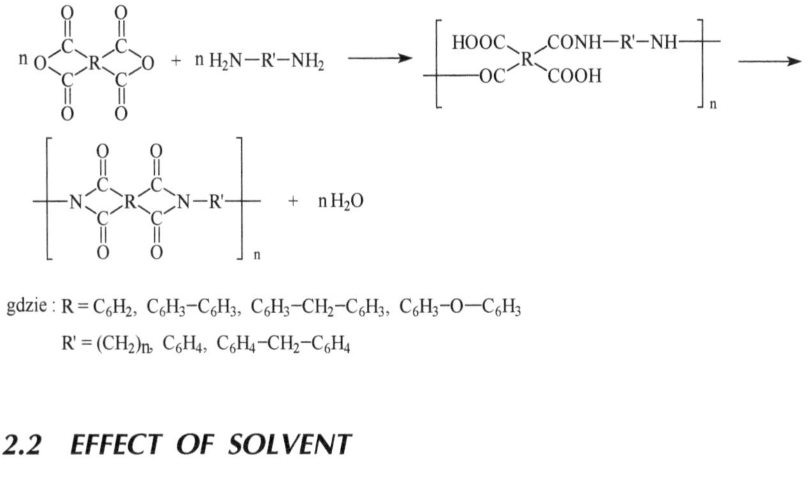

gdzie : R = C$_6$H$_2$, C$_6$H$_3$–C$_6$H$_3$, C$_6$H$_3$–CH$_2$–C$_6$H$_3$, C$_6$H$_3$–O–C$_6$H$_3$

R' = (CH$_2$)$_n$, C$_6$H$_4$, C$_6$H$_4$–CH$_2$–C$_6$H$_4$

5.2.2 EFFECT OF SOLVENT

The condensation polymerization reaction may alsobe carried out in solution.

Solvent plays an important role in this process. Due to the ionic nature of polycondensation, one mostly uses the polar solvents, which affect the kinetics of the reaction.

An example of such a reaction is the nonequilibriumpolycondensation of dicarboxylic acid chlorides with amines conducted in amide solvents.

It always passes through a stage of a reactive complex of chloroanhydride with amine. The process runs with a simultaneous dehydrochlorination by di-methylacetamide molecules with the formation of an appropriate hydrochloride:

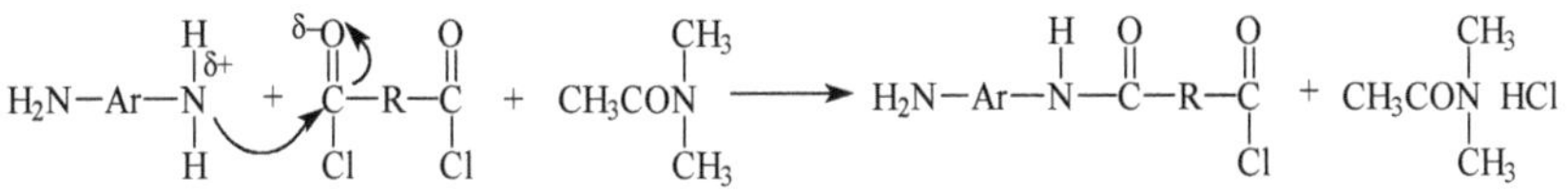

It was also found that during this reaction, intermediate adducts between the chloroanhydride and the solvent molecules are also formed. The formed adducts react with amines. The polycondensation process depends thus on the constant of the equilibrium balance of the following reaction:

$$R{-}NH_2 \bullet HCl \;+\; R'{-}\underset{CH_3}{\overset{\overset{\displaystyle O}{\|}\;\;CH_3}{C{-}N}} \;\rightleftharpoons\; R{-}NH_2 \;+\; R'{-}\underset{CH_3}{\overset{\overset{\displaystyle O}{\|}\;\;CH_3}{C{-}N}} \bullet HCl$$

In addition, the transamidation reaction may take place too, as illustrated by the following example:

$$CH_3{-}\underset{N(CH_3)_2}{\overset{\overset{\displaystyle O}{\|}}{C}} \;+\; ClCOAr \;\longrightarrow\; H_3C{-}C\underset{\underset{CH_3\,Cl^-}{\overset{|}{N^+{-}CH_3}}}{\overset{O{-}\overset{\overset{\displaystyle O}{\|}}{C}{-}Ar}{}} \;\longrightarrow\; \underset{H_3C}{\overset{H_3C}{{>}N}}{-}\overset{\overset{\displaystyle O}{\|}}{C}{-}Ar \;+\; CH_3\overset{\overset{\displaystyle O}{\|}}{C}{-}Cl$$

Application of the labeled 14C atoms technique showed that the growth rate of the polymer chain is 280 times larger than its termination. The activation energy of chain growth is of 6.27 kJ/mol.

The kinetic studies of the polyamide acid chlorides synthesis showed that it could be carried out in temperatures below 80 °C. At higher temperatures, the competitive reactions of transamidation and chain growth termination begin to outweigh.

5.2.3 EFFECT OF MONOMER CONCENTRATION

The process of polycondensation reaction takes place mostly as a second or thirdorder, therefore, the concentration of reactants has a significant impact on the speed of its course.

Changing the concentration of monomers in the balanced polycondensation reaction affects not only its course and rate but also the molecular weight of the resulting polymer.

In dilute solutions,often the competitive cyclization reaction predominates.

As an example, one can cite the polycondensation reaction of sebacic acid (dodecane acid) with hexamethylenediamine at equimolecular ratio and, in m-cresol environment. When the concentration of reactants is changed from 5 to 50%, the global reaction rate increases by about five–six times. The rate constants for this reaction, depending on concentration, are practically unchanged at a given temperature. The obtained experimental results are presented in Table 5.1.

TABLE 5.1   Effect of Concentration and Temperature on the Course of Polycondensation of SebacicAcid with Hexamethylenediamine in m-Cresol.

No	Monomer concentration [mol/kg]	Reaction rate constant [kg/(mol•min•102)]				Activation energy [kJ/ mol]
		433 K	443 K	456 K	463 K	
1	0.57	1.1	3.1	3.5	6.5	94.5
2	0.15	1.2	2.7	4.1	7.7	100.3
3	0.71	1.2	2.7	3.3	7.5	101.6
4	0.3	1.2	2.4	3.7	7.6	99.6
5	1.57	1.3	3.8	5.7	7.8	99.5

With the increasing monomer concentration up to a certain optimal value, one obtains polymers with higher molecular weight.

Further increase of the monomer concentration affects the molecular weight of the product, decreasing it. This is explained by the increase in viscosity of the medium. During the synthesis of polyamides, the best results are obtained with monomers at a concentration of 50%. In the case of polyesterification, it is preferable to use 20% solutions.

During the low-temperature polycondensation of chloroanhydride dicarboxylic acids with diamines, conducted at the phase boundary, the best solution is to use solutions with concentration equal to or slightly greater than 10%.

5.3   EFFECT OF CATALYSTS ON THE COURSE OF POLYCONDENSATION

The polycondensation reactions, depending on the starting monomers and the type of reaction, can be accelerated with cationic, anionic, or ion-coordination catalysts.

The actionmechanism of various types of catalysts can be summarized as follows:

- mechanism of cationic catalysis

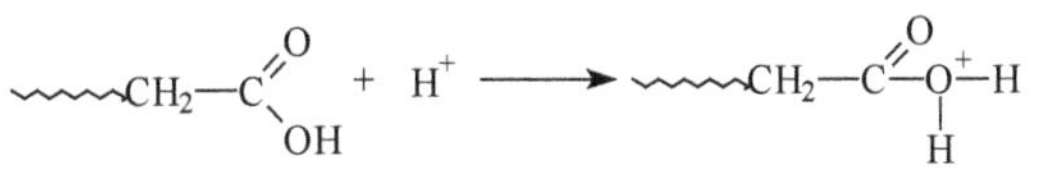

- mechanism of anionic catalysis

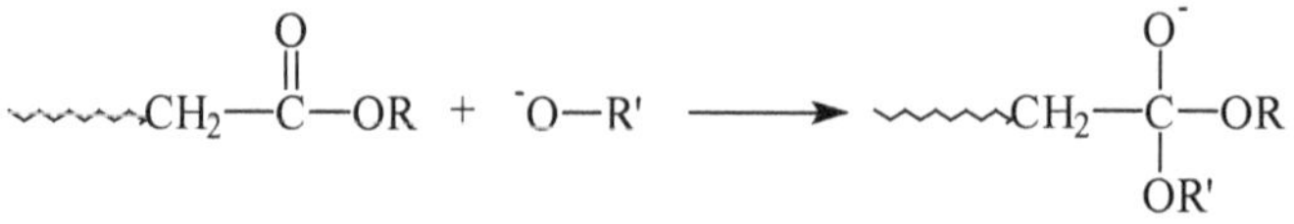

- mechanism of ion-coordination catalysis

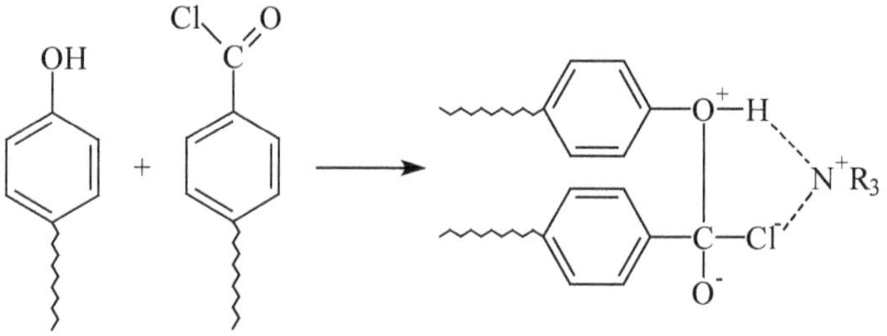

Now, there exists a number of polycondensation catalysts, derived from elements of practically all groups of the periodic table. Their catalytic activity depends on the particular type of polycondensation reaction.

Catalysts used in the synthesis of polyamides are generally tertiary amines or phosphines. The effect of basicity of amines on the course of synthesis of po ly(hexamethylenediaminesebacate)is shown in Table 5.2.

TABLE 5.2 Effect of the Basicity of Anionic Catalyst on the Course of Polycondensation of SebacicAcid with Hexamethylenediamine in the Presence of Triphenylphosphine and Carbon Tetrabromide.

No	Base	pK$_a$	Yield [%]	Intrinsic viscosity in H$_2$SO$_4$ at concentration 10g/dm^3 and at 303 K
1	Quinoline	4.81	58	0.17
2	Pyridine	5.25	62	0.27
3	3-picoline	5.63	79	0.42
4	2-picoline	5.94	21	0.13
5	4-picoline	6.03	64	0.3
6	2,6-lutidine	6.6	24	Insoluble
7	Morphine	8.33	0	–

Source: Adapted from Ogata, N. et al Polym. J. 1984, 16, 569.

Recently, more and more important are the interfacial catalysts used in the polycondensation reactions carried out at the interface. For this purpose, the most commonly used compounds are the quaternary ammonium salts and the crown ethers.

As a result of the polycondensation reaction of a,a'-dichloro-p-xylene in a mixture of tetrahydrofuran and dimethyl sulfoxide at the interface of aqueous solution of sodium hydroxide, a polymer with the following structure is formed:

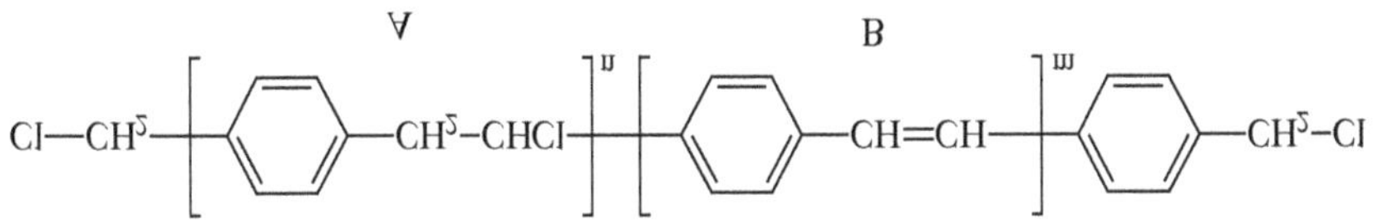

The influence of interphase ammonium catalysts on the course of this reaction, taking into account the emerging elements of structure A:B is presented in Table 5.3.

TABLE 5.3 Effect of Catalysts on the Course of Interphase Polycondensation of a,a'-dichloro-p-xylene in the Reaction with Sodium Hydroxide.

No	Catalyst	Concentra-tionNaOH [%]	Yield[%]	
			A	B
1	–	50	10	6
2	Benzyltriethylammonium chloride	50	21	12
3	Benzyltriethylammonium bromide	50	29	17
4	Benzyltriethylammoniumiodide	50	54	31
5	Tetrabutylammonium chloride	50	24	14
6	Tetrabutylammonium bromide	50	51	30
7	Cetyltrimethylammonium chloride	50	20	12
8	Benzyltrimethylammonium chloride	40	12	7
9	Benzyltrimethylammonium chloride	30	8	5

Source: Adapted from Yamazaki, N.; Imai, Y. Polym. J. 1983, 15, 905.
A—based on the xylidine structure.
B—based on the amount of reacted monomer.

Application of crown ethers as catalysts of the interface polycondensation made possible the development of a new method of synthesis of polycarbonates from alkylaryl halides and carbon dioxide or potassium carbonate.

The reactions proceeds according to the following schemes:

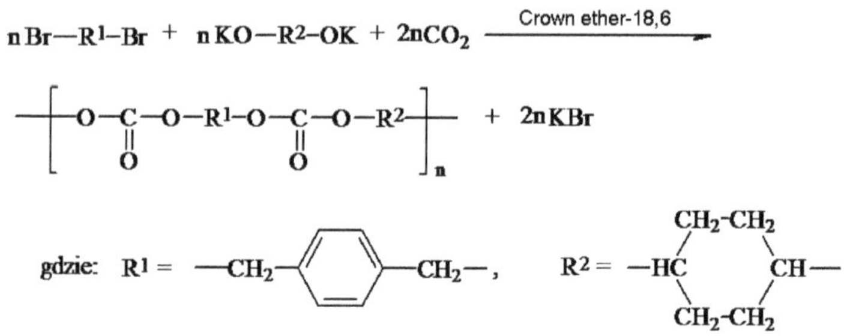

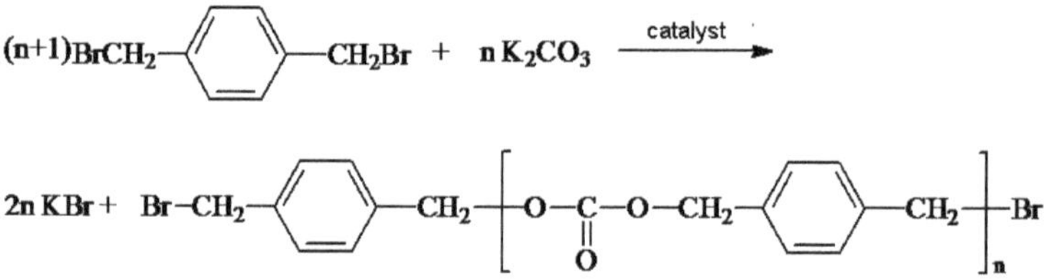

$$n\,Br—R^1—Br \;+\; n\,KO—R^2—OK \;+\; 2nCO_2 \xrightarrow{\text{Crown ether-18,6}}$$

$$\left[\!\!-O—\underset{\underset{O}{\|}}{C}—O—R^1—O—\underset{\underset{O}{\|}}{C}—O—R^2\!-\!\right]_n \;+\; 2n\,KBr$$

gdzie: $R^1 = $ —CH$_2$—⟨benzene ring⟩—CH$_2$—, $R^2 = $ —HC⟨CH$_2$-CH$_2$ / CH$_2$-CH$_2$⟩CH—

$$(n+1)BrCH_2\text{—⟨benzene ring⟩—}CH_2Br \;+\; n\,K_2CO_3 \xrightarrow{\text{catalyst}}$$

$$2n\,KBr + Br—CH_2\text{—⟨benzene ring⟩—}CH_2\left[\!\!-O—\underset{\underset{O}{\|}}{C}—O—CH_2\text{—⟨benzene ring⟩—}CH_2\!-\!\right]_n\!\!-Br$$

The obtained experimental results for this reaction are listed in Table 5.4.

TABLE 5.4   Characteristics of Polycondensation of a,a '-dibromo-p-xylene with Potassium Carbonate in the Presence of Crown Ether 18–6.

No	a,a'-dibromo-p-xylene [mmol]	K$_2$CO$_3$ [mm]	Crown ether [mmol]	Solvent	Temp. [°C]	Time [H]	Yield [%]
1	1.89	3.17	–	Benzene	60	45	0
2	3.03	5.55	0.38	Benzene	75	40	49.3
3	3.41	4.75	1.97	Benzene	60	44	66.8
4	3.79	4.04	1.89	Benzene	85	70	51.5
5	2.27	3.8	3.03	Dioxane	70	44	67.1
6	3.03	4.75	3.03	Dioxane	100	42	24.3

In a similar way, one can obtain polithiocarbonates. It is done by the catalyzed interface polycondensation of diphenols with thiophosgene (thiocarbonyl chloride) as shown in the following scheme:

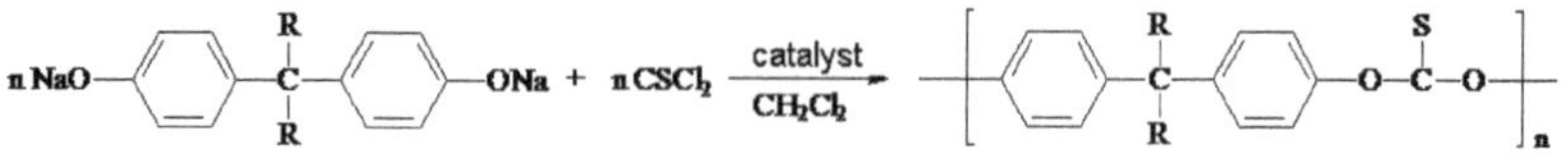

$$n\,NaO\text{—⟨ring⟩—}\underset{\underset{R}{|}}{\overset{\overset{R}{|}}{C}}\text{—⟨ring⟩—}ONa \;+\; n\,CSCl_2 \xrightarrow[\text{CH}_2\text{Cl}_2]{\text{catalyst}} \left[\!\!-\text{⟨ring⟩—}\underset{\underset{R}{|}}{\overset{\overset{R}{|}}{C}}\text{—⟨ring⟩—}O—\underset{\underset{O}{\|}}{\overset{\overset{S}{\|}}{C}}—O\!-\!\right]_n$$

As catalysts of this reaction, one can use tetrabutylammonium bromide, methyl cetyltrimethylammonium bromide, tributylmethylphosphonium bromide, methyltrioctylammonium chloride,or crown ethers.

The ways to catalyze the polycondensation by formation of charge transfer complexes with some polymers such as nitrated polystyrene during the polycondensation of 1,4-bis (methoxycarbonyl) piperazine are also known.

In addition to catalysts, in order to carry outthe polycondensation reaction effectively, one can also use the condensing means. An example of such a variant of synthesis is the polycondensation of p-aminobenzoic acid using the silicon tetrachloride:

$$2n\ H_2N\!-\!\!\bigcirc\!\!-\!COOH \ + \ n\ SiCl_4 \ \longrightarrow \ \left[\!-\!\bigcirc\!\!-\!\overset{O}{\underset{}{C}}\!-\!\overset{H}{\underset{}{N}}\!-\!\right]_{2n} + \ n\ SiO_2 \ + \ 4n\ HCl$$

5.4 EFFECT OF MONOMER STRUCTURE ON THE POLYCONDENSATION PROCESS

The course of polycondensation is affected by both the chemical structure of the reagents used as well as their quantitative ratio. As already mentioned, in the different types of polycondensation reactions, a dynamic equilibrium state is established, which is more or less shifted in favor of the nascent polymer. The use of a multifunctional monomer can lead to obtaining a cross-linked product, while the use of a single function monomer leads to the chain growth termination and the formation of low molecular weight compounds. A very important factor influencing the molecular weight of the resulting linear polymer is the quantitative ratio of used monomers. Excess of one of them causes a mutual blocking of functional groups, and thus the formation of a product with low molecular weight. In order to obtain a polymer with high molecular weight in the equilibrium polycondensation process, it is essential to preserve the stoichiometry of reacting monomers. A striking example of this may be the synthesis of Nylon (polyamide-6, 6). For the polycondensation reaction, one uses the salt formed from adipic acid and hexamethylenediamine, always keeping the molar ratio to be exactly 1:1.

In the nonequilibriumpolycondensation process, conducted at the interface, it is not necessary to use reagents in the molar ratio of 1:1, as a possible excess of one of them remains in a separate phase and does not take part in the reaction.

The dependence of the polymer molecular weight resulting from the excess of one of the monomers in equilibrium and nonequilibriumpolycondensation reactions is presented in Figure 5.1.

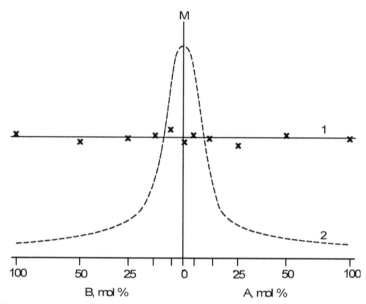

FIGURE 5.1   Dependence of the average molecular weight M of polyhexamethylenediamine on the excess of the components used: (1)nonequilibrium process (at interphase), (2) equilibrium process (in blend).

5.5   COPOLYCONDENSATION

The copolycondensation, also knownas condensation copolymerization, proceeds when different reagents with a similar chemical structureare introduced into the reaction environment. For example, in the polyesterification process of dicarboxylic acid with diol, various other dicarboxylic acids or glycols can be used.

In general, the copolycondensation process can be represented schematically by

$$HOOC-R_1-COOH + HO-R_{II}-OH \rightleftharpoons HOOC-R_1-COO-R_{II}-OH + H_2O$$

$$HOOC-R-COOH + HO-R_{II}-OH \rightleftharpoons HOOC-R-COO-R_{II}-OH + H_2O$$

At the end, a copolymer with the following structure is obtained:

$$HO-\left[OC-R-COO-R''-O\right]_n\left[OC-R'-COO-R''-O\right]_m H$$

where: $R= -(CH_2)_x-$, $-C_6H_4-$ $R'= -(CH_2)_y-$, $-C_6H_4-$ $R''= -(CH_2)_z-$

The rate of these reactions is proportional to the concentration of the groups R–COOH, R''–OH, and R'–COOHand can be described by the following equations:

$$\frac{-d[R-COOH]}{dt} = k_1[R-COOH][R''-OH]$$

$$\frac{-d[R'-COOH]}{dt} = k_2[R'-COOH][R''-OH]$$

After dividing these equations by themselves one gets

$$\frac{-d[R-COOH]}{d[R'-COOH]} = \frac{k_1[R-COOH]}{k_2[R'-COOH]} = k\frac{[R-COOH]}{[R'-COOH]}$$

The last equation shows that the composition of the reaction mixture will correspond to the copolymer composition only when $k = 1$

During the test of this process, it appeared that some secondary reactions can run too, such as acidolisis, alcoholysis, or transesterification. As result of these reactions, the previously formed homopolymer is converted into a copolymer and the composition of polycondensation product corresponds to the composition of reactants used. The condensation copolymers may be also formed by heating a mixture of polymers with similar structure, for example two polyamides. Similarly, by heating the block copolymer, one can obtain a normal statistical copolymer. This is possible by establishing a series of equilibrium states in the polycondensation process. The choice of appropriate reaction parameters makes possible the synthesis of copolycondensates with a more uniform structure, such as block copolymers.

5.6 BASIC TYPES OF POLYCONDENSATION REACTIONS

The most important examples of polycondensation reactions include:
- polyesterification

$$n\,HO-R-COOH \rightleftharpoons H\left[ORCO\right]_n OH + (n\text{-}1)H_2O$$

$$n\,HO-R-OH + n\,HOOC-R'-COOH \rightleftharpoons \left[OROOCR'CO\right]_n OH + (2n\text{-}1)H_2O$$

$$n\,HO-R-OH + n\,ClOC-R'-COCl \longrightarrow H\left[OROOCR'CO\right]_n Cl + (2n\text{-}1)\,HCl$$

gdzie : $R = -(CH_2)_x-$ $R' = -(CH_2)_y-$, $-C_6H_4-$

- polyamidation

$$n\,H_2N-R-COOH \rightleftharpoons H\left[HNRCO\right]_n OH + (n\text{-}1)H_2O$$

$$n\,H_2N-R-NH_2 + n\,HOOC-R'-COOH \rightleftharpoons H\left[NHRNHCOR'CO\right]_n OH + (2n\text{-}1)H_2O$$

$$n\,H_2N-R-NH_2 + n\,ClOC-R'-COOH \longrightarrow H\left[NHRNHCOR'CO\right]_n Cl + (2n\text{-}1)\,HCl$$

-polyanhydrization

$$2n\,HOOC-R-COOH \rightleftharpoons HO\left[\underset{O}{C}-R-\underset{O}{C}-O-\underset{O}{C}-R-\underset{O}{C}-O\right]_n H + (2n\text{-}1)H_2O$$

- polyamination   - polyacetylation

$$2n\,HO-R-OH + n\,R'-\overset{O}{\underset{H}{C}} \rightleftharpoons H\left[O-R-O-\underset{R'}{\overset{H}{C}}-OR\right]_n OH + (2n\text{-}1)H_2O$$

- polysulfuration

$$2n\,Cl-R-Cl + n\,Na_2S \longrightarrow Cl\left[R-S\right]_n Na + (2n\text{-}1)NaCl$$

- polyoxymethylation
(a) phenols

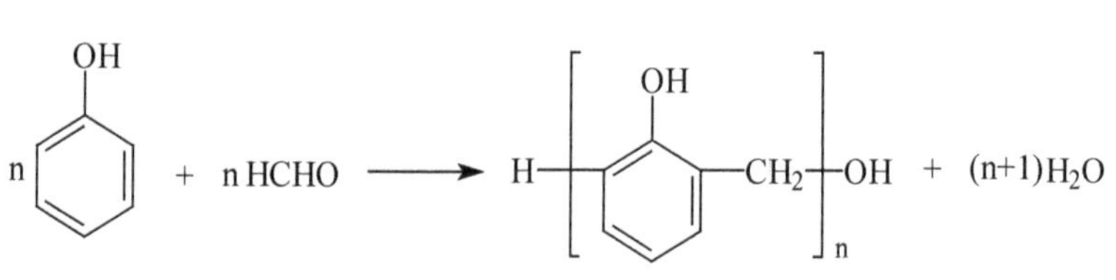

$$+ n\,HCHO \longrightarrow H\left[\text{...}CH_2\right]_n OH + (n+1)H_2O$$

(b) urea

$$n\,H_2N-\overset{O}{C}-NH_2 + n\,HCHO \longrightarrow H\left[HN-\overset{O}{C}-NH-CH_2\right]_n OH + (n\text{-}1)\,H_2O$$

With an excess of formaldehyde, the polyoxymethylation of phenol or urea continues and causes a crosslinking of the product.

-polybenzimidazolation

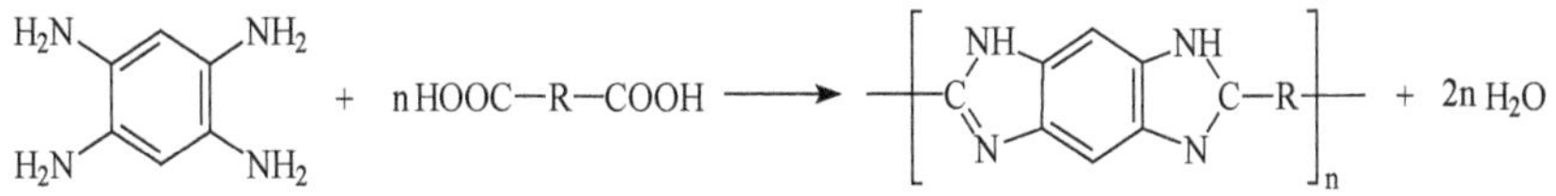

- polyimidation

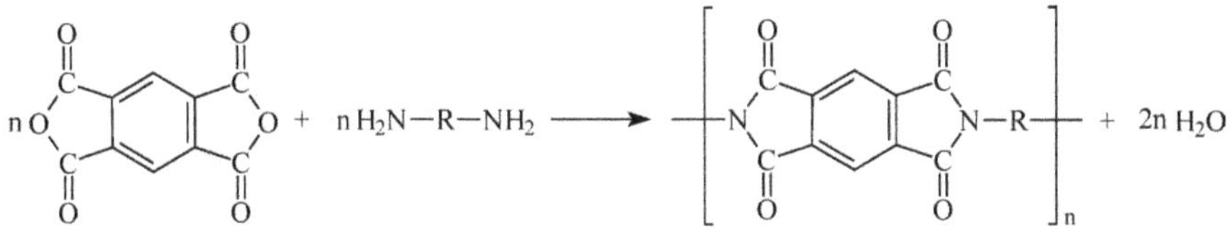

-polysiloxanation

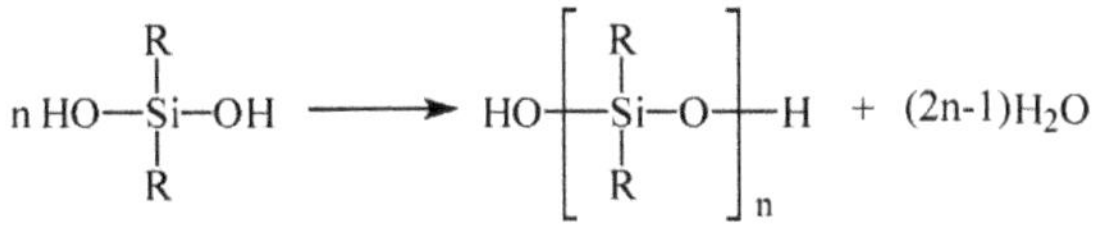

$$n\ HO-\underset{\underset{R}{|}}{\overset{\overset{R}{|}}{Si}}-OH \longrightarrow HO\left[\underset{\underset{R}{|}}{\overset{\overset{R}{|}}{Si}}-O\right]_n H + (2n-1)H_2O \qquad gdzie\ \ R = CH_3,\ C_2H_5,\ C_6H_5$$

- polycomplexation

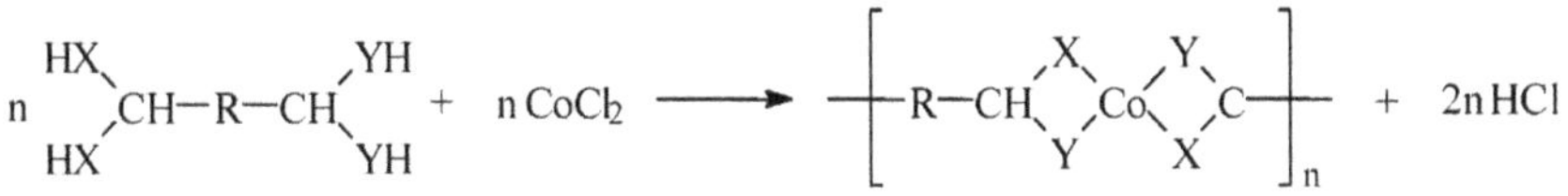

wher $Y = CN, (CH_3)_2N, C_5H_4N, X$ $= O, COO, S$

-polyhydrazidation

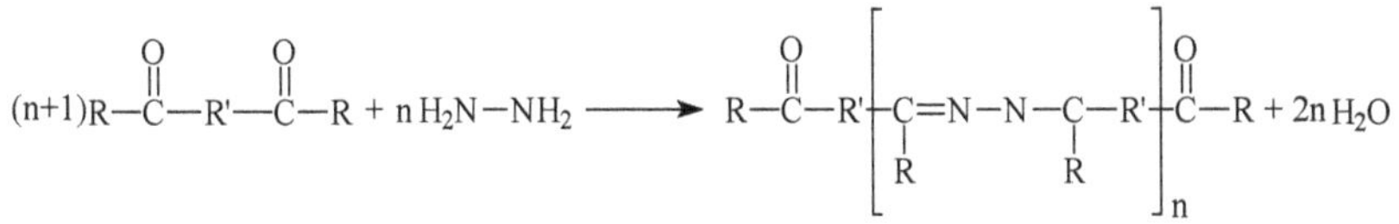

where R = CH3, (CH2)xCH3, R' = –(CH2)y–,
- dehydropolycondensation

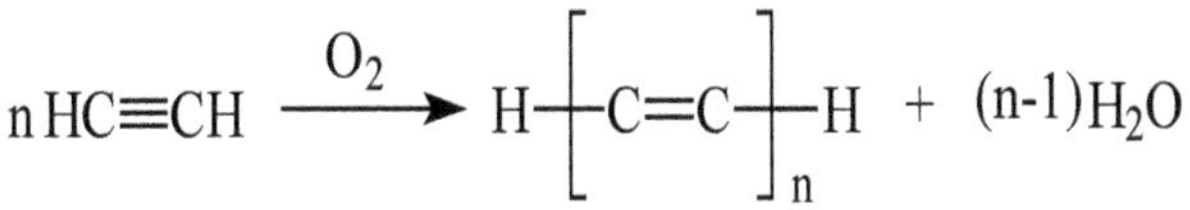

$$n\ HC{\equiv}CH \xrightarrow{\ O_2\ } H\left[C{=}C\right]_n H + (n{-}1)H_2O$$

The side reactions accompanying the polycondensation reaction:

The cyclization reaction is an important byproduct of the process competing with the homopolycondensation. This reaction takes place particularly easily when it is possible to form a permanent ring of five, six, or seven members. This reaction has an intramolecular character:

$$X-R-YH \longrightarrow R \overset{\frown}{\underset{\smile}{\quad}} Y + HX$$

Running the polycondensation reaction at low substrate concentration in the reaction mixture favors the cyclization process. Moreover, increasing the process temperature also favors the cyclization reaction. Despite the use of optimal parameters, a balance between both competitive reactions establishes and in the obtained product, in addition to the polymer, an amount of a low molecular mass cyclic compound is present. This compound is removed by extraction method, in a similar way such as removal of caprolactam from polyamide-6 or by distillation as in the case during the synthesis of polysiloxanes.

Another important group of processes accompanying the polycondensation reactions are the degradation reactions caused by the presence of bonds in polymer macromolecules, disintegrating under the influence of compounds present in the reaction mixture. Examples of such reactions may be the rupture of ester bond in the polyester molecule under the influence of alcohol (alcoholysis) or acid (acidolysis).

The greater the decrease in molecular weight of polycondensation products caused by the degradation reactions, the greater is the initial molecular weight of degraded polymer. The consequence of this is the equalization of chain lengths of individual macromolecules, thus by the same, decrease of polydispersity. For this reason, one introduces monofunctional organic acids during the synthesis of polyamide as well as the linearization processis also conducted during the synthesis of polyorganosiloxanes.

5.7   CROSSLINKING IN THE POLYCONDENSATION PROCESS

The use of tri-or more functional monomers in the process of polycondensation leads, in consequence, to crosslinking the formed polymer. The point at which the crosslinking reaction takes place is called gelation.

Flory has developed a statistical method for determining the gelation point, depending on the degree of conversion. He assumed that the probability of gelation depends on the branching factor α giving the probability of connecting two branching molecules.

The gelation point determines the value of critical branching factor ak, which depends on the functionality of the reacting monomers.

$$\alpha_k = \frac{1}{f-1},$$

where f is the monomer functionality

This formula shows that for trifunctional monomers (f = 3) ak = ½. It means that gelation will occur at 50% of reacted substrates.

One can also provide the dependence of ak coefficient on the degree of conversion p.

If in the reaction mixture, only trifunctional monomer molecules are present, such as

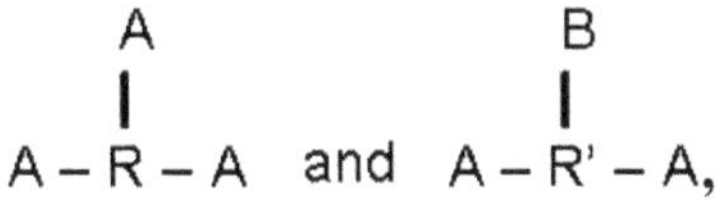

$$\begin{array}{cc} A & B \\ | & | \\ A-R-A & \text{and} & A-R'-A, \end{array}$$

then the probability of addition of one of them to the second one is equal to the probability of conversion of so many groups, namely

$$\alpha_k = p$$

If, however, difunctional and trifunctional monomer molecules are present in the reaction mixture

$$\begin{array}{cc} A & B \\ | & | \\ A-R-A & \text{and} & R'-B \end{array}$$

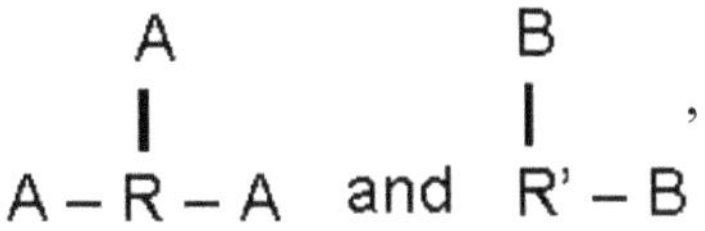

then assuming that the number of groups A is equivalent to the number of groups B, the probability of reaction for a molecule with group A terminated macromolecule will depend both on whether the group A will react with group B, as well as whether the other end of the monomer B–R'–B will react with the second group A of the macromolecular compound.

In this case,

$$\alpha_k = p$$

$$p^2 = \frac{1}{f-1} = \frac{1}{3-1} = 0.5$$

$$p = \sqrt{0.5} \approx 0.7$$

It follows from this that the point of gelation of equivalent mixture of difunctional and trifunctional monomers will take place at the conversion rate of 70%.

If the reaction mixture contains difunctional and trifunctionalmonomer molecules containing identical functional groups A and difunctional monomers reacting with groups A, then the coefficient αk is given by the following expression:

$$\alpha_k = \frac{rp^2\rho}{1-rp^2(1-p)},$$

where

$$
\begin{array}{c}
\text{A} \\
| \\
\text{A—R—A} \quad \text{A—R—A} \quad \text{B—R—B} \quad \text{- used monomers}
\end{array}
$$

$$r = \frac{N_0^A}{N_0^B} = \frac{\text{number of functional groups A}}{\text{number of functionalgroups B}}$$

$$\rho = \frac{N_{0rozg.}^A}{N_0^A} = \frac{\text{number of functional, groups A in a trifunctional monomer}}{\text{number of functional groups A}}$$

5.8   TECHNICAL METHODS OF POLYCONDENSATION PILOTING

Depending on the type of reaction, the polycondensation process can be carried out in a single or a heterogeneous phase system.

The reactions carried out in a single phase system include the polycondensation processes
- in solution,
- in melt.

The process of polycondensation in solution can be carried out for both the equilibrium and nonequilibrium reactions. Characteristic features of this process is that it takes place in a homogenous environment of relatively low temperature and the use of homogeneous catalysts is possible. The emitted low molecular weight byproducts are most frequently removed by distillation as exemplified by the azeotropic removal of water during thepolyesterification. Another way to remove the low molecular weight byproducts is the possibility of binding them by solvent molecules. A typical example of such a reaction is the synthesis of polycarbonate. The emitted in this reaction hydrogen chloride reacts with pyridine, used as a solvent, with formation of pyridine hydrochloride.

The course of polycondensation reaction in solution is influenced by the type and the amount of used solvent, catalyst, and the reaction temperature.

The advantages of this method are:

- easy evacuation of reaction heat,

- possibility of obtaining polymers of different molecular weight in the same reactor,

- possibility of influencing the reaction rate by changing the viscosity of the solution,

- relatively small number of side reactions,

- possibility of a direct use of the product in the form of a glue.

The primary disadvantage of this process, from an environment protection perspective, is the necessity to operate often with toxic solvents and the need to extract polymer from solution.

Polycondensation in the melt is performed at relatively high temperatures, usually above 200 °C.

The possibility of conducting polycondensation in the melt is, however, limited by the thermal resistance of both the monomers used, as well as the resulting product.

In order to inhibit undesirable side reactions such as the oxidation, the process is carried out in an inert atmosphere or under reduced pressure. The use of bothhigh temperature and the low pressure makes the removal of low molecular mass byproducts easy. The advantage of this method is the possibility of producing polymers with a relatively high molecular weight and fabrication of synthetic fibers directly from the melt.

Examples of such processes are the production of poly (ethylene terephthalate) and of polyamide 6.6.

Polycondensation in a heterogeneous system is performed:

- at phase boundary,

- in emulsion,

- in solid phase.

The polycondensation at the interface takes place when different substrates are soluble in various solvents, not miscible with each other. A typical example of such a reaction is the nonequilibriumpolycondensationof dicarboxylic acid chlorides with diamines or glycols.

As mentioned previously, in this case,the interfacial catalysts can be used and the addition of a binder to the aqueous phase makes it easy to remove the byproduct.

The advantages of polycondensation at the interface are:

- high reaction rate,

- possibility of conducting the process at room temperature,

- no need for a strict control of the monomers purity,
- possibility of using one of the substrates in excess,
- obtaining polymers with high molecular weight.

The disadvantages of the method is the need to use solvents and the evacuation of large quantities of sewage.

The polycondensation in emulsion is a variant of the method carried out at the interface. It allows a much larger contact area of reactants than it did at the interface as well as a partial or complete elimination of solvents. This process is performed in similar cases as the condensation on the phase boundary. As a result of the polycondensation of dicarboxylic acid dichlorides with diamines, a competitive reaction of hydrolysis takes place:

$$2\ H_2N-R-NH_2\ +\ ClOC-R'-COCl\ \xrightarrow{k_a}\ H_2N-R-NHCOR'CONHR-NH_2\ +\ 2\,HCl$$

$$ClOC-R'-COCl\ +\ 2H_2O\ \xrightarrow{k_r}\ HOOC-R'-COOH\ +\ 2\,HCl$$

:The prevalence of the amidation reaction rate over the hydrolysis reaction determines the following expression:

$$\frac{k_a}{k_r} = \frac{\lg\left(\dfrac{C_{A_0}}{C_{A_\infty}}\right)}{\lg\left(\dfrac{C_{(H_2O)_0}}{C_{(H_2O)_\infty}}\right)},$$

where

C_{A_0}, $C_{(H_2O)_\infty}$ —initial concentration of amine groups (or –COCl) and water,

C_{A_∞} —concentration of amine groups (or –COCl) after the time $\tau = \infty$

$C_{(H_2O)_\infty}$ —concentration of water after the time $\tau = \infty$

The degree of conversion in the amidation reaction is given by

$$\alpha = \frac{C_{A_0} - C_{A_\infty}}{C_{A_0}}$$

The obtained experimental results are presented in Table 5.5.

TABLE 5.5 Reaction Rate Constant for Amidation and Hydrolysis in the System: Tetrahydrofuran, Water, Na2CO3, NaOH by benzyl chloride (1–6) or aniline (7–11).

No	Monomer	a	k_a/k_r
1	Hexamethylenediamine	0.946	3170
2	p-phenylenediamine	0.954	4460
3	m-phenylenediamine	0.887	1050
4	o-phenylenediamine	0.832	646
5	Benzidine	0.877	1030
6	2,4-diaminophenol	0.714	260
7	2,6-naphthalene dicarboxylic acid dichloride	0.964	4800
8	Sebacylic acid dichloride	0.959	3470
9	Terephtalic acid dichloride	0.93	1920
10	Isophtalic acid dichloride	0.928	1630
11	Oxalyl dichloride	0.11	6.2

Source: Adapted from Sokołov, L. B.; Kogulova, W. I. Wysokomol. Soied. A. 1979, 21, 1050.

The method of polycondensation in the solid phase is practically applied in the process of crosslinking of polymers. Depending on the type of polymer and eventual catalyst used, it may be carried out at different temperatures.

This method is used mainly in the processing of plastics.

REFERENCES

Abusleme, J.; Giannetti, E. Emulsion Polymeryzation. *Macromolecules* 1991, *24*, 67–83.

Adesamya, I. Classification of Ionic Polymers. *Polym. Eng. Sci.* 1988, *28*, 1473.

Akbulut, U.; Toppare, L.; Usanmaz, A.; Onal, A. Electroinitiated Cationic Polymeryzation of Some Epoksides. *Makromol. Chem. Rapid. Commun.* 1983, *4*, 259.

Andreas, F.; Grobe, K. *Chemiapropylenu;* WNT: Warszawa, 1974.

Barg, E. I. *Technologiatworzywsztucznych;* PWT: Warszawa, 1957.

Billmeyer, F. W. *Textbook of Polymer Science*; J. Wiley and Sons: New York, 1984.

Borsig, E.; Lazar, M. Chemical Modification of Polyalkanes. *Chem. Listy* 1991, *85*, 30.

Braun, D.; Czerwiński, W.; Disselhoff, G.; Tudos, F. Analysis of the Linear Methods for Determining Copolymeryzation Reactivity Ratios. *Angew. Makromol. Chem.* 1984, *125*, 161.

Burnett, G. M.; Cameron, G. G.; Gordon G.; Joiner, S. N. Solvent Effects on the Free Radical Polymeryzation of Styrene. *J. Chem. Soc. Faraday Trans.*, *1* 1973, *69*, 322.

Challa, R. R.; Drew, J. H.; Stannet, V. T.; Stahel, E. P. Radiation-induced Emulsion Polymerization of Vinyl Acetate. *J. Appl. Polym. Sci.* 1985, *30*, 4261.

Candau, F. Mechamism and Kinetics of the Radical Polymeryzation of Acrylamide In Inverse Micelles. *Makromol. Chem.*, *MakromolSymp.* 1990, *31*, 27.

Ceresa, R. J. *Kopolimeryblokoweiszczepione*; WNT:Warszwa, 1965.

Charlesby, A. Chemiaradiacyjnapolimerów; WNT:Warszwa, 1962.

Chern, C. S. , Poehlein, G. W. : Kinetics of Cross-linking Vinyl Polymerization. *Polym. Plast. Technol. Eng.* 1990, 29, 577

Chiang, W. Y. , Hu, C. M. Studies of Reactions with Polymers. *J. Appl. Polym. Sci.* 1985, *30*, 3895–4045.

Chien, J. C.; Salajka, Z. Syndiospecific Polymeryzation of Styrene. *J. Polym. Sci. A*1991, *29*, 1243.

Chojnowski, J.; Stańczyk, W. E. Procesychlorowaniapolietylenu. *Polimery* 1972, *17*, 597.

Collomb, J.; Morin, B.; Gandini, A.; Cheradame, H. Cationic Polymerization Induced by Metal Salts. *Europ. Polym. J.* 1980, *16*, 1135.

Czerwiński, W. K. Solvent Effects on Free Radical Polymerization. *Makromol. Chem.* 1991, *192*, 1285–1297.

Dainton, F. S. *Reakcjełańcuchowe*; PWN:Warszwa, 1963.

Ding, J.; Price, C.; Booth, C. Use of Crown Ether in the Anionic Polymerization of Propylene Oxide. *Eur. Polym. J.* 1991, *27*, 895–901

Dogatkin, B. A. *Chemiaelastomerów*; WNT: Warszwa, 1976.

Duda, A.; Penczek, S. Polimeryzacjajonowaipseudojonowa. *Polimery*1989, *34*, 429.

Dumitrin, S.; Shaikk, A. S.; Comanita, E.; Simionescu, C. Bifunetional Initiators. *Eur. Polym. J.* 1983, *19*, 263.

Dunn, A. S. Problems of Emulsion Polymeryzation. *Surf. Coat. Int.* 1991, *74*, 50.

Eliseeva, W. I.; Asłamazova, T. R. Emulsionnajapolimeryzacja v otsustivieemulgatora. *Uspiechi-Chimii*1991, *60*, 398.

Erusalimski, B. L. Mechanism dezaktivacirastuszczichciepiej v procesachionnojpolimeryzacjivinylovychpolimerov. *Vysokomol. Soed. A.* 1985, *27*, 1571.

Fedtke, M. *Chimiczieskijereakcjipolimerrov;*Chimica:Moskwa, 1990.

Fernandez-Moreno, D.; Fernandez-Sanchez, C.; Rodriguez, J. G.; Alberola, A. Polymeryzation of Styrene and 2-vinylpyridyne Using AlEt-VClHeteroaromatic Bases. *Eur. Polym. J.* 1984, *20*, 109.

Fieser, L. F.; Fieser, M. *Chemiaorganiczna*; PWN:Warszwa,1962.

Flory, P. J. Nauka o makrocząsteczkach. *Polimery*1987, *32*, 346.

Frejdlina, R. Ch.; Karapetian, A. S. *Telomeryzacja*; WNT: Warszawa, 1962.

Funt, B. L.; Tan, S. R. The Fotoelectrochemical Initiation of Polymeryzation of Styrene. *J. Polym. Sci. A*1984, *22*, 605.

Gerner, F. J.; Hocker, H.; Muller, A. H.; Shulz, G. V. On the Termination Reaction in the Anionic Polymerization of Methyl Methacrylate in Polar Solvents. *Eur. Polym. J.* 1984, *20*, 349.

Ghose, A.; Bhadani, S. N. Electrochemical Polymerization of Trioxane in Chlorinated Solvents. *Indian J. Technol.* 1974, *12*, 443.

Giannetti, E.; Nicoletti, G. M.; Mazzocchi, R. Homogenous Ziegler–Natta Catalysis. *J. Polym. Sci. A*1985, *23*, 2117.

Goplan, A.; Paulrajan, S.; Venkatarao, K.; Subbaratnam, N. R. Polymeryzation of N,N'-Methylene Bisacrylamide Initiated by Two New Redox Systems Involving Acidic Permanganate. *Eur. Polym. J.* 1983, *19*, 817.

Gózdz, A. S.; Rapak, A. Phase Transfer Catalyzed Reactions on Polymers. *Makromol. Chem. Rapid Commun.* 1981, *2*, 359.

Greenley, R. Z. An Expanded Listing of Revized Q and e Values. *J. Macromol. Sci. Chem. A*1980, *14*, 427.

Greenley, R. Z. Recalculation of Some Reactivity Ratios. *J. Macromol. Sci. Chem. A*1980, *14*, 445.

Guhaniyogi, S. C.; Sharma, Y. N. Free-radical Modyfication of Polipropylene. *J. Macromol. Sci. Chem. A* 1985, *22*, 1601.

Guo, J. S.; El Aasser, M. S.; Vanderhoff, J. W. MicroemulsionPolymerization of Styrene. *J. Polym. Sci. A*1989, *27*, 691.

Gurruchaga, M.; Goni, I.; Vasguez, M. B.; Valero, M.; Guzman, G. M. An Approach to the Knowledge of the Graft Polymerization of Acrylic Monomers onto Polysaccharides using Ce (IV) as Initiator. *J. Polym. Sci. Part C*1989, *27*, 149.

Hasenbein, N.; Bandermann, F. Polymerization of Isobutene with $VOCl_3$ in Heptane. *Makromol. Chem.* 1987, *188*, 83.

Hatakeyama, H. et al. Biodegradable Polyurethanes from Plant Components. *J. Macromol. Sci. A.* 1995, *32*, 743.

Higashimura, T.; Law, Y. M.; Sawamoto, M. Living Cationic Polymerization. *Polym. J.* 1984, *16*, 401.

Hirota, M.; Fukuda, H. Polymerization of Tetrahydrofuran. *Nagoya-ski Kogyo KenkynshoKentyn-Hokoku*1984, *51*.

Hodge, P.; Sherrington, D. C. *Polymer-Supported Reactions in Organic Syntheses*;Nova Publisher: New York,2007;tłum. narosyjski Mir:Moskwa, 1983.

Inoue, S. Novel Zine Carboxylates as Catalysts for the Copolymeryzation of $CO_2$with Epoxides. *Makromol. Chem. Rapid Commun.* 1980, *1*, 775.

Janović, Z. *Polimerizacije I polimery*;IzdanjaKemije Ind. :Zagrzeb, 1997.

Jedliński, Z. J. : Współczesnekierunki w dziedziniesyntezypolimerów. *Wiadomości Chem.* 1977, *31*, 607.

Jenkins, A. D.; Ledwith, A. *Reactivity, Mechanism and Structure in Polymer Chemistry*; J. Wiley and Sons: London, 1974; tłum. na rosyjski, Mir:Moskwa, 1977.

Joshi, S. G.; Natu, A. A. Chlorocarboxylation of Polyethylene.*Angew. Makromol. Chem.* 1986, *140*, 99–115.

Kaczmarek, H. polimery, a środowisko. *Polimery*1997, *42*, 521.

Kang, E. T.; Neoh, K. G. Halogen Induced Polymerization of Furan. *Eur. Polym. J.* 1987, *23*, 719.

Kang, E. T.; Neoh, K. G.; Tan, T. C.; Ti, H. C. Iodine Induced Polymerization and Oxidation of Pryridazine. *Mol. Cryst. Lig. Cryst.* 1987, *147*, 199.

Karnojitzki, V. *Organiczieskijepierekisy*; I. I. L. :Moskwa, 1961.

Karpiński, K.; Prot, T. Inicjatoryfotopolimeryzacjiifotosieciowania. *Polimery*1987, *32*, 129.

Keii, T. Propene Polymerization with $MgCl_2$-Supported Ziegler Catalysts. *Makromol. Chem.* 1984, *185*, 1537.

Khanna, S. N.; Levy, M.; Szwarc, M. Complexes Formed by Anthracene with 'Living' Polystyrene. *Dormant Poly. Trans. Faraday Soc.* 1962,*58*, 747–761.

Koinuma, H.; Naito, K.; Hirai, H. Anionic Polymerization of Oxiranes. *Makromol. Chem.* 1982, *183*, 1383.

Korniejev, N. N.; Popov, A. F.; Krencel, B. A. *Komplieksnyjemetalloorganiczeskijekatalizatory*;K himia: Leningrad, 1969.

Korszak, W. W. Niekotoryjeproblemypolikondensacji. *Vysokomol. Soed. A*1979, *21*, 3.

Korszak, W. W. *Technologiatworzywsztucznych*; WNT: Warszawa, 1981.

Kowalska, E.; Pełka, J. Modyfikacjatworzywtermoplastycznychwłóknamicelulozowymi. *Polimery* 2001, *46*, 268.

Kozakiewicz, J.; Penczek, P. Polimeryjonowe. *Wiadomości Chem.* 1976, *30*, 477.

Kubisa, P. Polimeryzacjażyjąca. *Polimery* 1990, *35*, 61.

Kubisa, P.; Neeld, K.; Starr, J.; Vogl, O. Polymerization of Higher Aldehydes. *Polymer* 1980, *21*, 1433.

Kucharski, M. Nitroweiaminowepochodnepolistyrenu. *Polimery* 1966, *11*, 253.

Kunicka, M.; Choć, H. J. HydroliticDegradation and Mechanical Properties of Hydrogels. *Polym. Degrad. Stab.* 1998, *59*, 33.

Lebduska, J.; Dlaskova, M. Roztokovakopolymerace. *ChemickyPrumysl* 1990, *40*, 419–480.

Leclerc, M.; Guay, J.; Dao, L. W. Synthesis and Characterization of Poly(alkylanilines). *Macromolecules* 1989, *22*, 649.

Lee, D. P.; Dreyfuss, P. Triphenylphesphine Termination ofTetrahydrofuran Polymerization. *J. Polym. Sci. ,Polym. Chem. Ed.* 1980, *18*, 1627.

Leza, M. L.; Casinos, I.; Guzman, G. M.; Bello, A. Graft Polymerization of 4-vinylpyridineonto Cellulosic Fibres by the Ceric on Method. *Angew. Makromol. Chem.* 1990, *178*, 109–119.

Lopez, F.; Calgano, M. P.; Contrieras, J. M.; Torrellas, Z.; Felisola, K. Polymerization of Some Oxiranes. *Polym. Int.* 1991, *24*, 105.

Lopyrev, W. A.; Miaczina, G. F.; Szevaleevskii, O. I.; Hidekel, M. L. Poliacetylen. *Vysokomol. Soed.* 1988, *30*, 2019.

Mano, E. B.; Calafate, B. A. L. Electrolytically Initiated Polymerization of N-vinylcarbazole. *J. Polym. Sci. ,Polym. Chem. Ed.* 1983, *21*, 829.

Mark, H. F. *Encyclopedia of Polymer Science and Technology*; Concise 3rd ed. , WileyInterscience: Hoboken, New York, 2007.

Mark, H.; Tobolsky, A. V. *Chemiafizycznapolimerów*; PWN: Warszawa, 1957.

Matyjaszewski, K.; Davis, T. P. *Handbook of Radical Polymerization*; Wiley-Interscience: New York, 2002.

Matyjaszewski,K.; Mülle, A. H. E. 50 Years of Living Polymerization. *Prog. Polym. Sci.* **2006**, *31*, 1039–1040.

Mehrotra, R.; Ranby, B. Graft Copolymerization onto Starch. *J. Appl. Polym. Sci.* 1977, *21*, 1647, 3407–1978, 22, 2991.

Morrison, R. T.; Boyd, N. *Chemiaorganiczna*; PWN: Warszawa, 1985.

Morton, M. Anionic Polymerization. Academic Press: New York, 1983.

Munk, P. Introduction to Macromolecular Science. J. Wiley and Sons: New York, 1989,

Naraniecki. B. Cieczemikroemulsyjne. *Przem. Chem.* 1999, *78(4)*, 127.

Neoh, K. G.; Kang, E. T.; Tan, K. L. Chemical Copolymerization of Aniline with Halogen-Substituted Anilines. *Eur. Polym. J.* 1990, *26*, 403.

Nowakowska, M.; Pacha, J. Polimeryzacja olefin wobeckompleksówmetaloorganicznych. *Polimery* 1978, *23*, 290.

Ogata, N.; Sanui, K.; Tan S. Synthesis of Alifatic Polyamides by Direct Polycondensation with Triphenylphosphine. *Polym. J.* 1984, *16*, 569.

Onen, A.; Yagci, Y. Bifunctional Initiators. *J. Macromol. Sci. Chem.* *A* 1990, *27*, 743, *Angew. Makromol. Chem.* 1990, 181, 191

Osada, Y. Plazmiennajapolimerizaciaiplazmiennajaobrabotkapolimerov. *Vysokomol. Soed.* *A* 1988, *30*, 1815.

Pasika, W. M. Cationic Copolymerization of Epichlorohydrinwith Styrene Oxide and Cyclohexene Oxide. *J. Macromol. Sci. Chem. A*1991, *28*, 545.

Pasika, W. M. Copolymerization of Styrene Oxide And Cyklohexene Oxide. *J. Polym. Sci. Part A*1991, *29*, 1475.

Pielichowski, J.; Puszyński, A. *Preparatykamonomerów*; W. P. Kr. :Kraków, 2009, 79–96.

Pielichowski, J.; Puszyński, A. *Preparatykapolimerów*; TEZA WNT:Kraków, 2005.

Pielichowski, J.; Puszyński, A. *Technologiatworzywsztucznych*; WNT: Warszawa, 2003.

Pielichowski, J.; Puszyński, A. *Wybrane działy z technologiichemicznejorganicznej*; W. P. Kr. : Kraków.

Pistola, G.; Bagnarelli, O. Electrochemical Bulk Polymeryzation of Methyl Methacrylate in the Presence of Nitric Acid. *J. Polym. Sci. , Polym. Chem. Ed.* 1979, *17*, 1002.

Pistola, G.; Bagnarelli, O.; Maiocco, M. Evaluation of Factors Affecting The Radical Electropolymerization of Methylmethacrylate in the Presence of HNO$_3$. *J. Appl. Electrochem.* 1979, *9*, 343.

Połowoński, S. *Technikipomiarowo-badawcze w chemiifizycznejpolimerów*. W. P. L. :Łódź, 1975.

Porejko, S.; Fejgin, J.; Zakrzewski, L. *Chemiazwiązkówwielkocząsteczkowych*; WNT: Warszawa, 1974.

Pracazbiorowa: Chemiaplazmyniskotemperaturowej; WNT: Warszawa, 1983.

Puszyński, A. Chlorination of Polyethylene in Suspension. *Pol. J. Chem.* 1981, *55*, 2143.

Puszyński, A.; Godniak, E. Chlorination of Polyethylene in Suspension in the Presence of Heavy Metal Salts. *Macromol. Chem. Rapid Commun*1980, *1*, 617.

Puszyński, A.; Dwornicka, J. Chlorination of Polypropylene in Suspension. *Angew. Macromol. Chem.* 1986, *139*, 123.

Puszyński, J. A.; Miao, S. Kinetic Study of Synthesis of SiC Powders and Whiskers in the Presence KClO$_3$ and Teflon. *Int. J. SHS*1999, *8(8)*, 265.

Rabagliati, F. M.; Bradley, C. A. Epoxide Polymerization. *Eur. Polym. J.* 1984, *20*, 571.

Rabek, J. F. *Podstawyfizykochemiipolimerów*; W. P. Wr. :Wrocław, 1977.

Rabek, T. I. Teoretycznepodstawysyntezypolielektrolitówiwymieniaczyjonowych;PWN: Warszawa, 1962.

Regas, F. P.; Papadoyannis, C. J. Suspension Cross-linking of Polystyrene with Friedel-Crafts Catalysts. *Polym. Bull.* 1980, *3*, 279.

Rodriguez, M.; Figueruelo, J. E. On the Anionic Polymeryzation Mechanism of Epoxides. *Makromol. Chem.* 1975, *176*, 3107.

Roudet, J.; Gandini, A. Cationic Polymerization Induced by Arydiazonium Salts. *Makromol. Chem. Rapid Commun.* 1989, *10*, 277.

Sahu, U. S.; Bhadam, S. N. Triphenylphosphine Catalyzed Polimerization of Maleic Anhydride. *Makromol. Chem.* 1982, *183*, 1653.

Schidknecht, C. E. *Polimerywinylowe*; PWT: Warszawa, 1956.

Szwarc, M.; Levy, M.; Milkovich, R. Polymerization Initiated by Electron Transfer to Monomer. A New Method of Formation of Block Copolymers. *J. Am. Chem. Soc.* **1956**,**78**, 2656–2657.

Szwarc, M. 'Living' polymers. *Nature*1956, *176*, 1168–1169.

Sen, S.; Kisakurek, D.; Turker, L.; Toppare, L.; Akbulut, U. Electroinitiated Polymerization of 4-Chloro-2,6-Dibromofenol. *New Polym. Mater.* 1989, *1*, 177.

Sikorski, R. T. Chemicznamodyfikacjapolimerów, PraceNauk. *Inst. Technol. Org. Tw. Szt. P. Wr.* 1974, *16*, 33.

Sikorski, R. T. *Podstawychemiiitechnologiipolimerów*; PWN: Warszawa, 1984.

Sikorski, R. T.; Rykowski, Z.; Puszyński, A. Elektronenempfindliche Derivate von Poly-methylmethacrylat. *Plaste und Kautschuk*1984, *31*, 250.

Simionescu, C. I.; Geta, D.; Grigoras, M. Ring-Opening Isomerization Polymeryzation of 2-Methyl-2-Oxazoline Initiated by Charge Transfer Complexes. *Eur. Polym. J.* 1987, *23*, 689.

Sheinina, L. S.; Vengerovskaya, Sh. G.; Khramova, T. S.; Filipowich, A. Y. Reakcjipiridinoviichczietvierticznychsoliej s epoksidnymisoedineniami. *Ukr. Khim. Zh.* 1990, *56*, 642.

Soga, K.; Toshida, Y.; Hosoda, S.; Ikeda, S. A Convenient Synthesis of a Polycarbonate. *Makromol. Chem.* 1977, *178*, 2747.

Soga, K.; Hosoda, S.; Ikeda, S. A New Synthetic Route to Polycarbonate. *J. Polym. Sci. ,Polym. Lett. Ed.* 1977, *15*, 611.

Soga, K.; Uozumi, T.; Yanagihara, H.; Siono, T. Polymeryzation of Styrene with Heterogeneous Ziegler-Natta Catalysts. *Makromol. Chem. Rapid. Commun.* 1990, *11*, 229.

Soga, K.; Shiono, T.; Wu, Y.; Ishii, K.; Nogami, A.; Doi, Y. Preparation of a Stable Supported Catalyst for Propene Polymeryzation. *Makromol. Chem. Rapid Commun.* 1985, *6*, 537.

Sokołov, L. B.; Logunova, W. I. Otnositielnajareakcionnosposobnostmonomeroviprognozirovanicpolikondensacji v emulsionnychsistemach. *Vysokomol. Soed. A.* 1979, *21*, 1075.

Soler, H.; Cadiz, V.; Serra, A. Oxirane Ring Opening with Imide Acids. *Angew. Makromol. Chem.* 1987, *152*, 55.

Spasówka, E.; Rudnik, E. Możliwościwykorzystaniawęglowodanów w produkcjibiodegradowalnychtworzywsztucznych. *Przemysł Chem.* 1999, *78*, 243.

Stevens, M. P. *Wprowadzenie do chemiipolimerów*; PWN: Warszawa, 1983.

Stępniak, I.; Lewandowski, A. Elektrolitypolimerowe. *Przemysł Chem.* 2001, *80*(9), 395.

Strohriegl, P.; Heitz, W.; Weber, G. Polycondensation using Silicon Tetrachloride. *Makromol. Chem. Rapid Commun.* 1985, *6*, 111.

Szur, A. M. *Vysokomolekularnyjesoedinenia*;VysszajaSzkoła:Moskwa, 1966.

Tagle, L. H.; Diaz, F. R.; Riveros, P. E. Polymerization by Phase Transfer Catalysis. *Polym. J.* 1986, *18*, 501.

Takagi, A.; Ikada, E.; Watanabe, T. Dielectric Properties of Chlorinated Polyethylene and Reduced Polyvinylchloride, Memoirs of the Faculty of Eng. *Kobe Univ.* 1979, *25*, 239.

Tani, H. Stereospecific Polymerization of Aldehydes and Epoxides. *Adv. Polym. Sci.* 1973, *11*, 57.

Vandenberg, E. J. CoordynationCopolymeryzation of Tetrahydrofuran and Oxepane with Oxetanes and Epoxides. *J. Polym. Sci. Part A*1991, *29*, 1421.

Veruovic, B. *Navody pro laboratornicviceni z chemiepolymeru*; SNTL: Praha, 1977.

Vogdanis, L.; Heitz, W. Carbon Dioxide as a Mmonomer. *Makromol. Chem. Rapid Commun.* 1986, *7*, 543.

Vollmert, B. *Grundriss der Makromolekularen Chemic*; Springer-Verlag: Berlin, 1962.

Wei, Y.; Tang, X.; Sun, Y. A Study of the Mechanism of Aniline Polymerization. *J. Polym. Sci. Part A*1989, *27*, 2385.

Wei, Y.; Sun, Y.; Jang, G. W.; Tang, X. Effects P-Aminodiphenylamine on Electrochemical Polymerization of Aniline. *J. Polym. Sci. ,Part C*1990, *28*, 81.

Wei, Y.; Jang, G. W. Polymerization of Aniline and Alkyl Ringsubstituted Anilines in the Presence of Aromatic Additives. *J. Phys. Chem.* 1990, *94*, 7716.

Wiles, D. M. Photooxidative Reactions of Polymers. *J. Appl. Polym. Sci. , Appl. Polym. Symp.* 1979, *35*, 235.

Wilk, K. A.; Burczyk, B.; Wilk, T. Oil in Water Microemulsions as Reaction Media. 7th International Conference Surface and Colloid Science, Compiegne, 1991.

Winogradova, C. B. Novoe v obłastipolikondensacji. *Vysokomol. Soed. A*1985, *27*, 2243.

Wirpsza, Z. *Poliuretany*; WNT: Warszawa, 1991.

Wojtala, A. Wpływwłaściwościorazotoczeniapoliolefinnaprzebiegichfotodegradacji. *Polimery*2001, *46*, 120.

Xue, G. Polymerization of Styrene Oxide with Pyridine. *Makromol. Chem. Rapid Commun.* 1986, *7*, 37.

Yamazaki, N.; Imai, Y. Phase Transfer Catalyzed Polycondensation of Bishalomethyl Aromatic compounds. *Polym. J.* 1983, *15*, 905.

Yasuda, H. *Plasma Polymerization*; Academic Press Inc. : Orlando, 1985.

Yuan, H. G.; Kalfas, G.; Ray, W. H. Suspension Polymerization. *J. Macromol. Sci. Rev. Macromol. Chem. Phys.* 1991, *C 31*, 215.

Zubakowa, L. B.; Tevlina, A. C.; Davankov, A. B. *Sinteticzeskijeionoobmiennyjemateriały*;Khim ia:Moskwa, 1978.

Zubanov, B. A. : *Viedienie v chimiupolikondensacionnychprocesov*;Nauka:Ałma Ata, 1974.

Żuchowska, D. *Polimerykonstrukcyjne*; WNT: Warszawa, 2000.

Żuchowska, D. *Strukturaiwłaściwościpolimerówjakomateriałówkonstrukcyjnych*;W. P. Wr. :Wrocław, 1986.

Żuchowska, D.; Steller, R.; Meisser, W. Structure and Properties of Degradable Polyolefin–Starch Blends. *Polym. Degrad. Stab.* 1998, *60*, 471.

CHAPTER 6

CONDENSATION POLYMERS: A REVIEW

G. E. ZAIKOV

Russian Academy of Sciences, Moscow, Russia

CONTENTS

6. 1 PHENOPLASTS

Phenoplasts (phenolics) are synthetic resins formed by the condensation polymerization reaction of phenol or its derivatives with aldehydes. The most commonly used are phenol formaldehyde resins (PF).

The polycondensation reaction of phenol with aldehydes is catalyzed by both the hydrogen and the hydroxide ions. Because of the favorable equilibrium states the reaction can be carried out in an aqueous solution.

In the first stage of the reaction, an electrophilic binding of aldehyde molecule in ortho- or paraposition to the phenolic group with a simultaneous rearrangement of the hydrogen atom to the oxygen atom takes place.

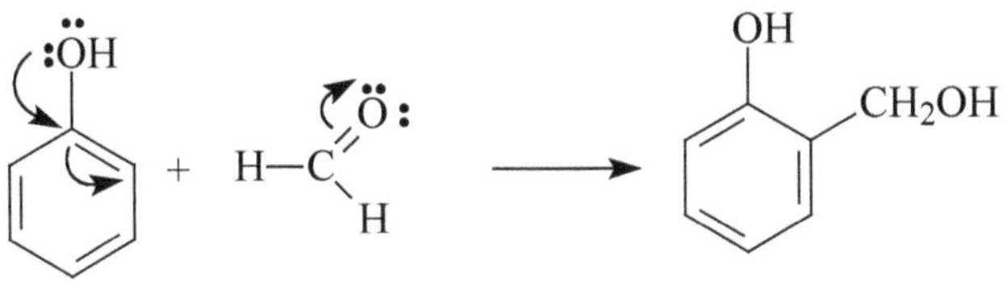

In the acidic environment, in which there is no stabilizing effect of hydrogen bonds, the rate of condensation reaction of hydroxymethylene groups is very large. In the case of trifunctional phenol it results in an immediate cross-linking. To avoid this, the polycondensation reaction in an acidic environment is performed in excess of phenol. In practice, the maximum acceptable molar ratio of formaldehyde to phenol is 10:12. The optimum ratio giving the possibility of safe operation of the process and obtaining the best properties of the resin is the ratio of 26. 5–27. 5 g of formaldehyde per 100 g of phenol. It corresponds approximately to the molar ratio 6:7. The resins obtained in these conditions, called novolacs, are thermoplastic and are soluble in organic solvents, which indicates their linear structure:

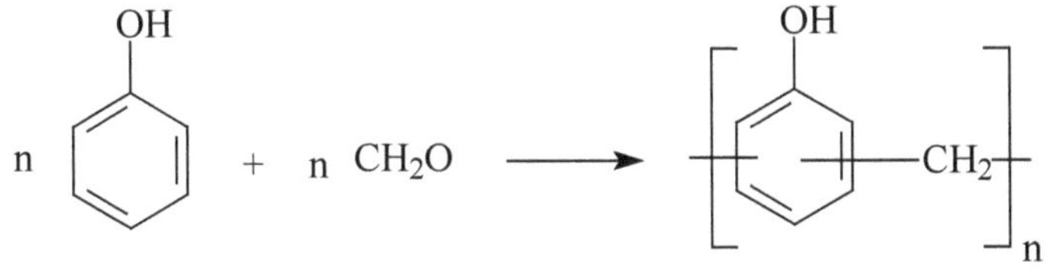

Novolacs can be cross-linked by heating with polyoxymethylene or urotropin.

During the polycondensation of phenol with formaldehyde in alkaline environment an excess of formaldehyde is used. Since the rate of this reaction is much lower than in an acidic environment a soluble resin with linear structure can be isolated. This resin is calledresol resin.

During the polycondensation of phenol with formaldehyde the following reactions take place:

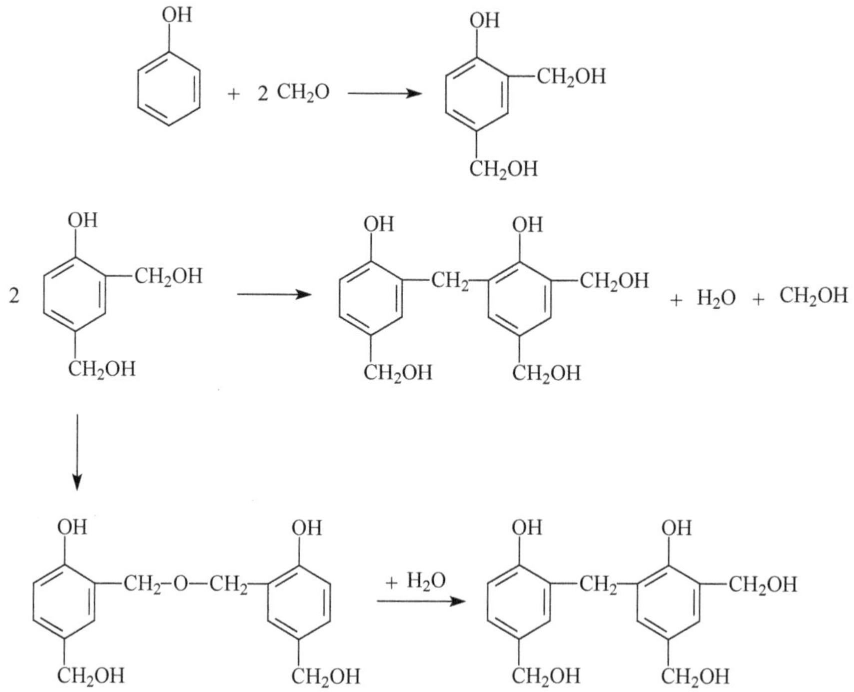

Repetitionof above processes lead to the formation of macromolecules, in which the aromatic rings are connected by methylene and dimethylenether bridges.

Under heating, the resols transform into partially cross-linked resitols, and then into the insoluble and infusible resites.

The reactive hydroxymethyl groups allow a chemical modification of these resins. The phenol formaldehyde resins, etherified by butanol or esterified by fatty acids, are more soluble and more flexible, allowing their use as lacs for paintingpurposes.

Phenol formaldehyde resins (phenoplasts) (PF) produced during the condensation of phenol with other aldehydes such as acetaldehyde and furfuralare alsoknown.

6. 2   AMINOPLASTS (CARBAMIDE RESIN)

Aminoplasts mean generally the macromolecular reaction products of compounds containing amino groups in the molecule or amide groups with aldehydes. Main types of aminoplasts are products of polycondensation reaction of formaldehyde with urea, melamine, dicyanodiamide, guanidine, thiourea, etc.

As a result of the reaction of amine compounds with aldehydes, the aldehyde molecule is attached to the nitrogen atom of amino group with a simultaneous rearrangement of the hydrogen atom to the oxygen atom of aldehyde group. While using formaldehyde,compounds containing hydroxymethylamino groupsareformed,which are easy to separate. With an excess of formaldehyde it is possible to replace all the hydrogen atoms of amine groups by methylol groups.

$$R-NH_2 + CH_2O \longrightarrow R-NH-CH_2OH$$

$$R-NH-CH_2OH + CH_2O \longrightarrow R-N(CH_2OH)_2$$

The reaction for obtaining methylol (hydroxymethyl) derivatives runs in the aquatic environment and is catalyzed both by hydrogen and hydroxyl ions. The pH value also determines the rate of reaction of urea with formaldehyde. The process is the slowest one in the pH range of 4–8. Increasing pH above 8 or reducing it below 4 speeds up the processsignificantly. The resulting methylol compounds undergo polycondensation with the formation of macromolecules linked by methylene and dimetylene ether bridges.

$$-NH-CH_2OH + H_2N- \longrightarrow -NH-CH_2-NH- + H_2O$$

$$-NH-CH_2OH + HOCH_2-NH- \longrightarrow -NH-CH_2-O-CH_2-NH- + H_2O$$

The dimetylene ether bridges arise when the polycondensation reactions are carried out in a neutral or alkaline environment. They may also occur in slightly acidic environment with excess of large molar formaldehyde.

The polycondensation reaction of multifunctional amine compounds with formaldehyde is conducted in an inert environment in order to avoid a premature crosslinking of product. The obtained urea or melamine–formaldehyde-derived polycondensates are brought to pH = 7 and in this state they remain stable for about 6 months. Crosslinking of these resins is accelerated by hydrogen ions and by elevated temperature. The resulting products are used as adhesives, binders for molding and laminates and as impregnating materials.

The urea or melamine–formaldehyde modified with butyl alcohol resins are used forchemically setting varnishes. In these products, a part of methylol group undergoes an etherification by butyl alcohol. The butyl chains cause inner plastification of the formed resins.

6. 3   POLYESTERS

Polyesters are formed by polycondensation of multifunctional acids with alcohols with evolution of water molecules. In this reaction, the state of equilibrium establishes rapidly and the resulting water acts as a hydrolysing factor. Therefore, in order to shift the equilibrium state, it is necessary to remove water from the reaction environment by carrying out the process at high temperature (220–250 °C) or by azeotropic distillation of water from toluene or xylene. The reaction proceeds according to the scheme:

$$n\ HO-R-OH + n\ HOOC-R'-COOH \rightleftharpoons H\left[O-R-O-\overset{\overset{O}{\|}}{C}-R'-\overset{\overset{O}{\|}}{C}\right]_n OH + (2n-1)\,H_2O$$

Catalysts for this reaction are acids, such as sulfuric acid and p-toluenesulfonic acid. The use of elevated temperatures lead to a series of side reactions, and to eliminate them the process is conducted in nitrogen atmosphere or in a solvent whose use allows to lower the reaction temperature.

Among the other methods of obtaining polyesters are:

(a) homopolycondensation of hydroxyacids in which the number of methylene groups exceeds five. It ensures a complete elimination of cyclization:

$$n\ HO-R-COOH \rightleftharpoons H\left[O-R-\overset{\overset{O}{\|}}{C}\right]_n OH + (n-1)\,H_2O$$

(b) alcoholysis reaction of dimethyl esters of dicarboxylic acid with glycols, catalyzed by acid or acetates of cadmium, zinc, and lead.

$$n\ HO-R-OH + n\ H_3CO-\overset{\overset{O}{\|}}{C}-R-\overset{\overset{O}{\|}}{C}-OCH_3 \rightleftharpoons H\left[O-R-O-\overset{\overset{O}{\|}}{C}-R-\overset{\overset{O}{\|}}{C}\right]_n OCH_3$$

$$+ (2n\text{-}1)\ CH_3OH$$

(c) tpolycondensation reaction at the interface of glycols with carboxylic acid dichlorides:

$$n{+}1\ HO{-}R{-}OH\ +\ n\ Cl{-}\overset{O}{\overset{\|}{C}}{-}R'{-}\overset{O}{\overset{\|}{C}}{-}Cl\ \longrightarrow\ H{-}\!\!\left[O{-}R{-}O{-}\overset{O}{\overset{\|}{C}}{-}R'{-}\overset{O}{\overset{\|}{C}}\right]_n\!\!{-}O{-}R{-}OH$$

Polyesters, formed as result of the above-mentioned reaction, can be divided into linear polyesters, alkyd resins, unsaturated polyesters, and polyarylates.

Linear polyesters are obtained from dicarboxylic acids and glycols. Aliphatic polyesters are characterized by a low melting temperature (below 80 °C) and are well soluble in many organic solvents, such as chloroform, benzene, acetone, tetrahydrofuran, and ethyl acetate. The ester bindings of aliphatic polyester are easily hydrolyzed at room temperature under the influence of the aqueous solution of base. The resistance of aliphatic polyesters to hydrolysis increases with the increase of the number of methylene groups separating ester bonds. The linear polyesters derived from aromatic dicarboxylic acids are characterized by a greater degree of crystallization and higher melting temperatures. The biggest importance in this group of polymers has poly (ethylene terephthalate) (PET), which is obtained by the reaction of dimethyl terephthalate with ethylene glycol. This polymer is used for packing purposes (bottles for beverages) and also to produce foils and synthetic fibers (Dacron, Polyester, Terylene, and Elana).

The alkyd resins are polyesters obtained from aliphatic polyols and phthalic acids. In the first stage of reaction,linear polyesters, which can be then cross-linked are obtained. These resins are often modified with fatty acids in order to improve their flexibility, solubility, and resistance to water.

The unsaturated polyester resins are obtained by polycondensation of glycols with fumaric acid or maleic acid anhydride. Here, it is possible to use other carboxylic acids as a comonomer, such as adipic, phthalic, and hexachloroendomethylenetetrahydrophthalic (so-called HET acid).

The introduction of chlorine, bromine, phosphorus, or boron atoms to the polyester molecule reduces the flammability of the product. The obtained linear polyesters, containing double bonds, can be cross linked by radical copolymerization with vinyl or allyl compounds. The most frequently used crosslinking agent is styrene.

Poliarylates are products of polycondensation of diphenols with aromatic dicarboxylic acids. The characteristic chemical structure of these polymers, whose main chains are composed almost exclusively of aromatic rings linked together by chemical bonds, gives them some specific properties. Above all they are characterized by a high softening temperature, thermal resistance, and resistance to chemical agents.

6.4 POLYCARBONATES

Polycarbonates are linear polyesters of carbonic acid. Due to special methods,specific physicochemical properties, and thermoplasticity they constitute a distinct group of polymers.

Depending on the structure of the polymer chain, polycarbonates are divided into aliphatic, aliphatic–aromatic, and aromatic. The greatest practical importance represent the aromatic polycarbonates.

Polycarbonates are obtained by condensation polymerization reaction of bifunctional alcohols or phenols with phosgene or carbonic acid esters. The reaction may be carried out at the interface or in a solvent.

The reaction at the interface is carried out in two-phase system composed of alkaline aqueous solution of biphenolate and phosgene in an organic solvent (usually in methylene chloride). The process to obtain a polycarbonate is a two-step process. In the first stage, the reaction proceeds in the presence of excess of phosgene and leads to the formation of polycarbonates with molecular weight of 5000–10,000 and terminal chloromethyl groups:

$$n \, NaO-Ar-ONa + (n+1) \, COCl_2 \longrightarrow Cl-\overset{O}{\overset{\|}{C}}\left[O-Ar-O-\overset{O}{\overset{\|}{C}}\right]_n Cl + n \, NaCl$$

An excess of phosgene is necessary as a part of it is hydrolyzed by an undesirable side reaction.

$$COCl_2 + 4 \, NaOH \longrightarrow Na_2CO_3 + 2 \, NaCl + H_2O$$

The second phase of the reaction is carried out with the excess of sodium hydroxide. Then follows the hydrolysis of terminal chloroformate groups to phenolic and a further polycondensation by the reaction of chloroformate groups with phenolic ones:

$$Cl-\overset{O}{\overset{\|}{C}}\left[O-Ar-O-\overset{O}{\overset{\|}{C}}\right]_n Cl + 5 \, NaOH \longrightarrow Na\left[O-Ar-O-\overset{O}{\overset{\|}{C}}\right]_n OH$$

$$+ \, 2 \, NaCl + Na_2CO_3 + 2 \, H_2O$$

$$Na\left[O-Ar-O-\overset{O}{\overset{\|}{C}}\right]_n OH + Cl-\overset{O}{\overset{\|}{C}}\left[O-Ar-O-\overset{O}{\overset{\|}{C}}\right]_n Cl \xrightarrow[- \, NaCl]{NaOH} Na\left[O-Ar-O-\overset{O}{\overset{\|}{C}}\right]_{n+m} OH$$

The reaction is catalyzed by a small amount of tertiary amines and quaternary ammonium bases.

The synthesis of polycarbonates can also be carried out in a solution in presence of a substance binding the emitted hydrochloride. Tertiary amines and pyridineare most commonly used for this purpose.

Another method for obtaining polycarbonates is the ester exchange reaction of biphenols with the alkyl or aryl carbonates.

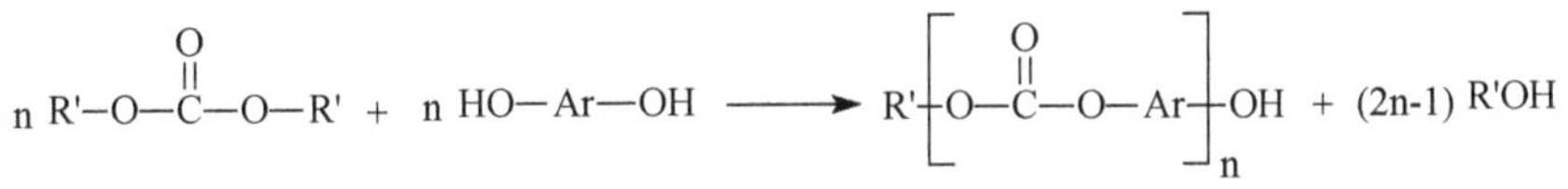

$$n \; R'-O-\overset{\overset{O}{\|}}{C}-O-R' \; + \; n \; HO-Ar-OH \longrightarrow R'\left[O-\overset{\overset{O}{\|}}{C}-O-Ar\right]_n OH \; + \; (2n\text{-}1) \; R'OH$$

In this reaction, an equilibrium state establishes and it is necessary to remove its byproducts,alcohol, or phenolfrom the reaction environment. Therefore, the process is carried out in the second stage at553–573 K (280–300 °C) and under a reduced pressure. As products, one obtains polycarbonates with molecular weight not exceeding 50,000.

The ester exchange catalysts are oxides, hydroxides, hydrides, and alkoxides of alkali metals and alkaline earths (oxides of Ca, Sr, Ba, and Ra) as well as oxides, hydroxides, and salts of heavy metals.

In this process, the side reactions may occur associated with the rearrangement and cross-linking of the product.

The greatest practical importance exhibit polycarbonates obtained by reacting phosgene with 4,4'-ddihydroxy diphenylpropane, called Dian (bisphenol A).

The reaction proceeds according to the scheme:

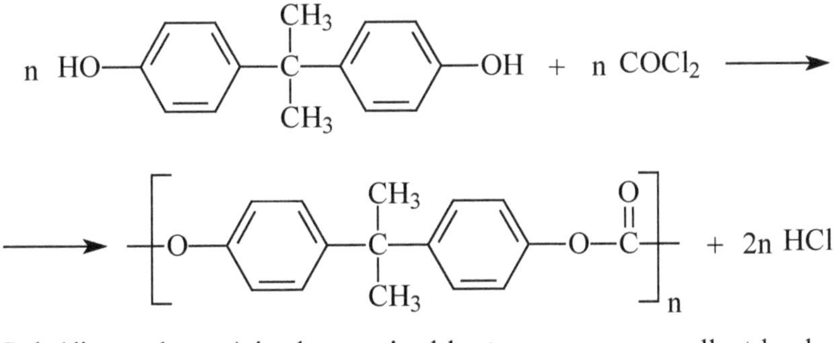

Poly(dian carbonate) is characterized by transparency, excellent hardness, and resistance to heat in a very wide range. It is used in the manufacture of bulletproof glass in cars, in diplomatic representations and as windows in banks, to cover the shieldings of precious stained glass windows, screens for the audience at the hockey rinks, cabs in supersonic jets, diving masks, and shields used by the antiterrorist brigade. The possibility of sterilizing the polycarbonate plastic by steam also made it an interesting material to manufacture parts for medical equipment, safety glasses, infant feeding bottles, containers, and packaging.

6. 5 POLY(ETHER-KETONES) (PEEK)

Poly(ether–ketones) are obtained by polycondensation of diphenols with 4,4'-difluoracetophenon.

The reaction of hydroquinone with 4,4'-difluoracetophenon is carried out in diphenylsulfon and in the presence of anhydrous potassium carbonate at temperature of 300–350 °C.

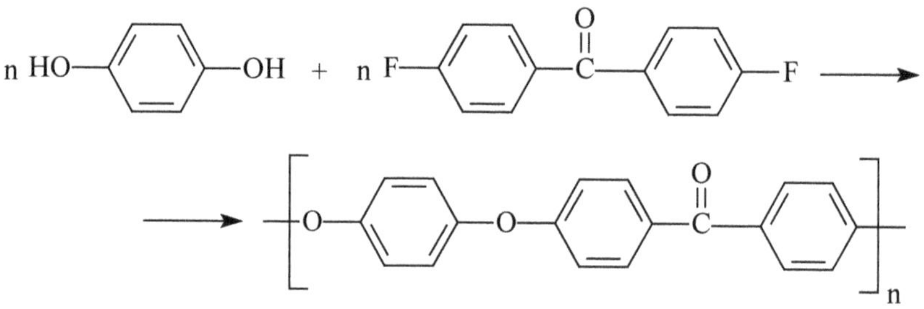

The poly (ether-ether-ketone) (PEEK) obtained in this way contains 30–40% of crystallites with low flammability (oxygen index = 35). It can be used up to the temperature of 280 °C. The polymer is produced since 1982 by an English company ICI under the trade name of Victrex.

In 1990, an US company DuPont Co. began to produce poly(ether–ketone–ketone (PEKKA) by the reaction of diphenyl oxide with terephthalic acid dichloride:

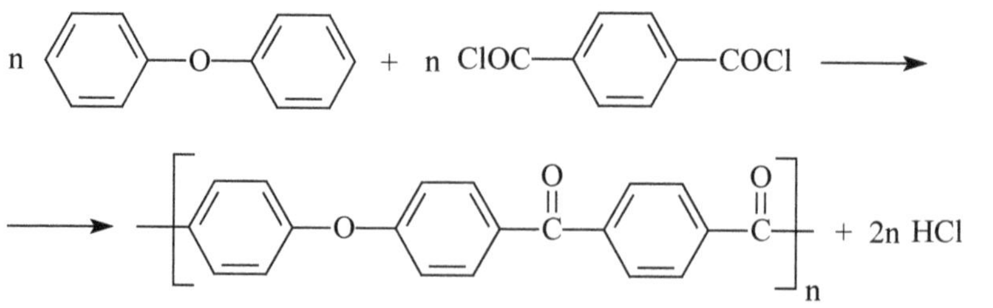

The emitted hydrogen chloride is neutralized by the addition of a weak base. Replacing a part terephthalic acid dichloride by an equivalent amount of izoftaloil dichloride reduces the degree of crystallinity and improves the processing properties of polymer.

6. 6   POLYAMIDES

Polyamides are polymers with a chain structure in which the individual hydrocarbon segments are connected by amide –CO–NH– bonds.

These connections can be obtained by three methods:
- polycondensation of dicarboxylic acids with diamines,
- polycondensation of ω amino acids,
- polymerization of lactams

In 1936, in laboratory of W. H. Carothers at Du Pont Co. the synthesis of poly(adipate hexamethylendimine), calledpolyamide 6. 6. , was elaborated. It proceeds the following scheme:

$$n\ H_2N(CH_2)_6NH_2\ +\ n\ (CH_2)_4(COOH)_2 \longrightarrow H\!-\!\!\left[HN(CH_2)_6NHCO(CH_2)_4CO\right]_n\!\!-\!OH$$

$$+\ (2n\text{-}1)\ H_2O$$

The industrial production of this polymer known as "nylon" (NY—New York, LON—London) started in 1940 (1938?). At the beginning, it was used for production of thin fibers designed for stockings. The outbreak of war caused the use of production of nylon in the manufacture of parachutes. For this reason, polyamides influenced the course of the war and contributed significantly to the increase of research on polymers.

Another important polyamide is polycaprolactam, called polyamide 6. It is obtained from caprolactam (caproate acid lactam), which is subjected initially to hydrolysis into aminocaproic acid and then maintained at temperature of 240–250 °C with a simultaneous distillation of the water. The polycondensation reaction of caprolactam is an equilibrium reaction, and 8–12% of unreacted caprolactam remains in the product. For this reason, the obtained polyamide 6 cannot be used directly for production of fibers as it is the case of polyamide 6. 6 but must be fragmented and subjected to extraction of unreacted caprolactam with water ormatanol. After extraction the product is remelted and processed into fibers or granulates. The production and the processing of polyamides at high temperatures must take place in a neutral atmosphere like nitrogen, because the product in melt phase oxidizes easily and darkens.

Polycaprolactam can be also obtained by anionic polymerization of caprolactam. The obtained polymer is characterized by a very high molecular weight and a low liquidity, which prevents its thermal processing. The product is suitable for machining only.

Polyamides are used in the manufacture of synthetic fibers and as construction materials processed by injection molding.

6. 7   AROMATIC POLYAMIDES

Introduction to the polyamide chain of alicyclic or aromatic rings results in improving the thermal resistance of the product.

Among the various aliphatic–aromatic polyamides polyhexamethylene terephtalamideis noteworthy, called polyamide 6T. It is obtained by polycondensation of salts of terephthalic acid with hexamethylenediamine. The resulting salt is easy to clean crystalline compound. Its use in polycondensation, instead of the mixture of diamine with dicarboxylic acid, ensures the maintenance of equimolar ratio of functional groups during the process.

The fiber from polyamide 6T is formed from sulfuric acid solutions in a wet way or from 15 to 25% solution of polymer in trifluoroacetic acid by a method involving evaporation of solvent. The newly formed fibers are stretched by 200–400% at the temperature of 160–180 °C. It provides them the strength of 0. 2–0. 5 N/tex and the elongation at break of 10–40%. An important advantage of products made from these fibers is to maintain the shape at elevated temperatures. They have also a good resistance to the action of bases. Dyeing of this type of fibers is difficult. A good coloring provide only the dispersed dyes.

A growing interest attract aromatic polyamides derived from dicarboxylic acids and aromatic diamines. The highest importance from this polyamide group has the m-phenylene isophthalamide, produced under the trade name "Nomex" in US and "Feniton" in Russia, respectively.

The aromatic polyamides are obtained by reacting diamines with dicarboxylic acid chlorides:

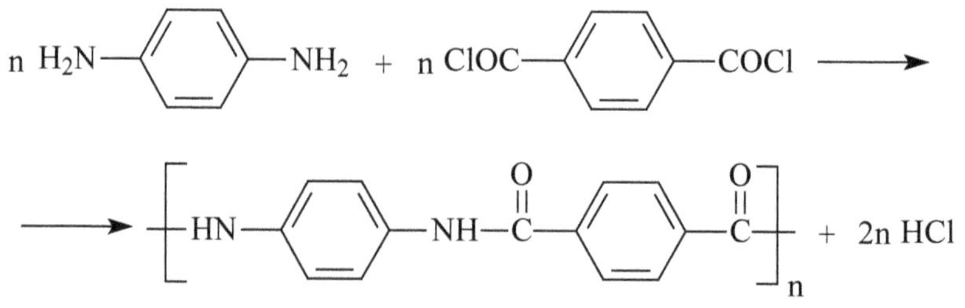

This reaction can be carried out by polycondensation at the interface, emulsion, or in solution. In industrial practice, the solution method is mainly used. It has the advantage that with the formation of polymer, one obtains the spinning solution with sufficient concentration from which fibers can be produceddirectly. Dimethylacetamide is used as solvent. It has the advantage to be acceptor of hydrogen chloride which is secreted during the reaction. Nevertheless, the calcium hydroxide is added to the polymer solution, which neutralizes the reaction medium. The calcium chloride formed during the reaction increases the stability of the spinning solution. The addition of alkali metal salts to the solution, in which the polycondensation takes place, affects positively the course of the reaction ensuring the formationof a polymer with higher molecular weight.

The fibers made from aromatic polyamides can be formed by dry evaporation of the solvent or by wet precipitation of polymer in form of fibers in a coagulating solution. The aqueous solvent (dimethylformamide or alcohol) solutions are used as the coagulation bath.

6.8   POLYIMIDES

Polyimides are products of polycondensation of tetracarboxylic acids or their dianhydrades with diamines. In the synthesis,pyromellitic acid anhydride or its derivatives and various diamines are most frequently used. The polycondensation reaction is carried out in the solution. Dimethylformamide, dimethylacetamide, dimethylsulfoxide, and N-methylpyrrolidone are used as solvents.

In the first stage of the synthesis, carried out at 253–343 K(−20–70 °C),polyamide acidsareformed. Heated at temperatures above 473 K (200 °C), under reduced pressure or in the presence of dehydrating factors such acetic anhydride and ketenes, the polyamide acids undergo a cyclization process to polyimides.

The reaction proceeds according to the scheme:

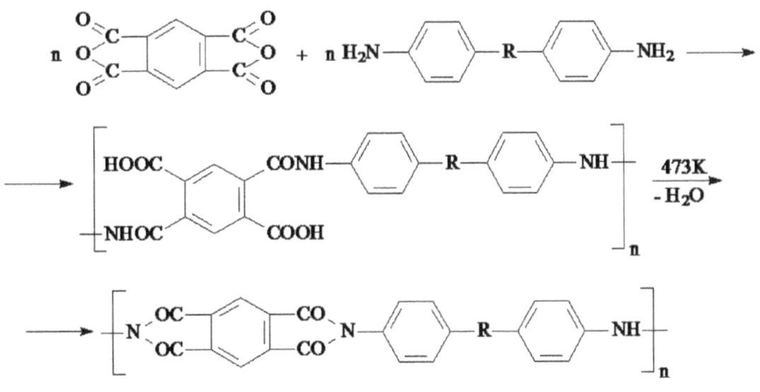

R = CH_2, O, S, etc.

Polyimides are thermostable polymers becausethey do not melt and soften, and ignite, but undergo a slow pyrolysis. Through the thick polyimide fabric flame penetrates after 15 min only.

Polyimides are not soluble and do not swell in most known organic solvents. The prolonged action of steam causes a slight decrease in fiber strength of polyimide only. However, the action of concentrated acids and bases causes a harm to the material through its hydrolysis.

Because of their good properties,polyimide areused to produce heat- and flame-resistant fibers.

Polyimide fibers are obtained from the solution polyamidoacid through its coagulation, or by evaporation of the solvent.

In the case of wet processing, the content of polymer in the spinning solution should not exceed 10%. The solution is introduced into a coagulation bath consisting of a mixture of solvent and aqueous solution of inorganic salts (e. g. , calcium thiocyanate).

The formed fiber is dried and subjected to a staged heat treatment with the task of providing the fibers with adequate physicomechanical properties. During the heat treatment, the polymer undergoes cyclization and the fiber is stretched by 170–200%.

Polyimide fibers are manufactured under the brand names: Arimid (Russia) and Kapton (USA).

6. 9   POLYSULFONES

Polysulfones are polymers containing the main chain of the sulfone groups – SO_2-.

Bindings of this type can be obtained as a result of the polycondensation reaction of chlorinated derivatives of diphenylo sulfone with sodium salts of diphenyls or by copolymerization of sulfur dioxide with vinyl monomers.

A practical importance in the polymer technology only the first method has found.

On industrial scale polysulfones are obtained by the reaction of 4,4'-dichlorodiphenylosulfone with bisphenol A (dian)or with 1,4-dihydroxybenzene:

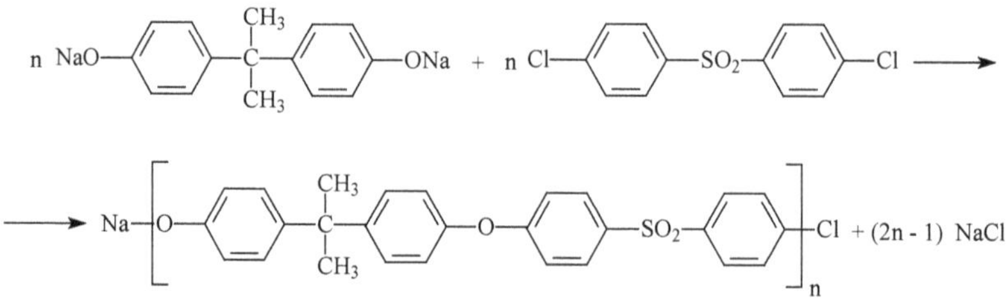

In the commercial grades polysulfones, this type of polycondensation degree n ranges usually from 50 to 80. Polysulfones exhibit excellent utility qualities in a wide range of temperatures: from -100 to $+150\ °C$.

Polysulfones resist the action of inorganic acids, alkalis, alcohols, and aliphatic hydrocarbons. They are soluble in chlorinated hydrocarbons and are partially soluble in esters, ketones, and aromatic hydrocarbons.

Thecommercial products are obtained from polysulfones by the injection method. Polysulphones can alsobe processed into panels characterized by a good transparency.

6. 10 POLYSULFIDES

The earliest known polymers containing sulfur in the main chain are thioplasts (polysulfide rubbers), called tiocols (polysulfide resines). They are obtained by the reaction of alkyl dichloride with sodium polysulfide. An example of such a reaction is the synthesis of thiocol A with 1,2-dichloroethane:

$$n \ ClCH_2CH_2Cl \ + \ n \ Na_2S_4 \longrightarrow \left(CH_2-CH_2-\underset{\underset{S}{\|}}{S}-\underset{\underset{S}{\|}}{S} \right)_n + 2n \ NaCl$$

The polycondensation process proceeds easily at 80 °C in an aqueous medium and in the presence of magnesium hydroxide suspension. The resulting latex is subjected to coagulation process with acids and the isolated polymer is separated from the residue, washed with water, dried, and pressed into blocks.

As a result of the action of bases on thiocols one can remove the lateral sulfur atoms with formation of a disulfide polymer with chemical formula:

$$\left(CH_2-CH_2-S-S \right)_n$$

Thioplasts are obtained also from the dichlorodiethyl ether (Thiocol D) and from glycerol dichlorohydrin (Vulcaplas).

Thioplasts have found application as synthetic rubbers and can be vulcanized using the zinc oxide. During the vulcanization process a slight tear gas is emitted. It hinders the technology process and necessitates cooling of forms in the press before removing the vulcanizates from them. Thioplasts can alsobe cross-linked by mixing them with polyester, epoxy, and phenolic resins. The mechanical properties of thioplast vulcanizates are worse than those of rubber. Moreover, thiocols have an unpleasant smell and tends to hardening during storage, which limits their application.

An important advantage of thioplasts is their very low solubility in oils and in many other organic solvents, low gas permeability, and resistance to ozone. Therefore, thiocols are used in production of gaskets and hoses resistant to gasoline and oils, as well as a guard for cables instead of lead. They are also used as binders in the manufacture of solid rocket propellants due to their excellent combustibility.

Polysulfides are also obtained during the opening of three-, four-and five-member cyclic sulfides, in the presence of cationic, anionic, and coordination catalysts. An example of this type of reaction is the polymerization of the ethylene sulfide.

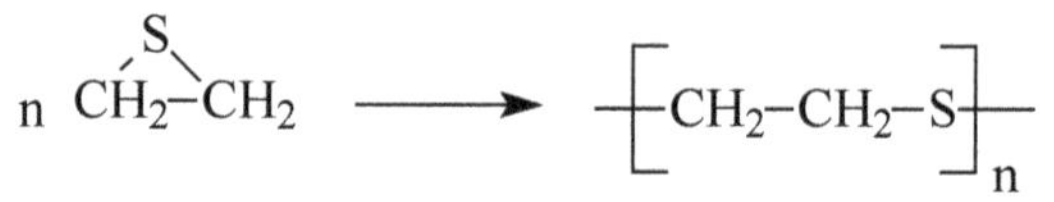

$$n \ \overset{S}{\overset{\diagup\diagdown}{CH_2-CH_2}} \longrightarrow \left[CH_2-CH_2-S \right]_n$$

Good results are also obtained during the reaction of dithiols with dialkenes. If this reaction is carried out by the radical mechanism then it runs in disagreement with the Markovnikov's rule and the formed polymer does not contain side groups.

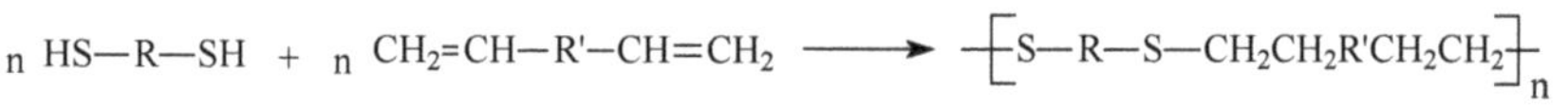

$$n \ HS-R-SH \ + \ n \ CH_2=CH-R'-CH=CH_2 \longrightarrow \left[S-R-S-CH_2CH_2R'CH_2CH_2 \right]_n$$

During the ionic reaction catalyzed by hydrogen or hydroxyl ions the resulting polymer contains the side methyl groups:

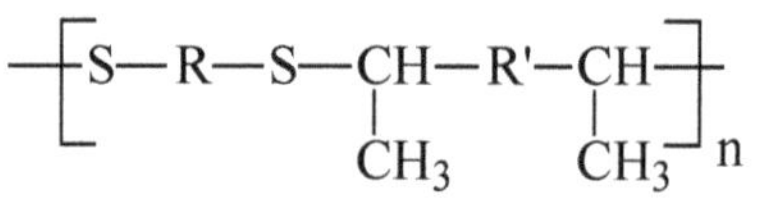

$$\left[S-R-S-\underset{\underset{CH_3}{|}}{CH}-R'-\underset{\underset{CH_3}{|}}{CH} \right]_n$$

Polysulfides with a high molecular weight (20,000–60,000) are obtained in reaction of dithiols with alkynes through the stage of formation of a reactive vinyl sulfide. This reaction is usually initiated by γ rays.

$$n+1 \ HS-R-SH \ + \ n \ HC{\equiv}CR' \longrightarrow HS \left[R-S-CH_2-\underset{\underset{R'}{|}}{CH}-S \right]_n RSH$$

The polymers obtained in this way are used as additives for adhesives and putty.

Dithiols react with aldehydes and ketones forming the polythioacetals which are more resistant to hydrolysis:

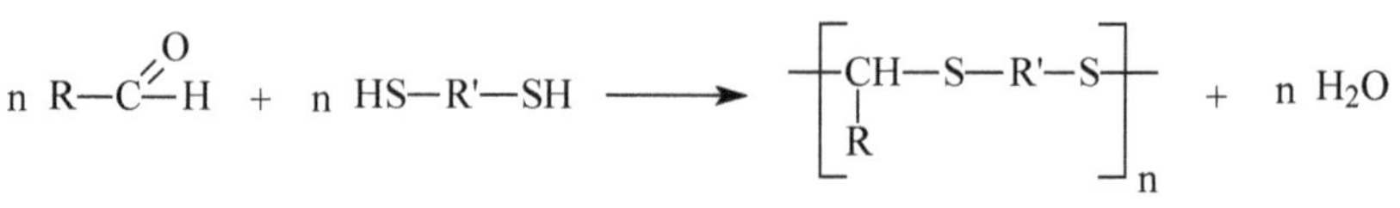

$$n \ R-\overset{O}{\overset{\diagup\diagdown}{C}}-H \ + \ n \ HS-R'-SH \longrightarrow \left[\underset{\underset{R}{|}}{CH}-S-R'-S \right]_n \ + \ n \ H_2O$$

Aromatic polysulfidesare alsoknown for their property of self-distinguishing, characterized by a good thermal and chemical resistance. .

They are formed by the reaction of p-dichlorobenzene with sodium sulfide or by the alkaline hydrolysis of p-chlorothiophenol:

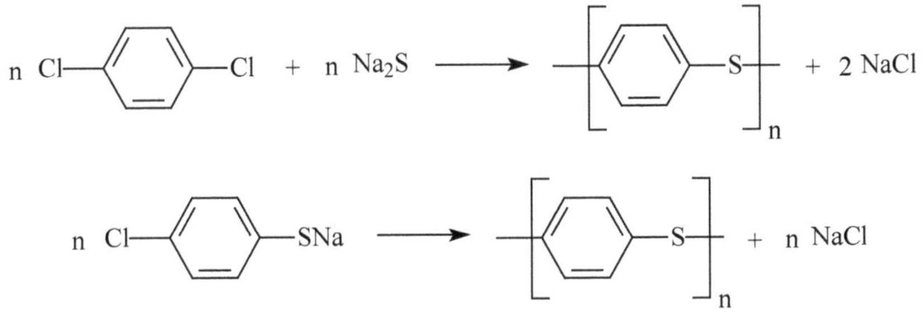

During the condensation of diphenylether with sulfur chlorides a polymer-containing ether and sulfide groups areformed.

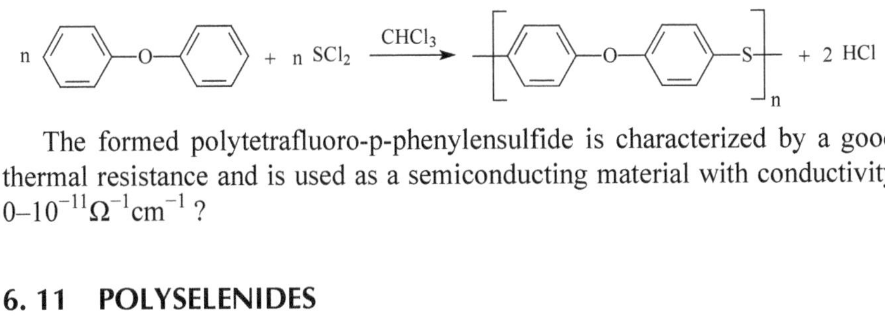

The formed polytetrafluoro-p-phenylensulfide is characterized by a good thermal resistance and is used as a semiconducting material with conductivity $0-10^{-11}\Omega^{-1}cm^{-1}$?

6. 11   POLYSELENIDES

Recently, a method was developed for the synthesis of polyselenides by a reaction of aromatic chlorine-containing derivativeswith sodium selenide (metallic sodium and black selenium) in dimethylformamide or in pyridine at the temperature of 120 °C.

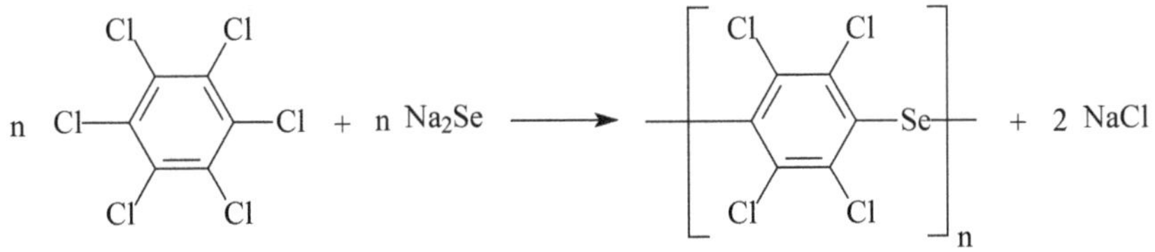

The obtained polymer has a glass transition temperature of 104 °C, thermal resistance up to 320 °C and conductivity of $1.5 \times 10^{-11}\Omega^{-1}$ cm^{-1}.

6. 12   SILICONES

Silicones are macromolecular organosilicon compounds which, owing to their excellent properties, find increasing industrial applications. They are obtained from alkyl- or arylochlorosilanes by hydrolysis, followed by polycondensation of multifunctional monomers. At present, the arylochlorosilanes are obtained by reaction of alkyl halogens with silicon alloy with copper.

The reaction mixture of synthesis products is separated on distillation columns. In a similar way the phenylchlorosilanesare obtained.

By hydrolysis of dichlorosilanes and neutralization with soda offormed siloxanes one obtains products with low molecular weights, applicable as silicone oils. In order to obtain a specific grade of the oil, the polymerization process is led in an appropriate proportion mixture of single and bifunctional monomers with a catalyst (H_2SO_4), at room or elevated temperature with a simultaneous control of viscosity increase. The silicone oils can be converted into emulsions, pastes, sealants, and lubricants.

Silicone rubbers, produced from bifunctional monomers with a high degree of purityare of practical importance. One obtains polysiloxanes with consistency of a syrup and molecular weight in the order of 300,000–800,000. These polymers are mixed in a crusher with fillers, pigments, and stabilizers. The obtained premix can be stored for several months until the vulcanization. The vulcanization may be carried out with organic peroxides or sulfur, together with accelerators. The vulcanization of silicone rubber with peroxides is carried out in the ordinary and transfer molding press for 20 min at temperature of 100–180 °C and under pressure of 50–100 atm. The silicone rubbers after vulcanization require an annealing for 2–10 h at elevated temperatures (200–250 °C). Silicones have also found application in paint industry.

REFERENCES

Abusleme, J.; Giannetti, E. . Emulsion Polymeryzation. *Macromolecules*1991,24.

Adesamya, I. Classification of Ionic Polymers. *Polym. Eng. Sci.* 1988, *28*, 1473.

Akbulut, U.;Toppare, L.;Usanmaz, A.; Onal, A. Electroinitiated Cationic Polymeryzation of Some Epoksides. *Makromol. Chem. Rapid. Commun.* 1983, *4*, 259.

Andreas, F.; Grobe, K. *Chemia propylenu*; WNT:Warszawa, 1974.

Barg, E. I. *Technologia tworzyw sztucznyc*; PWT:Warszawa, 1957.

Billmeyer, F. W. *Textbook of Polymer Science*; John Wiley and Sons: New York, 1984.

Borsig, E.; Lazar, M. Chemical Modyfication of Polyalkanes. *Chem. Listy*1991, *85*, 30.

Braun, D.; Czerwiński, W.; Disselhoff, G.; Tudos, F. Analysis of the Linear Methods for Determining Copolymeryzation Reactivity Ratios. *Angew. Makromol. Chem.* 1984, *125*, 161.

Burnett, G. M.; Cameron, G. G.; Gordon, G.; Joiner, S. N. Solvent Effects on the Free Radical Polymerization of Styrene. *J. Chem. Soc. Faraday Trans. 1*1973, *69*, 322.

Challa, R. R.; Drew, J. H.; Stannet, V. T.; Stahel, E. P. Radiationindueed Emulsion Polymeryzation of Vinyl Acetate. *J. Appl. Polym. Sci.* 1985, *30*, 4261.

Candau, F. Mechamism and Kinetics of the Radical Polymeryzation of Acrylamide In Inverse Micelles. *Makromol. Chem. , Makromol Symp.* 1990, *31*, 27.

Ceresa, R. J. *Kopolimery blokowe i szczepione*; WNT: Warszawa, 1965.

Charlesby, A. *Chemia radiacyjna polimerów*; WNT: Warszawa,1962.

Chern, C. S; Poehlein, G. W. Kinetics of Cross-Linking Vinyl Polymerization. *Polym. Plast. Technol. Eng.* 1990, *29*, 577.

Chiang, W. Y.; Hu, C. M. Studies of Reactions with Polymers. *J. Appl. Polym. Sci.* 1985, *30*, 3895–4045.

Chien, J. C.; Salajka, Z. Syndiospecific Polymeryzation of Styrene. *J. Polym. Sci. A*1991, *29*, 1243.

Chojnowski, J.; Stańczyk, W. E. Procesy chlorowania polietylenu. *Polimery*1972, *17*, 597.

Collomb, J.; Morin, B.; Gandini, A.; Cheradame, H. Cationic Polymeryzation Induced by Metal Salts. *Europ. Polym. J.* 1980, *16*, 1135.

Czerwiński, W. K. Solvent Effects on Free-Radical Polymerization. *Makromol. Chem.* 1991, *192*, 1285–1297.

Dainton, F. S. *Reakcje łańcuchowe*; PWN:Warszawa, 1963.

Ding, J.; Price, C.; Booth, C. Use of Crown Ether in the Anionic Polymeryzation of Propylene Oxide. *Eur. Polym. J.* 1991, *27*, 895–901.

Dogatkin, B. A. *Chemia elastomerów*; WNT:Warszawa, 1976.

Duda, A.; Penczek, S. Polimeryzacja jonowa i pseudojonowa. *Polimery*1989, *34*, 429.

Dumitrin, S.; Shaikk, A. S.; Comanita, E.; Simionescu, C. Bifunetional Initiators. *Eur. Polym. J.* 1983, *19*, 263.

Dunn, A. S. Problems of Emulsion Polymeryzation. *Surf. Coat. Int.* 1991, *74*, 50.

Eliseeva, W. I.; Asłamazova, T. R. Emulsionnaja polimeryzacja v otsustivie emulgatora. *Uspiechi Chimii*1991, *60*, 398.

Erusalimski, B. L. Mechanism dezaktivaci rastuszczich ciepiej v procesach ionnoj polimeryzacji vinylovych polimerov. *Vysokomol. Soed. A.* 1985, *27*, 1571.

Fedtke, M. *Chimiczieskijereakcjipolimerrov*;Khimia:Moskwa, 1990.

Fernandez-Moreno, D.; Fernandez-Sanchez, C.; Rodriguez, J. G.; Alberola, A. Polymeryzation of Styrene and 2-vinylpyridyne Using AlEt-VCl Heteroaromatic Bases. *Eur. Polym. J.* 1984, *20*, 109.

Fieser, L. F.; Fieser, M. *Chemia organiczna*; PWN: Warszawa, 1962.

Flory, P. J. Nauka o makrocząsteczkach. *Polimery*1987, *32*, 346.

Frejdlina, R. Ch.; Karapetian, A. S. *Telomeryzacja*; WNT: Warszawa, 1962.

Funt, B. L.; Tan, S. R. The Fotoelectrochemical Initiation of Polymeryzation of Styrene. *J. Polym. Sci. A*1984, *22*, 605.

Gerner, F. J.; Hocker, H.; Muller, A. H.; Shulz, G. V. On the Termination Reaction In The Anionic Polymeryzation of Methyl Methacrylate in Polar Solvents. *Eur. Polym. J.* 1984, *20*, 349.

Ghose, A.; Bhadani, S. N. Elektrochemical Polymeryzation of Trioxane in Chlorinated Solvents. *Indian J. Technol.* 1974, *12*, 443.

Giannetti, E.; Nicoletti, G. M.; Mazzocchi, R. Homogenous Ziegler-Natta Catalysis. *J. Polym. Sci. A*1985, *23*, 2117.

Goplan, A.; Paulrajan, S.; Venkatarao, K.; Subbaratnam, N. R. Polymeryzation of N,N'- Methylene Bisacrylamide Initiated by Two New Redox Systems Involving Acidic Permanganate. *Eur. Polym. J.* 1983, *19*, 817.

Gózdz, A. S.; Rapak, A. Phase Transfer Catalyzed Reactions on Polymers. *Makromol. Chem. Rapid Commun.* 1981, *2*, 359.

Greenley, R. Z. An Expanded Listing of Revized Q and e Values. *J. Macromol. Sci. Chem. A*1980, *14*, 427.

Greenley, R. Z. Recalculation of Some Reactivity Ratios. *J. Macromol. Sci. Chem. A*1980, *14*, 445.

Guhaniyogi, S. C.; Sharma, Y. N. Free-Radical Modyfication of Polipropylene. *J. Macromol. Sci. Chem. A*1985, *22*, 1601.

Guo, J. S.; El Aasser, M. S.; Vanderhoff, J. W. Microemulsion Polymerization of Styrene. *J. Polym. Sci. A*1989, *27*, 691.

Gurruchaga, M.; Goni, I.; Vasguez, M. B.; Valero, M.; Guzman, G. M. An Approach to the Knowledge of The Graft Polymerization of Acrylic Monomers Onto Polysaccharides Using Ce (IV) as Initiator. *J. Polym. Sci. Part C*1989, *27*, 149.

Hasenbein, N.; Bandermann, F. Polymeryzation of Isobutene With $VOCl_3$ in Heptane. *Makromol. Chem.* 1987, *188*, 83.

Hatakeyama, H. et al. Biodegradable Polyurethanes from Plant Components. *J. Macromol. Sci. A.* 1995, *32*, 743.

Higashimura, T.; Law, Y. M.; Sawamoto, M. Living Cationic Polymerization. *Polym. J.* 1984, *16*, 401.

Hirota, M.; Fukuda, H. Polymerization of Tetrahydrofuran. *Nagoya-ski Kogyo Kenkynsho Kentyn Hokoku*,1984, 51.

Hodge, P.; Sherrington, D. C. *Polymer-Supported Reactions in Organic Syntheses*;
J. Wiley and Sons: New York,1980; tłum. na rosyjski Mir: Moskwa, 1983.

Inoue, S. Novel Zine Carboxylates as Catalysts for the Copolymeryzation of CO_2 With Epoxides. *Makromol. Chem. Rapid Commun.* 1980, *1*, 775.

Janović, Z. *Polimerizacije I polimery*; Izdanja Kemije Ind. : Zagrzeb, 1997.

Jedliński, Z. J. Współczesne kierunki w dziedzinie syntezy polimerów. *Wiadomości Chem.* 1977, *31*, 607.

Jenkins, A. D. , Ledwith, A. , Ed. ;*Reactivity, Mechanism and Structure in Polymer Chemistry*; John Wiley and Sons: London, 1974; tłum. na rosyjski, Mir: Moskwa, 1977.

Joshi, S. G.; Natu, A. A. Chlorocarboxylation of Polyethylene. *Angew. Makromol. Chem.* 1986, *140*, 99–115.

Kaczmarek, H. polimery, a środowisko. *Polimery* 1997, *42*, 521.

Kang, E. T.; Neoh, K. G. Halogen Induced Polymerization of Furan. *Eur. Polym. J.* 1987, *23*, 719.

Kang, E. T.; Neoh, K. G.; Tan, T. C.; Ti, H. C. Iodine Induced Polymerization and Oxidation of Pryridazine. *Mol. Cryst. Lig. Cryst.* 1987, *147*, 199.

Karnojitzki, V. Organiczieskije pierekisy, I. I. L. :Moskwa, 1961.

Karpiński, K.; Prot, T. Inicjatory fotopolimeryzacji i fotosieciowania. *Polimery*1987, *32*, 129.

Keii, T. Propene Polymerization with $MgCl_2$-Supported Ziegler Catalysts. *Makromol. Chem.* 1984, *185*, 1537.

Khanna, S. N.; Levy, M.; Szwarc, M. Complexes formed by Anthracene with 'Living' Polystyrene. *Dormant Polymers. Trans Faraday Soc.* 1962,*58*, 747–761.

Koinuma, H.; Naito, K.; Hirai, H. Anionic Polymerization of Oxiranes. *Makromol. Chem.* 1982, *183*, 1383.

Korniejev, N. N.; Popov, A. F.; Krencel, B. A. *Komplieksnyje metalloorganiczeskije katalizatory*;Khimia: Leningrad, 1969.

Korszak, W. W. Niekotoryje problemy polikondensacji. *Vysokomol. Soed. A*1979, *21*, 3.

Korszak, W. W. *Technologia tworzyw sztucznych*; WNT: Warszawa, 1981.

Kowalska, E.; Pełka, J. Modyfikacja tworzyw termoplastycznych włóknami celulozowymi. *Polimery*2001, *46*, 268.

Kozakiewicz, J.; Penczek, P. Polimery jonowe. *Wiadomości Chem.* 1976, *30*, 477.

Kubisa, P. Polimeryzacja żyjąca. *Polimery*1990, *35*, 61.

Kubisa, P. Neeld, K.; Starr, J. ;Vogl, O. Polymerization of Higher Aldehydes. *Polymer*1980, *21*, 1433.

Kucharski, M. Nitrowe i aminowe pochodne polistyrenu. *Polimery*1966, *11*, 253.

Kunicka, M.; Choć, H. J. Hydrolitic Degradation and Mechanical Properties of Hydrogels. *Polym. Degrad. Stab.* 1998, *59*, 33.

Lebduska, J.; Dlaskova, M. Roztokova kopolymerace. *Chemicky Prumysl*1990, *40*, 419–480.

Leclerc, M.; Guay, J.; Dao, L. W. Synthesis and Characterization of Poly(alkylanilines). *Macromolecules*1989, *22*, 649.

Lee, D. P.; Dreyfuss, P. Triphenylphesphine Termination of Tetrahydrofuran Polymerization. *J. Polym. Sci. ,Polym. Chem. Ed.* 1980, *18*, 1627.

Leza, M. L.; Casinos, I.; Guzman, G. M.; Bello, A. Graft Polymerization of 4-vinylpyridineOnto Cellulosic Fibres by the Ceric on Method. *Angew. Makromol. Chem.* 1990, *178*, 109–119.

Lopez, F.; Calgano, M. P.; Contrieras, J. M.; Torrellas, Z.; Felisola, K. Polymerization of Some Oxiranes. *Polym. Int.* 1991, *24*, 105.

Lopyrev, W. A.; Miaczina, G. F.; Szevaleevskii, O. I.; Hidekel, M. L. Poliacetylen, *Vysokomol. Soed.* 1988, *30*, 2019.

Mano, E. B.; Calafate, B. A. L. Electrolytically Initiated Polymerization of N-vinylcarbazole. *J. Polym. Sci. ,Polym. Chem. Ed.* 1983, *21*, 829.

Mark, H. F. *Encyclopedia of Polymer Science and Technology*; Concise 3rd ed.; WileyInterscience: Hoboken, New York, 2007.

Mark, H.; Tobolsky, A. V. *Chemia fizyczna polimerów*; PWN: Warszawa, 1957.

Matyjaszewski, K.; Davis, T. P. *Handbook of Radical Polymerization*; WileyInterscience: New York, 2002.

Matyjaszewski,K. ;Mülle, A. H. E. 50 years of Living Polymerization. *Prog. Polym. Sci.* **2006**, *31*, 1039–1040.

Mehrotra, R.; Ranby, B. Graft CopolymerizationOnto Starch. *J. Appl. Polym. Sci.* 1977, *21*, 1647.

Morrison, R. T.; Boyd, N. *Chemia organiczna*; PWN:Warszawa,1985.

Morton, M. *Anionic Polymerization*; Academic Press: New York, 1983.

Munk, P. *Introduction to Macromolecular Science*; John Wiley and Sons: New York, 1989.

B. Ciecze mikroemulsyjne. *Przem. Chem.* 1999, *78(4)*, 127.

Neoh, K. G.; Kang, E. T.; Tan, K. L. Chemical Copolymerization of Aniline with Halogen-Substituted Anilines. *Eur. Polym. J.* 1990, *26*, 403.

Nowakowska, M.; Pacha, J. Polimeryzacja olefin wobec kompleksów metaloorganicznych. *Polimery*1978, *23*, 290.

Ogata, N.; Sanui, K.; Tan, S. Synthesis of Alifatic Polyamides by Direct Polycondensation with Triphenylphosphine. *Polymer J.* 1984, *16*, 569.

Onen, A.; Yagci, Y. Bifunctional Initiators. *J. Macromol. Sci. Chem.* A1990, *27*, 743, *Angew. Makromol. Chem.* 1990, *181*, 191.

Osada, Y. Plazmiennaja polimerizacia i plazmiennaja obrabotka polimerov. *Vysokomol. soed.* A1988, *30*, 1815.

Pasika, W. M. Cationic Copolymerization of Epichlorohydrin With Styrene Oxide And Cyclohexene Oxide. *J. Macromol. Sci. Chem.* A1991, *28*, 545.

Pasika, W. M. Copolymerization of Styrene Oxide and Cyklohexene Oxide. *J. Polym. Sci. Part* A1991, *29*, 1475.

Pielichowski, J.; Puszyński, A. Preparatyka monomerów, W. P. Kr. Kraków.

Pielichowski, J.; Puszyński, A. Preparatyka polimerów, TEZA WNT 2005 Kraków.

Pielichowski, J.; Puszyński, A. Technologia tworzyw sztucznych, WNT 2003, Warszawa.

Pielichowski, J.; Puszyński, A. Wybrane działy z technologii chemicznej organicznej. W. P. Kr. Kraków.

Pistola, G.; Bagnarelli, O. Electrochemical Bulk Polymeryzation of Methyl Methacrylate in the Presence of Nitric Acid. *J. Polym. Sci. , Polym. Chem. Ed.* 1979, *17*, 1002.

Pistola, G.; Bagnarelli, O.; Maiocco, M. Evaluation of Factors Affecting The Radical Electropolymerization of Methylmethacrylate in the Presence of HNO_3. *J. Appl. Electrochem.* 1979, *9*, 343.

Połowoński, S. Techniki pomiarowo-badawcze w chemii fizycznej polimerów; W. P. L. : Łódź, 1975.

Porejko, S.; Fejgin, J.; Zakrzewski, L. *Chemia związków wielkocząsteczkowych*; WNT: Warszawa 1974.

Praca zbiorowa: Chemia plazmy niskotemperaturowej; WNT: Warszawa, 1983.

Puszyński, A. Chlorination of Polyethylene in Suspension. *Pol. J. Chem.* 1981, *55*, 2143.

Puszyński, A.; Godniak, E. Chlorination of Polyethylene in Suspension in the Presence of Heavy Metal Salts. *Macromol. Chem. Rapid Commun.* 1980, *1*, 617.

Puszyński, A.; Dwornicka, J. Chlorination of Polypropylene in Suspension. *Angew. Macromol. Chem.* 1986, *139*, 123.

Puszyński, J. A.; Miao, S. Kinetic Study of Synthesis of SiC Powders and Whiskers In the Presence $KClO_3$ and Teflon. *Int. J. SHS*1999, *8(8)*, 265.

Rabagliati, F. M.; Bradley, C. A. Epoxide Polymerization. *Eur. Polym. J.* 1984, *20*, 571.

Rabek J. F. : Podstawy fizykochemii polimerów, W. P. Wr. , Wrocław 1977

Rabek T. I.: *Teoretyczne podstawy syntezy polielektrolitów i wymieniaczy jonowych*; PWN: Warszawa 1962.

Regas, F. P.; Papadoyannis, C. J. Suspension Cross-linking of Polystyrene With Friedel-Crafts Catalysts. *Polym. Bull.* 1980, *3*, 279.

Rodriguez, M.; Figueruelo, J. E. On the Anionic Polymeryzation Mechanism of Epoxides. *Makromol. Chem.* 1975, *176*, 3107.

Roudet, J.; Gandini, A. Cationic Polymerization Induced by Arydiazonium Salts. *Makromol. Chem. Rapid Commun.* 1989, *10*, 277.

Sahu, U. S.; Bhadam, S. N. Triphenylphosphine Catalyzed Polimerization of Maleic Anhydride. *Makromol. Chem.* 1982, *183*, 1653.

Schidknecht, C. E. *Polimery winylowe*; PWT: Warszawa 1956.

Szwarc, M.; Levy, M.; Milkovich, R. Polymerization Initiated by Electron Transfer to Monomer. A New Method of Formation of Block Copolymers. *J. Am. Chem. Soc.* **1956**,78, 2656–2657.

Szwarc, M. 'Living' Polymers. *Nature*1956, *176*, 1168–1169.

Sen, S.; Kisakurek, D.; Turker, L.; Toppare, L.; Akbulut, U. Elektroinitiated Polymerization of 4-Chloro-2,6-Dibromofenol. *New Polym. Mater.* 1989, *1*, 177.

Sikorski, R. T. Chemiczna modyfikacja polimerów, Prace Nauk. *Inst. Technol. Org. Tw. Szt. P. Wr.* 1974, *16*, 33.

Sikorski, R. T. *Podstawy chemii i technologii polimerów*; PWN: Warszawa, 1984.

Sikorski, R. T.; Rykowski, Z.; Puszyński, A. Elektronenempfindliche Derivate von Poly-methylmethacrylat. *Plaste und Kautschuk*1984, *31*, 250.

Simionescu, C. I.; Geta, D.; Grigoras M. : Ring-Opening Isomerization Polymeryzation of 2-Methyl-2-Oxazoline Initiated by Charge Transfer Complexes, *Eur. Polym. J.* 1987, *23*, 689.

Sheinina, L. S.; Vengerovskaya, Sh. G.; Khramova, T. S.; Filipowich, A. Y. Reakcji piridinov i ich czietvierticznych soliej s epoksidnymi soedineniami. *Ukr. Khim. Zh.* 1990, *56*, 642.

Soga K. , Toshida Y. , Hosoda S. , Ikeda S. : A Convenient Synthesis of a Polycarbonate, Makromol. Chem. 1977, 178, 2747

Soga K. , Hosoda S. , Ikeda S. : A New Synthetic Route to Polycarbonate, J. Polym. Sci. ,Polym. Lett. Ed. 1977, 15, 611

Soga, K.; Uozumi, T.; Yanagihara, H.; Siono, T. Polymeryzation of Styrene with Heterogeneous Ziegler-Natta Catalysts. *Makromol. Chem. Rapid. Commun.* 1990, *11*, 229.

Soga, K.; Shiono, T.; Wu, Y.; Ishii, K.; Nogami, A.; Doi, Y. Preparation of a Stable Supported Catalyst for Propene Polymeryzation. *Makromol. Chem. Rapid Commun.* 1985, *6*, 537.

Sokołov, L. B.; Logunova, W. I. Otnositielnaja reakcionnosposobnost monomerov i prognozirovanic polikondensacji v emulsionnych sistemach. *Vysokomol. Soed. A.* 1979, *21*, 1075.

Soler, H.; Cadiz, V.; Serra, A. Oxirane Ring Opening with Imide Acids. *Angew. Makromol. Chem.* 1987, *152*, 55.

Spasówka, E.; Rudnik, E. Możliwości wykorzystania węglowodanów w produkcji biodegradowalnych tworzyw sztucznych. *Przemysł Chem.* 1999, *78*, 243.

Stevens, M. P. *Wprowadzenie do chemii polimerów*; PWN: Warszawa, 1983.

Stępniak, I.; Lewandowski, A. Elektrolity polimerowe. *Przemysł Chem.* 2001, *80(9)*, 395.

Strohriegl, P.; Heitz, W.; Weber, G. Polycondensation Using Silicon Tetrachloride. *Makromol. Chem. Rapid Commun.* 1985, *6*, 111.

Szur, A. M. *Vysokomolekularnyje soedinenia*; Vysshaja Szkoła: Moskwa, 1966.

Tagle, L. H.; Diaz, F. R.; Riveros, P. E. Polymeryzation by Phase Transfer Catalysis. *Polym. J.* 1986, *18*, 501.

Takagi, A.; Ikada, E.; Watanabe, T. Dielectric Properties of Chlorinated Polyethylene and Reduced Polyvinylchloride. *Memoir Faculty Eng. ,Kobe Univ.* 1979, *25*, 239.

Tani, H. Stereospecific Polymeryzation of Aldehydes and Epoxides. *Adv. Polym. Sci.* 1973, *11*, 57.

Vandenberg, E. J. Coordynation Copolymeryzation of Tetrahydrofuran and Oxepane with Oxetanes and Epoxides. *J. Polym. Sci. Part A*1991, *29*, 1421.

Veruovic, B. *Navody pro laboratorni cviceni z chemie polymer*; SNTL: Praha, 1977.

Vogdanis, L.; Heitz, W. Carbon Dioxide as a Mmonomer. *Makromol. Chem. Rapid Commun.* 1986, *7, 543*.

Vollmert, B. *Grundriss der Makromolekularen Chemic*; Springer-Verlag: Berlin, 1962.

Wei, Y.; Tang, X.; Sun, Y. A study of the Mechanism of Aniline Polymeryzation. *J. Polym. Sci. Part A* 1989, *27*, 2385.

Wei, Y.; Sun, Y.; Jang, G. W.; Tang, X. Effects P-Aminodiphenylamine on Electrochemical Poly-meization of Aniline. *J. Polym. Sci. ,Part C* 1990, *28*, 81.

Wei, Y.; Jang, G. W. Polymerization of Aniline and Alkyl Ringsubstituted Anilines in the Presence of Aromatic Additives. *J. Phys. Chem.* 1990, *94*, 7716.

Wiles, D. M. Photooxidative Reactions of Polymers. *J. Appl. Polym. Sci. , Appl. Polym. Symp.* 1979, *35*, 235.

Wilk, K. A.; Burczyk, B.; Wilk, T. Oil in Water Microemulsions as Reaction Media. 7th International Conference Surface and Colloid Science, Compiegne, 1991.

Winogradova, C. B. Novoe v obłasti polikondensacji. *Vysokomol. Soed. A* 1985, *27*, 2243.

Wirpsza, Z. *Poliuretany*; WNT: Warszawa, 1991.

Wojtala, A. Wpływ właściwości oraz otoczenia poliolefin na przebieg ich fotodegradacji. *Polimery* 2001, *46*, 120.

Xue, G. Polymerization of Styrene Oxide with Pyridine. *Makromol. Chem. Rapid Commun.* 1986, *7*, 37.

Yamazaki, N.; Imai, Y. Phase Transfer Catalyzed Polycondensation of Bishalomethyl Aromatic compounds. *Polym. J.* 1983, *15*, 905.

Yasuda, H. *Plasma Polymerization*; Academic Press, Inc. : Orlando, 1985.

Yuan, H. G.; Kalfas, G.; Ray, W. H. Suspension polymerization. *J. Macromol. Sci. Rev. Macromol. Chem. Phys.* 1991, C 31, 215.

Zubakowa, L. B.; Tevlina, A. C.; Davankov, A. B. *Sinteticzeskije ionoobmiennyje materiały*; Khimia: Moskwa, 1978.

Zubanov, B. A. *Viedienie v chimiu polikondensacionnych procesov*; Nauka: Ałma Ata, 1974.

Żuchowska, D. Polimery konstrukcyjne. WNT: Warszawa, 2000.

Żuchowska, D. *Struktura i właściwości polimerów jako materiałów konstrukcyjnych*; W. P. Wr. : Wrocław, 1986.

Żuchowska, D.; Steller, R.; Meisser, W. Structure and Properties of Degradable Polyolefin—Starch Blends. *Polym. Degrad. Stab.* 1998, *60*, 471.

CHAPTER 7

TRENDS IN CHEMICAL REACTIONS OF POLYMERS

G. E. ZAIKOV

Russian Academy of Sciences, Moscow, Russia

CONTENTS

7.1 GENERAL CHARACTERISTICS

During the storage or the use of polymers and the plastics derived from them, they may be subject to a number of different chemical reactions, both intentional and incidental as a result of actions by various physical or chemical agents.

The intended chemical reactions carried out on polymers change their properties and allow obtaining a large number of new varietiesfrom one polymer. The latter are quite different from the original, broadening in this way the assortment of plastics and possibilities of their applications. The research area dealing with this issue is called the chemical modification of polymers. It is now widely expanded and further developed. The knowledge of routes of these reactions allows the synthesis of new polymers with specific chemical structure and intended for pre-planned applications.

The chemical reactions carried out on polymers can be divided into several types, depending on whether they run on individual polymer moleculeswith or without the participation of other small molecule reagents, or as an intermolecular reaction occurring between the macromolecules of the same or different polymers.

The most important types of reactions of chemical modification of polymers include:
- introduction of functional groups to polymer molecules,
- transformation of functional groups in polymer macromolecules,
- intramolecularcyclization,
- oxidation and reduction,
- polymer grafting,
- crosslinking,
- degradation of polymers.

The chemical reactions carried out on polymer macromolecules are characterized by a defined specificity. In his kinetic studies, Flory has shown that the reactivity of carboxyl and hydroxyl groups during the synthesis of polyester does not depend on chain length of the nascent polymer. On this basis, he concluded that the reactivity of functional groups in polymer molecules do not differ from the reactivity of the same functional groups in small molecule compounds. This suggestion was confirmed when testing a variety of reactions in dilute solutions of polymers and small molecule compounds. However, it is important to keep in mind the complex structure of polymers, whose chains are tangled, and in the solid state, they are often combined with other chains by the Van der Waals forces. A lot of them form crystalline areas. All these factors make the access of a reactive particle (e.g., a radical) to a specific location in the chain of macromolecules difficult or even impossible. For this reason, the chemical modification of polymers with long chains runs in accessible places and the resulting product has usually the structure of a statistical copolymer. It means that it is not substituted in each mer.

The course of chemical reactions in polymers, enabling the introduction of new functional groups or transformation of functional groups in the macromolecules, is affected by the following factors:

- availability of functional groups,
- chain length (degree of entanglement of macromolecules),
- effect of neighboring groups,
- effect of configuration,
- electrostatic effects,
- cooperativeeffects,
- effect of inhomogeneous activity,
- solution concentration,
- supermolecular effects in heterogeneous reactions.
- the effect of mechanical stress.

7.1.1 AVAILABILITY OF FUNCTIONAL GROUPS

For an effective course of chemical modification of polymer, it is important that its functional groups exhibit the same reactivity. Large macromolecules, contained in solution, are partiallytangled. In the solid, they are associated with other macromolecules via intermolecular forces.

This macrostructure of the polymer molecules makes that in the reaction environment the functional groups in polymer are not equally sensitive to the reactive agents. Especially in a heterogeneous system, diffusion factor plays an important role. Functional groups present in the outer layers of polymer ball react relatively easily, while inside the reaction depends on the possibility of reactive particles to penetrate in the part of polymer.

Functional groups inaccessible for steric reasons remain unchanged in the reaction product.

An example of the impact of the availability of functional groups on the reaction rate can be the amidation reactions with aniline of various types of copolymers of maleic anhydride in dimethylformamide. The rate of amidation of the maleic anhydride copolymer with ethylene is, for example, two orders of magnitude higher than the rate of amidation of maleic anhydride copolymer with norbornene.

7.1.2 INFLUENCE OF THE CHAIN LENGTH

Typical macromolecular reactions, not observed in the chemistry of small molecule compounds, are the degradation processes (chain decay into smaller fragments) and the intramolecular cyclization.

As the polymer chain length grows, the viscosity of its solutions or blends increases. It hampers the mobility of molecules and thus reduces their reactivity. Also entanglement of polymer molecules increases with the chain length growth.

The course of reactions of functional groups in polymers of different molecular weights depends on the macrostructure of molecules, which is associated with the availability of functional groups.

On the basis of experimental studies, it was shown that the effect of the chain length of macromolecules on the reactivity of functional groups may vary, depending on the nature and structure of the polymer. For example, it was found that during the reaction of phenylisocyanate with polyesters of different molecular weight, the reaction rate decreases with the increasing chain length of polyester.

Similarly, during the esterification of maleic anhydride copolymers with polypropylene, with molecular weights of 42,000, 51,000, 91,000, and 103,000, the reaction rate decreases with increasing average molecular weight.

Quite different results were obtained during the amidation of styrene maleic anhydride copolymers with average molecular weights of 24,000 and 46,000. In this case, the reaction rate practically does not depend on the chain length.

The results of these reactions are presented in Table 7.1.

TABLE 7.1   Influence of Molecular Weight of Maleic Anhydride Copolymer with Styrene on the AmidationReaction.

No	Temperature [°C]	Amine used	$k \times 10^2$ [dm3/mol s]	
			$M = 24,000$	$M = 46,000$
1	60	Aniline	1.45	1.38
2	45	p-toluidine	2.37	2.1
3	30	p-anisidine (p-methoxyaniline)	1.75	1.8
4	45	m-chloroaniline	0.68	0.71

Source: Adapted from Rätzsch, M.; Phien, V. Faserforsch. u. Textiltech. 1975,26, 165.

7.1.3 EFFECT OF NEIGHBORING GROUPS

The neighboring groups, both the nearest and the further lying away from the functional group formed in the reaction center, impact its electronic structure and thus its reactivity. The effect of neighboring groups may lead both to a reduction and to an increase in the reaction rate.

The hydrolysis of poly(vinyl acetate) to poly(vinyl alcohol) is a characteristic example of a reaction, which reflects the influence of the neighboring groups on the reactivity of the substituent in the polymer molecule. With the progressing process of hydrolysis, the reaction rate increases, because the hydroxyl groups activate the neighboring acetoxyl groups. Therefore, the reactivity of various groups increases in the following order:

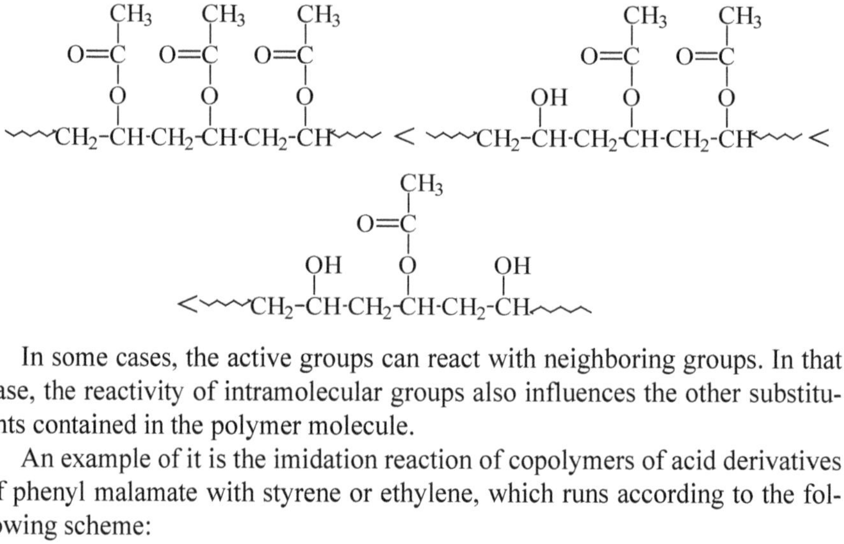

In some cases, the active groups can react with neighboring groups. In that case, the reactivity of intramolecular groups also influences the other substituents contained in the polymer molecule.

An example of it is the imidation reaction of copolymers of acid derivatives of phenyl malamate with styrene or ethylene, which runs according to the following scheme:

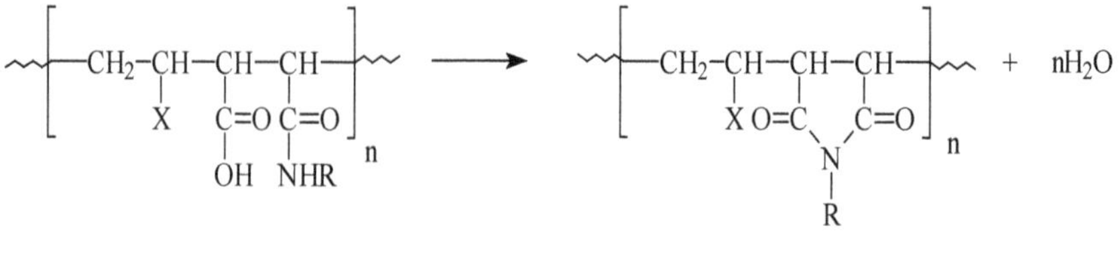

where X = H, C6H5

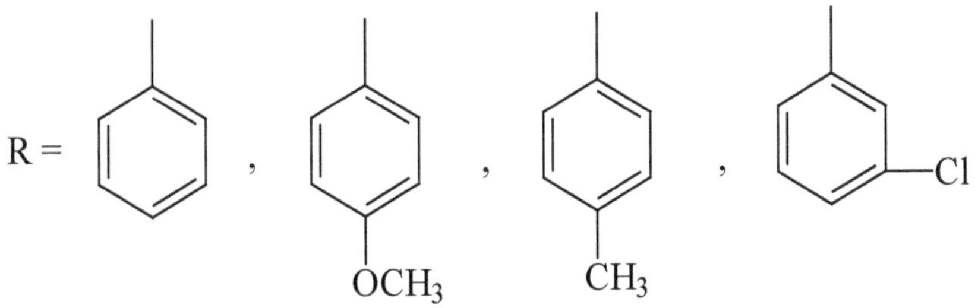

R =

OCH₃   CH₃   , —Cl

With a decrease of the basicity of amide substituent,the cyclization reaction rate decreases. The imidation rate of copolymers with styrene is much higher than that with ethylene. It shows the influence of phenyl substituent on the course of this reaction.

7.1.4 CONFIGURATION EFFECTS

Configuration is called the distribution of atoms in molecule, which is characteristic for a particular stereoisomer.

The substituents present in polymer chains can be arranged either statistically (atactic polymers) or regularly (syndiotactic and isotactic polymers).

The reactivity of functional groups in polymers change with a change of spatial configuration of macromolecules.

The configuration effects in the chemical reactivity of polymers depend on the kind of functional groups present in macromolecules and on the type of chemical reaction.

During the study of the hydrolysis of poly(methyl methacrylate), it was found that the isotactic polymer reacts with a high rate, whereas the atactic and the syndiotactic polymers P react slowly. This phenomenon is explained by the fact that the ester groups in the isotactic polymer form complexes with the neighboring carboxyl groups, so that the hydrolysis reaction proceeds more easily.

During the similar study done with the labeled carbon atoms, it was observed that the isotactic poly(methyl methacrylate) reacts 16 times faster than the atactic and 43 times faster than the syndiotactic polymer, respectively.

Some tactic polymers, such as polyethylene and polypropylene, are characterized by a much larger degree of crystallinity than the atactic polymers. During a chemical reaction in the solid state, the reaction takes place in the amorphous areas of polymer, while the crystalline areas are weakly reactive.

An example of such a reaction is the chlorination. Atactic polypropylene is much easier to chlorinate than the isotactic polypropylene with a high degree of crystallinity. Dissolving polymer in a solvent or conducting the process above the melting point of polymer crystallites enables its more uniform substitution.

7.1.5 CONFORMATIONAL EFFECTS

Conformations are the possibilities of mutual transformation of different spatial arrangements of atoms in molecules, as a result of the rotation around single bonds. There are two basic types of conformational structures: opposite conformation (called also eclipsed conformation) and alternate conformation (called also staggered conformation). Several intermediate conformations occurring between the two are called skew conformations.

Conformational effects that take place during the reaction of functional groups in macromolecules overlap often with the other discussed effects.

The phenomenon occurs primarily during the reactions carried out in solutions. During the dissolution,the polymer molecules pass into solution in the form of coils, which, depending on the impact on them of suitable solvent molecules, are more or less developed.

In good solvents and in dilute solutions, the availability of functional groups is much greater, which affects the reaction rate. However, the reaction rate changes clearly when the polarity of the solvent differs significantly from the polarity of the resulting modified polymer. For this reason, it is advisable to carry out the reaction in a solvent (and not in a mixture) to facilitate the attainability of polymer functional groups at least at the beginning of the reaction. If a mixture of solvents is used, such as a solvent for the output polymer and for the product after the modification, and possibly for the small molecule compound then one obtains a compound with various degrees of substitution.

The conformational effects play a particularly important role in the intramolecular cyclization reaction carried out in dilute solutions. The defined conformation and the distance between the chain ends influence the course of formation of stable rings with 5–12 members.

7.1.6 ELECTROSTATIC EFFECTS

The electrostatic effects are observed during the reactions with the use of ionic polymers, in comparison with similar reactions carried out on small molecule compounds.

As an example, the use of protonated poly(4-vinylpyridine) as a catalyst for the hydrolysis of sodium 4-acetoxy-3-nitrobenzenesulfonate is shown below. In this reaction,poly(4-vinylpyridine) shows a much higher catalytic activity than 4-methylpyridine.

The mechanism of this reaction is shown in the following scheme:

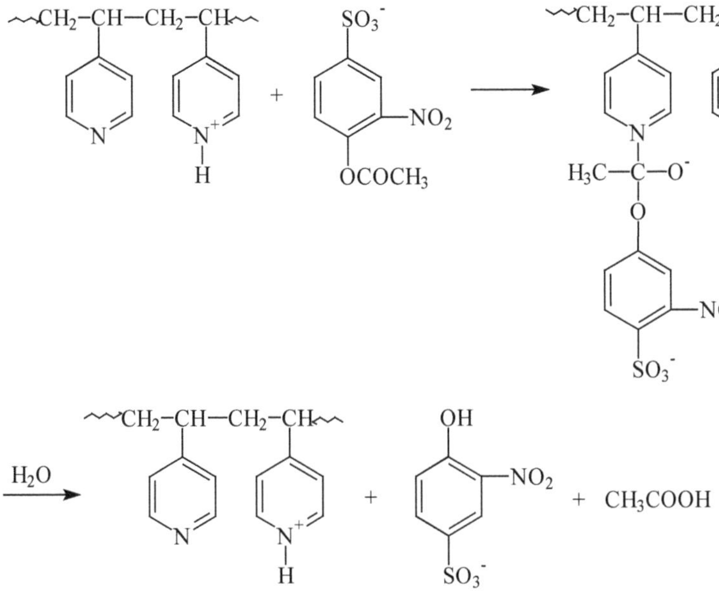

The electrostatic interactions also depend on the solvent polarity, pH, and on the type of neighboring groups.

7.1.7 COOPERATIVE EFFECTS

The cooperative effects in chemical reactivity of polymers may be subject to the influence of neighboring groups, association processes, and catalysts presence. They play an important role in many biochemical processes involving proteins and nucleic acids, and in particular enzymes. The cooperative action in course of chemical reactions involving macromolecules can be presented schematically as follows:

The functional group A in the polymer chain becomes an active site during a reaction with functional group M of a small molecule compound. At the same time, a mutual interaction between the group A' and another functional group N of the reagent takes place. It facilitates the main reaction by stabilizing the transition compound. Groups A and A' in the polymer chain may be the same or different.

7.1.8 EFFECT OF INHOMOGENEOUS ACTIVITY

This effect is related to the possible formation of hydrogen bonds in polymer as well as to the hydrophobic interactions, which influence the reactivity of different functional groups. During the amidation reaction of maleic anhydride copolymers with aliphatic amines at a stoichiometric molar ratio, the substitution at the polymer chain does not take place completely, but only to half when the reaction carried out on small molecule compounds is running to the end.

7.1.9   INFLUENCE OF CONCENTRATION

The course of the reaction of chemical modification of polymer in solution depends on the concentration of polymer, the actual concentration of active groups, and the concentration of small molecule reagent. During the substitution or replacement of functional groups, one uses usually a 2.5-fold excess of small molecule reagent.

In very dilute solutions, the cyclization reactions are possible. A significant increase in the concentration of polymer solution leads to an increase of viscosity and to a series of side reactions, thereby increasing the heterogeneity of the product.

During the chemical modification of polymers, carried out in a homogenous environment, the processes typically run as second-order reactions. The rate of this reaction depends on both solvent polarity and on solvation effects.

7.1.10   SUPRAMOLECULAR EFFECTS IN HETEROGENEOUS REACTIONS

The influence of supramolecularpolymer structures plays an important role during their chemical modification. On the reactions running in solutions impose at the same time conformational effects which are difficult to distinguish from supramolecular.

A typical example of the impact of supramoleculareffects on the process takes place during the acetylation of cellulose. The course of reaction depends on the supramolecular structure of cellulose. During the acetylation of natural cellulose fibers for 24 h, one obtains 26.3–27.7% substitution degree of acetyl groups in cellulose. Acetylation in the same conditions of viscose fibers leads to the substitution degree of order of 14–16% only.

Similarly, the thermal dehydration of poly(vinyl alcohol) depends on the ordering degree of polymer chains. Samples of poly(vinyl alcohol), containing oriented molecules, are more resistant to the action of temperature.

Chemical reactions of polymers in the solid phase can run easily in amorphous areas or on the surface of amorphous or crystalline areas. Exploiting this phenomenon, one can obtain polyethylene with a given degree of crystallinity by the destruction of the amorphous areas with fuming nitric acid. In the reaction conditions, the crystalline polyethylene does not react with nitric acid.

7.1.11 EFFECT OF MECHANICAL STRESS

The mechanical stress in polymer molecules can affect the activation of chemical bonds in macromolecule, something that plays a particularly important role in such reactions as the degradation of polymers, crosslinking, and grafting. The mechanochemistry of polymers is dealing with these issues. The most important external factors that cause the effect of mechanical strain are the forcing processes, such asextrusion,calendering, injection molding, rolling, grinding, hammering, crushing, osmotic pressure, phase transitions, and the impact of electrohydraulic as well as ultrasound actions.

Another example of the effect of mechanical stress is the process ofdeacetylationof polyvinylacetylimidazole in the presence of 3-nitroterephthalic monoester. Depending on the type of ester used, different results are obtained. The rate of this *deacetylation* reaction in the presence of octadecyl monoester is 100–400 times larger than when using an analogous ethyl ester. The reason for this is the interpenetration of molecules and the mutual interactions of aliphatic chains and acid groups in polymer molecule.

7.2 MODIFICATION REACTIONS OF POLYMERS

7.2.1 INTRODUCTION OF NEW FUNCTIONAL GROUPS

Introduction of new functional groups into the polymer molecules may be carried out by the substitution reaction of hydrogen atom or other functional groups as well as by the addition reaction to double bond when using the unsaturated polymers. A model for this type ofprocess is the chlorination reaction. It may follow different mechanisms, depending on the type of used chlorinating agent and the process conditions. Chlorination of polymers by free-radical mechanism may take place by using chlorine or sulfuryl chloride in the presence of initiators such as organic peroxides, dinitrylazoisobutyric acid, or by exposure to ultraviolet light. The reaction can be carried out in solution, suspension, and in fluidized bed.

During the chlorination of polymers in solution, the substitution of polymer chains by chlorine proceeds initially statistically, and then the influence of substituted chlorine atoms on the kinetics of the further chlorination reaction is observed.

The advantage of the method of chlorination in solution is the destruction of the crystalline structure of polymer as a result of its dissolution and the solvation of individual polymeric ball (more or less coiled chains). Thus, the chlorinating factor have an easier access to various areas of polymer chain. This allows a relatively uniform chlorine substitution. The mechanism of the free-radical chlorination reaction is as following:

$$R-N-N-R \longrightarrow 2\,R\bullet\ +\ N_2$$

$$R\bullet\ +\ Cl_2 \longrightarrow RCl\ +\ Cl\bullet$$

or

$$Cl_2 \xrightarrow{\ hv\ } 2\,Cl\bullet$$

$$\sim\!\!\sim\!\!CH_2\!\sim\!\!\sim\ +\ Cl\bullet \longrightarrow \sim\!\!\sim\!\!\overset{\bullet}{C}H\!\sim\!\!\sim\ +\ HCl$$

$$\sim\!\!\sim\!\!\overset{\bullet}{C}H\!\sim\!\!\sim\ +\ Cl_2 \longrightarrow \sim\!\!\sim\!\!\underset{\underset{Cl}{|}}{C}H\!\sim\!\!\sim\ +\ Cl\bullet$$

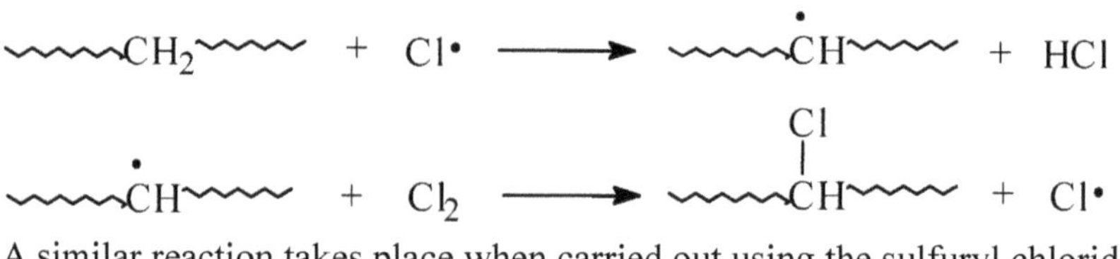

A similar reaction takes place when carried out using the sulfuryl chloride:

$$R\bullet\ +\ SO_2Cl_2 \longrightarrow RCl\ +\ \overset{\bullet}{S}O_2Cl$$

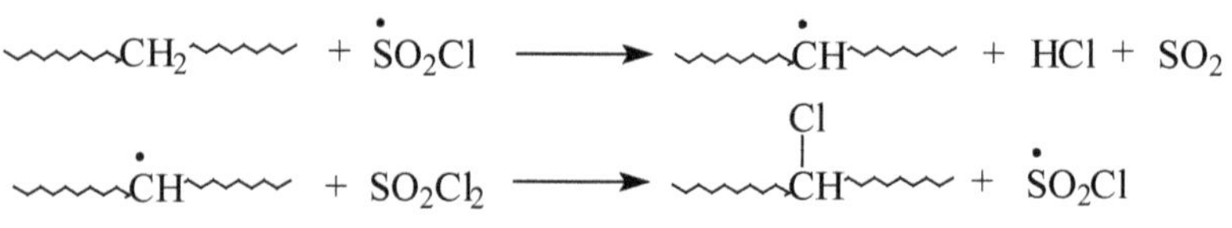

$$\sim\!\!\sim\!\!CH_2\!\sim\!\!\sim\ +\ \overset{\bullet}{S}O_2Cl \longrightarrow \sim\!\!\sim\!\!\overset{\bullet}{C}H\!\sim\!\!\sim\ +\ HCl\ +\ SO_2$$

$$\sim\!\!\sim\!\!\overset{\bullet}{C}H\!\sim\!\!\sim\ +\ SO_2Cl_2 \longrightarrow \sim\!\!\sim\!\!\underset{\underset{Cl}{|}}{C}H\!\sim\!\!\sim\ +\ \overset{\bullet}{S}O_2Cl$$

The disadvantage of chlorination reactions in solution is the necessity of often using the flammable and toxic solvents and the difficulty with polymer extraction.

Unlikely runs the process of chlorination carried out in a fluidized bed or in suspension. Polymers containing crystalline areas chlorinate then only in the amorphous areas. Only the destruction of the crystalline structure, by carrying out the reaction at a temperature above the melting point of crystallites, allows a more uniform process of chlorination. There are also difficulties with penetration of chlorinating agent in the internal layers of polymer. It involves the influence of diffusion factors on the course of this reaction. It follows from the above considerations that the chlorination products of polymers in solution and in solid phase differ in both the micro-and the macrostructure. The products chlorinated in the solid phase are characterized by substitution in certain areas of the molecules only and retain the original crystalline structure. They are also characterized by a poorer solubility than polymers chlorinated in solution.

An important parameter influencing the course of chlorination reaction is temperature at which the chlorination process is conducted. Figure 7.1 shows the influence of temperature on the course of chlorination process of a low-density polyethylene, in suspension, by chlorine produced in the reaction environment by dispensing a solution of potassium chlorate to the solution of hydrochloric acid.The increase of reaction temperature increases the rate in the initial period of the course, which is associated with better accessibility of chlorinating agent to outer layers.

In the next stage of the reaction, the diffusion processes dominate and the reaction rate decreases.

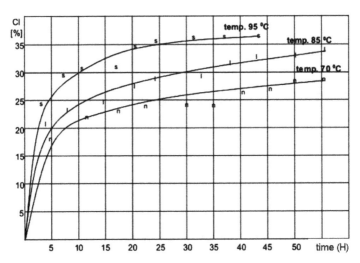

FIGURE 7.1   Influence of temperature on the course of chlorination of polyethylene in suspension.
Source: Adapted from Puszyński, A.Pol. J. Chem. 1981, 55, 2143.

During the chlorination of polymers in suspension, the reaction medium also affects the process. The ions in solution, in which the polymer powder was dispersed, act on the polymer matrix, causing acceleration or hinder the chlorination reaction. This applies, first of all, to the polymers containing crystalline groups.

Figure 7.2shows, as example, the impact of salt solutions at concentration of 0.5 mol/dm^3 on the course of the chlorination process of isotactic polypropylene dispersed in an aqueous solution of hydrochloric acid.

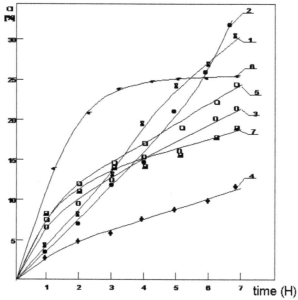

FIGURE 7.2   Influence of the reaction environment on the course of the chlorination process of polypropylene in an aqueous solution of hydrochloric acid and in the presence of salts at a concentration of 0.5 mol/dm3:(1)without salt, (2) CoCl2, (3) (NH4)2MoO4, (4) CuCl2, (5) KCl, (6)NaCl, (7)NaCl with concentration of 1.5 mol/dm3.
Source: Adapted from Puszyński, A. J. Dwornicka: Angew. Makromol. Chem. 1986, 139, 123.

It follows from the shown graph that most salts increase the effectiveness of chlorination reactions in the first period, while the copper cations show a clear inhibitive effect.

Known are also chlorination methods of polymers through the cationic mechanism by using Lewis acid-type catalysts, for example, the aluminum chloride.

In the case of chlorination in these conditions of aliphatic polymers, one often obtains cross-linked products.

Impact of the mechanism on the type of products formed during the reaction can be illustrated by the example of polystyrene chlorination. During this process, running according to the free-radical mechanism, the substitution takes place to the polymer chain, while in the course of ionic polymerization, the substitution reaction takes place in the ring.

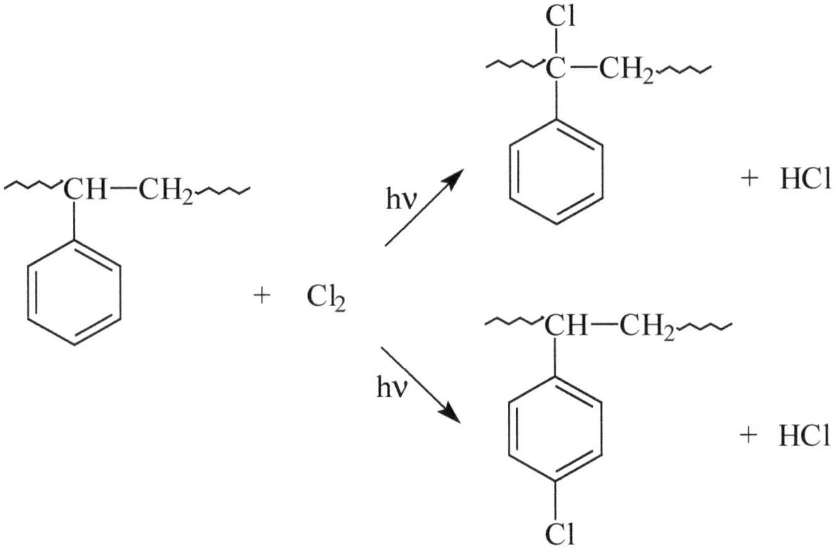

Similar results are obtained during the bromination of polymers.

The exchange of hydrogen atoms in the polymer molecule to other functional groups is carried out using conventional methods of organic preparation. Especially easy to run are electrophilic substitution reactions in aromatic rings. During the nitration of polystyrene, one obtains poly(p-nitrostyrene). The substitution taking place almost exclusively in para position is due to the steric factors. Similarly, during the sulfonation of polystyrene or its derivatives, polymeric sulfonic acids are formed, which found application as soluble polyelectrolytes and ion exchangers in the case of cross-linked polymers.

The method of substitution can be also used to modify the polymer molecules in finished products. An example of this process is the fluorination of polyethylene bottles. It is possible to obtain a fluorinated polymer coating with a thickness of 0.1 mm.

7.2.2 TRANSFORMATION OF FUNCTIONAL GROUPS

Conversion of one functional group into another one is mainly used to obtain polymers, which are difficult to synthesize by other methods or cannot be obtained by a direct polymerization.

A classic example of such a reaction is obtaining poly(vinyl alcohol). Vinyl alcohol does not exist, because it rearranges immediately to acetaldehyde. For this reason, poly(vinyl alcohol) is obtained by hydrolysis of poly(vinyl acetate) or poly(vinyl formate). The hydrolysis reaction can be carried out in an acidic or alkaline environment. However, the formed product always contains a certain amount of acetyl or formyl groups and actually is a copolymer of vinyl alcohol with initial ester. Very good results can be obtained by running the hydrolysis reaction in the presence of crown ethers as its catalysts.

During the hydrolysis of polyacrylonitrile, poly(acrylic acid) is formed:

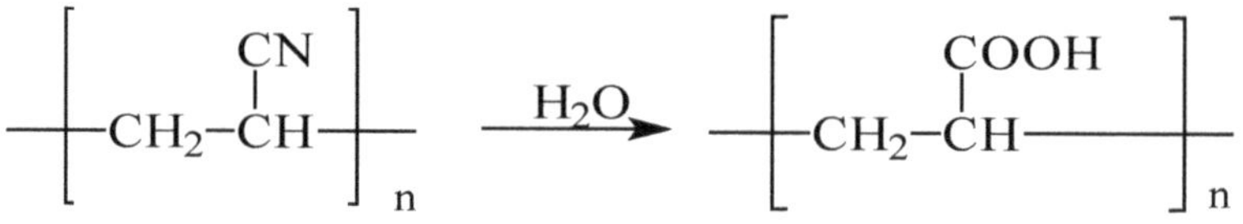

$$\left[-CH_2-\underset{\underset{CN}{|}}{CH}-\right]_n \xrightarrow{\;H_2O\;} \left[-CH_2-\underset{\underset{COOH}{|}}{CH}-\right]_n$$

Syndiotactic and isotactic poly(methacrylic acid) is obtained by the hydrolysis of poly(trimethylsilyl methacrylate):

$$\left[-CH_2-\underset{\underset{OSi(CH_3)_3}{\overset{C=O}{|}}}{\overset{|}{\underset{}{C}}}-\underset{CH_3}{\overset{|}{}}\right]_n + n\,H_2O \xrightarrow{\;OH^-\;} \left[-CH_2-\underset{\underset{OH}{\overset{C=O}{|}}}{\overset{|}{\underset{}{C}}}-\underset{CH_3}{\overset{|}{}}\right]_n + n\,(CH_3)_3SiOH$$

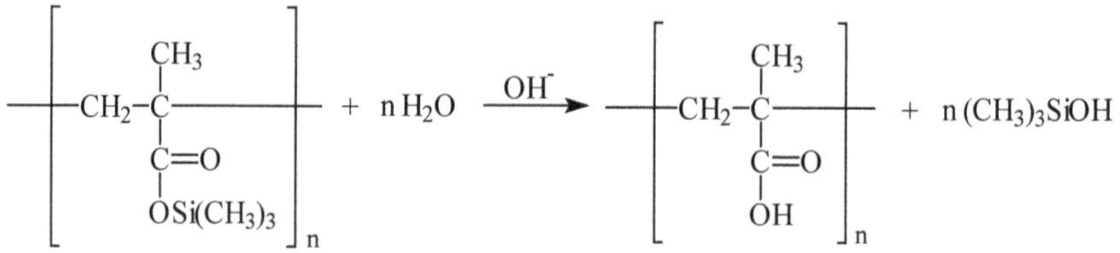

Syndiotactic poly(trimethylsilyl methacrylate) is obtained by anionic polymerization carried out in tetrahydrofuran, while the isotactic polymer is formed in toluene.

Another example of substitution reaction used to synthesize new polymers is the use of Hofmann method to produce polyamines from an easily accessible polyacrylamide.

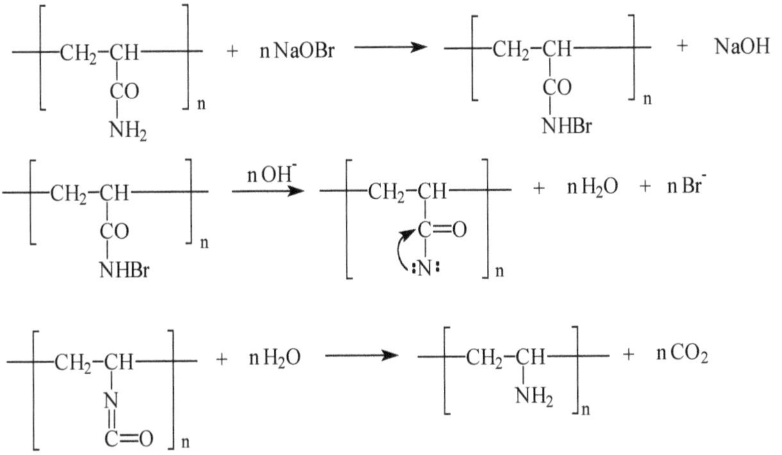

$$\left[-CH_2-CH-\underset{\underset{NH_2}{\overset{|}{CO}}}{}\right]_n + n\,NaOBr \longrightarrow \left[-CH_2-CH-\underset{\underset{NHBr}{\overset{|}{CO}}}{}\right]_n + NaOH$$

$$\left[-CH_2-CH-\underset{\underset{NHBr}{\overset{|}{CO}}}{}\right]_n \xrightarrow{n\,OH^-} \left[-CH_2-CH-\underset{C=O}{}\right]_n + n\,H_2O + n\,Br^-$$

$$\left[-CH_2-CH-\underset{\underset{C=O}{\overset{||}{N}}}{}\right]_n + n\,H_2O \longrightarrow \left[-CH_2-CH-\underset{NH_2}{}\right]_n + n\,CO_2$$

The chloromethylated polystyrene, formed by the reaction of polystyrene with chloromethyl ether, is an important raw material for further modification. It can be used to obtain a series of derivatives with the following structures:

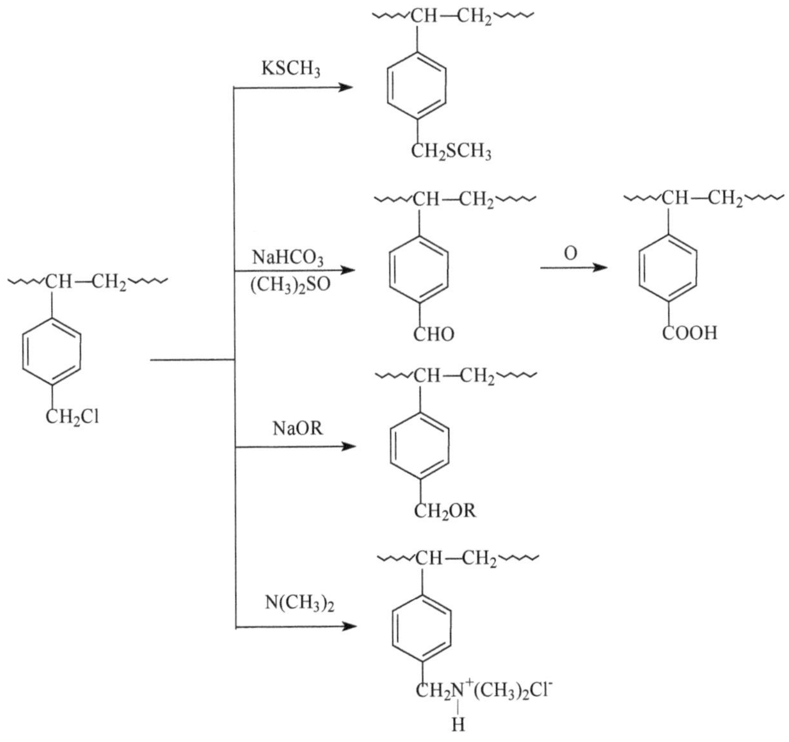

A series of polystyrene derivatives can be also obtained through the stage oflithium polystyrene:

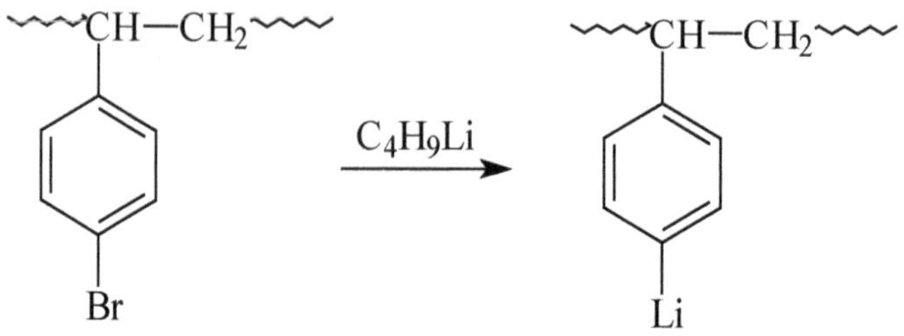

Lithium polystyrene is obtained by the reaction of n-polybromostyrene with n-butyllithium. It can react with such compounds like chlorodialkyl phosphine, boric acid ester, carbon dioxide, or dimethyldisulfide with formation of products with the following structure:

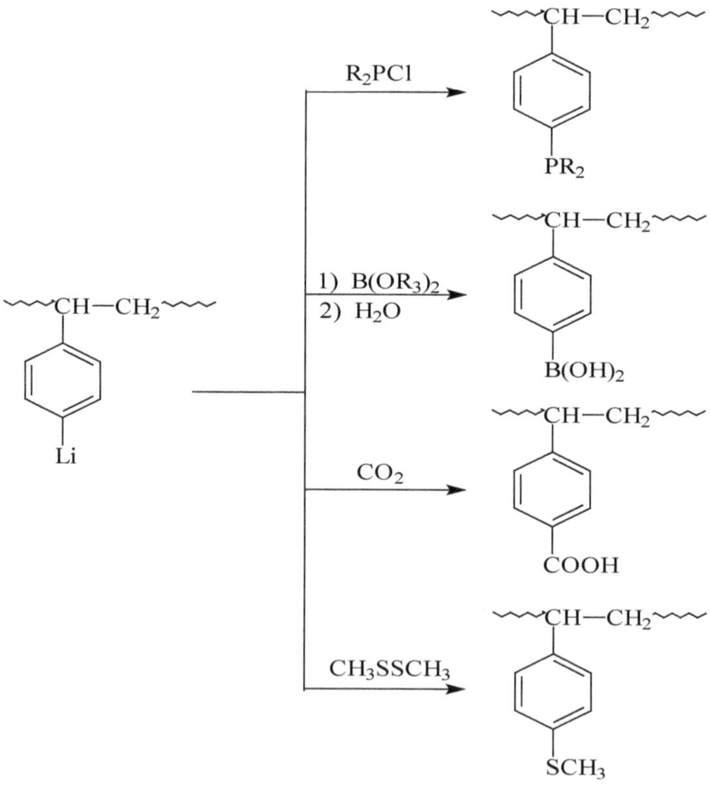

Introduction to the polymer molecule of segments containing primary aromatic amines or phenolic groups enables the synthesis of polymeric dyes. For this purpose, one runs the diazotization reaction on polymer containing primary amino groups at the aromatic carbon atom and then the coupling reaction with phenols or aromatic amines. It is also possible to run the coupling reaction of diazonium compounds with polymers containing hydroxyl or amine groups attached to the benzene ring. The course of these processes provide the following reactions:

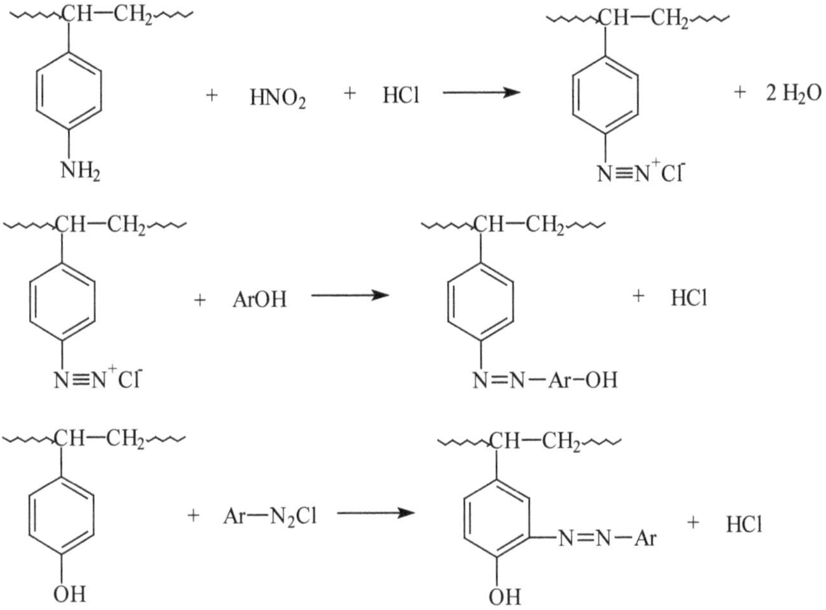

The synthesis of copolymers in which one component contains reactive groups provides further opportunities for chemical modification. An example of this can be the modification of copolymers of maleic anhydride, which can be carried out into suitable esters or imides.

The reaction with aniline proceeds through the stage of phenyl malamate acid derivative and leads to the formation of phenylmaleimide copolymer.

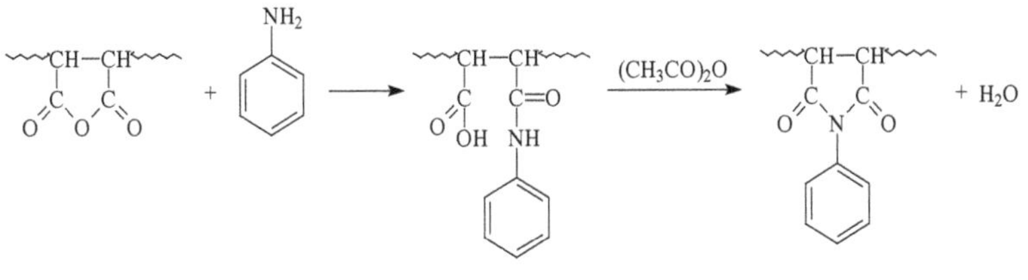

The chemical modification combinedwith the transformation or splitting off of functional groups can sometimes take place in the plastics processing industry as an adverse reaction. An example of such a reaction is the dehydrochlorination of poly(vinyl chloride).

The above examples show that due to the high reactivity of polymers, the possibilities of their chemical modification are very high. Some of these reactions have already found practical application on industrial scale.

The previously discussed synthesis of poly(vinyl alcohol), polyvinyloacetals, copolymers of ethylene with vinyl alcohol, cellulose nitrate (known also as nitrocellulose), cellulose acetate, ethyl cellulose, carboxymethyl cellulose, chitosan, chlorinated rubber, butoxylatedmelamine–formaldehyde resin, ion exchangers, and many other economically useful productscan serveas an example of the synthesis of chemically modified polymers.

7.2.3 INTRAMOLECULAR CYCLIZATION

The cyclic systems are introduced into the polymer chains mainly in order to increase their rigidity by inhibiting the free rotation of individual segments of the chain or to provide them the ladder structure that increases the heat resistance of polymer.

The cyclization of an intramolecular reaction takes place under the influence of catalysts or elevated temperatures.

The cyclization of polydienes takes place under the influence of protonic acids, tin tetrachloride, and titanium tetrachloride. The cyclization reaction of rubber is an ionic reaction and runs through the stage of formation of carbonion ion, the last arising as a result of the proton or titanium tetrachloride attachment to the double bond:

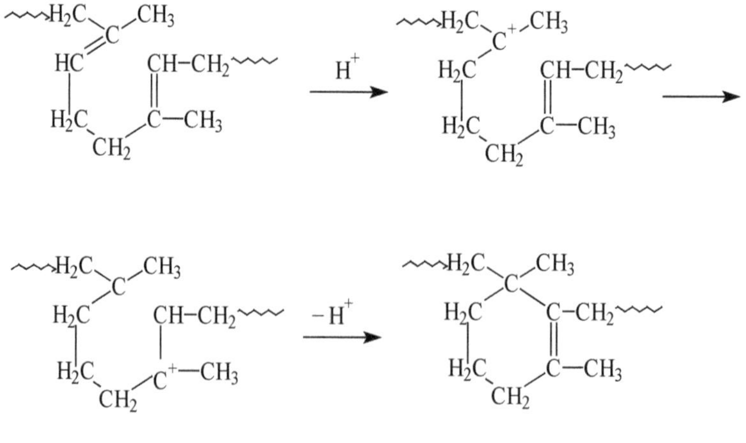

The cyclization of rubber is accompanied by a decrease of the unsaturation degree of polymer. Catalysts are not part of the cyclization products, but affect significantly the structure of the cyclized rubber.

An example of the synthesis of a ladder polymer is the cyclization of poly-acrylonitrile at temperature of 200 °C with the formation of a heterocyclic heat-resistant polymer with semiconductor properties:

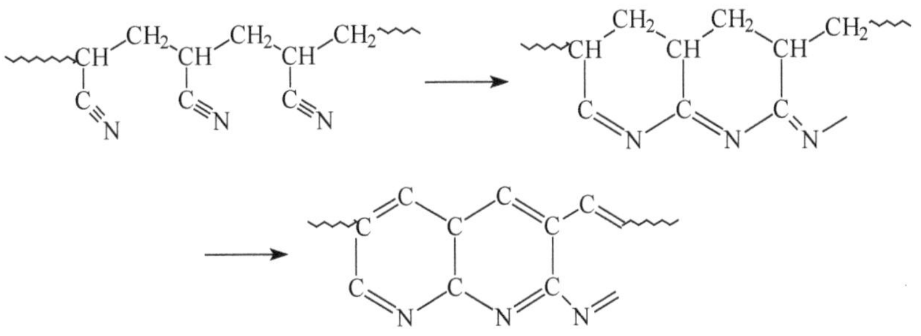

The ladder structures are also formed during the intramolecular aldol con-densation of poly(methyl vinyl ketone) in the presence of an alkaline catalyst.

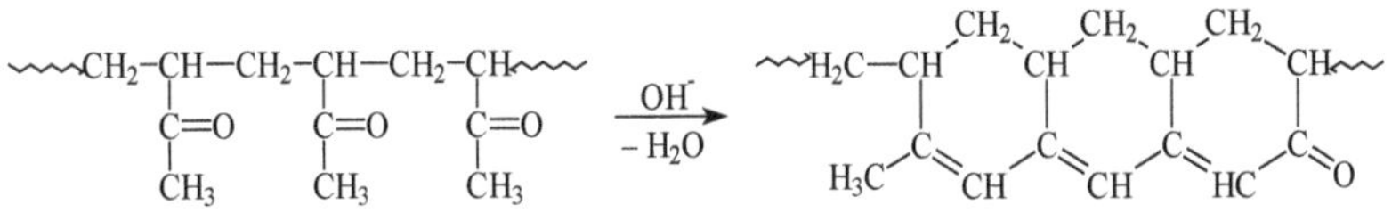

During the dehydration of polyacrylic acid, a correspondingpolyanhydride is formed:

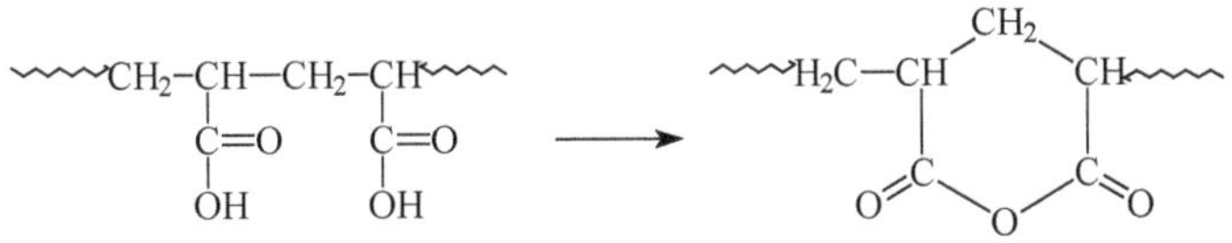

By attaching the aldehyde and phenylhydroxylamine molecules to unsatu-rated polymers, polymers with isooxazine rings are formed:

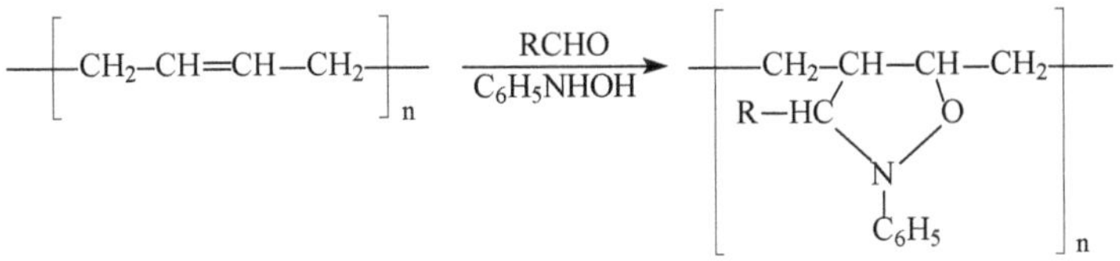

7.2.4 OXIDATION OF POLYMERS

The total oxidation of polymer leads to its decay with liberation of carbon dioxide, water, and possibly other degradation products. In less dramatic terms, this process is often linked to the degradation of macromolecules.

By selecting an appropriate oxidizing agent and using proper reaction parameters, it is possible to run a selective oxidation enabling the chemical modification of the polymer.

One of the most important reactions of this type is the formation in unsaturated polymers of oxirane (epoxy) units.

Selective oxidizing agents are, in this case, the carboxylic peracids such as the peracetic acid.

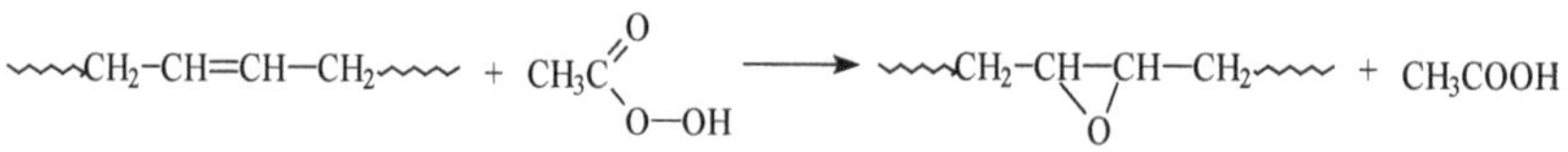

$$\text{\~\~CH}_2\text{-CH=CH-CH}_2\text{\~\~} + CH_3C\underset{O-OH}{\overset{O}{\diagup}} \longrightarrow \text{\~\~CH}_2\text{-CH-CH-CH}_2\text{\~\~} + CH_3COOH$$

It is possible, in a similar way, to oxidize poly(2-vinylpyridine) to a corresponding N-oxide, which found application as a biologically active polymer in the treatment of silicosis:

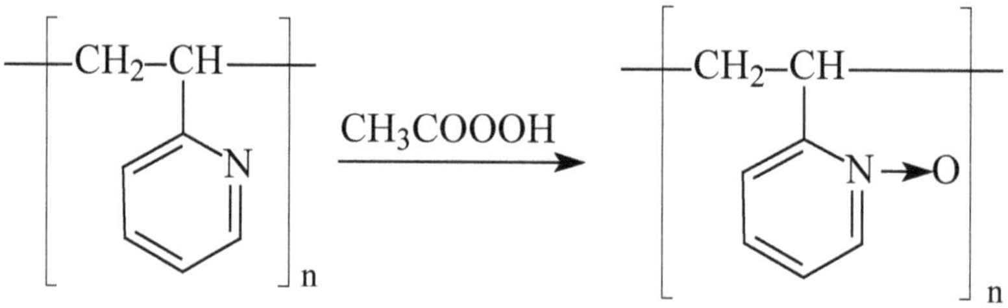

Application of a very mild oxidizing agent, such as pyridine dichromate, allows the oxidation of primary alcohol groups in poly(methylallyl alcohol) to poly(methyl acroleine):

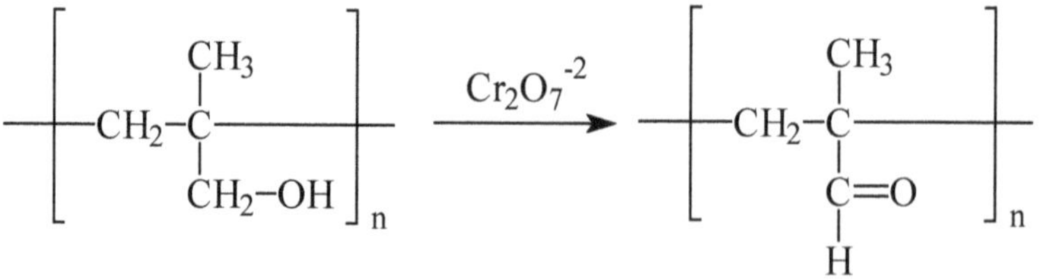

The reaction of oxidation of polymers is also often connected with the cross-linking process. For example, by mixing a solution of poly(vinyl alcohol) with a solution of potassium dichromate, a photosensitive mixture is formed.

Covering the polyamide grid with this mixture and evaporating the solvent in dark permits to obtain a photosensitive coating, which in exposed places tends to be insoluble. This procedure allows to obtain matrices for screen printing of fabrics.

7.2.5   REDUCTION OF POLYMERS

The reduction method can also be used to modify polymers. It enables the formation of a polymer molecule of amine groups by reducing the nitro groups:

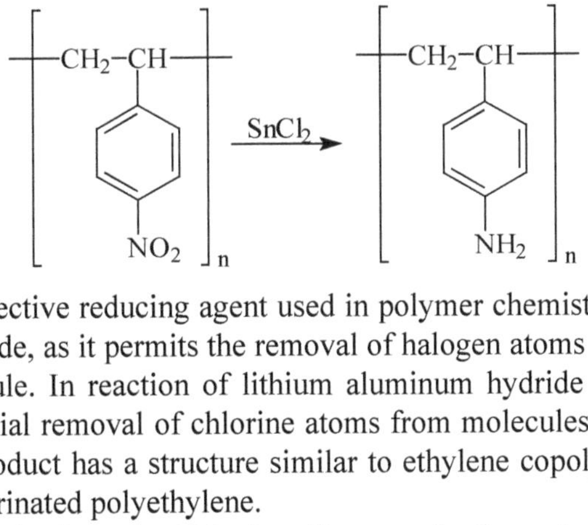

A highly selective reducing agent used in polymer chemistry is the lithium aluminum hydride, as it permits the removal of halogen atoms contained in the polymer molecule. In reaction of lithium aluminum hydride with poly(vinyl chloride), a partial removal of chlorine atoms from molecules takes place and the resulting product has a structure similar to ethylene copolymer with vinyl chloride or chlorinated polyethylene.

The lithium aluminum hydride also allows a selective reduction of the ester group to alcohol group. For example, during the reduction of poly(methyl methacrylate) with lithium aluminum hydride in dioxane, depending on the polymer concentration in solution, one can obtain the poly(methyl allyl alcohol) or its copolymers with methyl methacrylate. Some results of this process are listed in Table 7.2.

TABLE 7.2   Effect of Concentration of Poly(Methyl Methacrylate) (PMMA) in Dioxane on the Course of its Reduction with Lithium Aluminum Hydride.

No	PMMA concentration [g/dm^3]	Yield [%]	Saponification number [mgKOH/g]	Conversion [% mol]
1	10	89.4	349	37.5
2	6.9	77.2	372	33.4

TABLE 7.2 (*Continued*)

3	4	84	0	100
4	2	78	46	91.8

Source: Adapted fromSikorski, R. T.; Rykowski, Z.;Puszyński, A.Plast u. Kautschuk1984, 31, 250.

The reaction proceeds according to the scheme:

$$\left[-CH_2-\underset{\underset{COOCH_3}{|}}{\overset{\overset{CH_3}{|}}{C}}-\right]_n \xrightarrow{LiAlH_4} \left[-CH_2-\underset{\underset{CH_2OH}{|}}{\overset{\overset{CH_3}{|}}{C}}-\right]_n$$

In the same way, one can get poly(allyl alcohol) by reduction of poly(ethyl acrylate)

7.2.6 GRAFTING OF POLYMERS

Grafting of polymers is an important method of synthesis of copolymers by chemical modification of monomers.

The principle of this process consists on formation of an active center along the polymer chain, which starts polymerization of another monomer. The formed active centers could be either radical or ionic. This process can be presented by the following scheme:

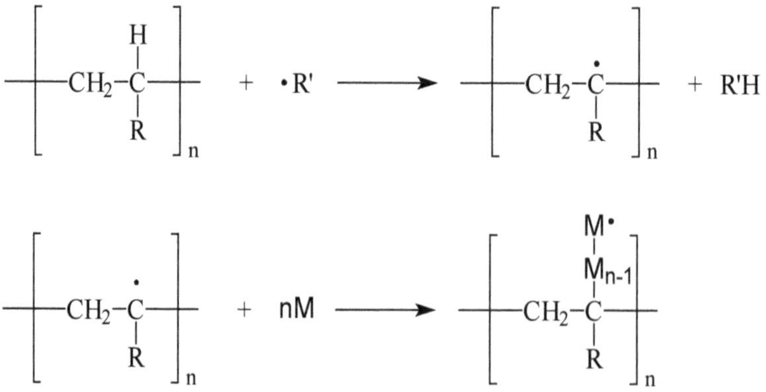

$$\left[-CH_2-\underset{\underset{R}{|}}{\overset{\overset{H}{|}}{C}}-\right]_n + \cdot R' \longrightarrow \left[-CH_2-\underset{\underset{R}{|}}{\overset{\cdot}{C}}-\right]_n + R'H$$

$$\left[-CH_2-\underset{\underset{R}{|}}{\overset{\cdot}{C}}-\right]_n + nM \longrightarrow \left[-CH_2-\underset{\underset{R}{|}}{\overset{\overset{\overset{M\cdot}{|}}{\underset{M_{n-1}}{|}}}{C}}-\right]_n$$

where M represents the monomer molecule.

Initiation of a grafting reaction can be carried out by the ozonization or oxidation method:

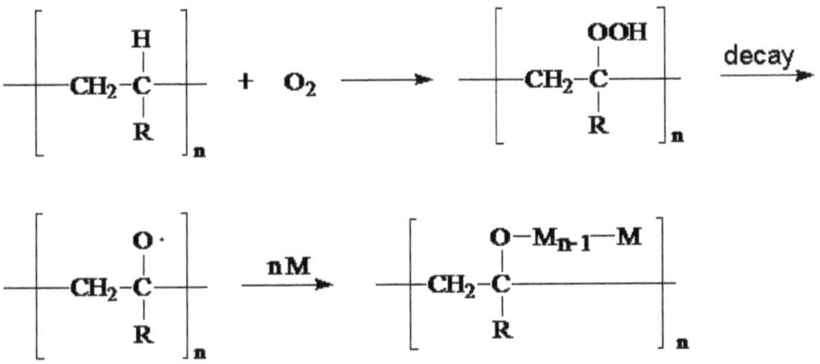

$$\left[-CH_2-\underset{\underset{R}{|}}{\overset{\overset{H}{|}}{C}}-\right]_n \; + \; O_2 \longrightarrow \left[-CH_2-\underset{\underset{R}{|}}{\overset{\overset{OOH}{|}}{C}}-\right]_n \xrightarrow{\text{decay}}$$

$$\left[-CH_2-\underset{\underset{R}{|}}{\overset{\overset{O\cdot}{|}}{C}}-\right]_n \xrightarrow{nM} \left[-CH_2-\underset{\underset{R}{|}}{\overset{\overset{O-M_{n-1}-M}{|}}{C}}-\right]_n$$

The free-radical grafting reactions of polymers are often initiated by cerium (IV) ions, which are reduced under the influence of sunlight with formation of free radicals:

$$Ce^{+4} + OH^- \longrightarrow Ce^{+3} + \cdot OH$$

These ions can also initiate a competitive homopolymerization reaction, because they generate free radicals of $M\cdot$ type:

$$M + Ce^{+4} \longrightarrow Ce^{+3} + \cdot M + H^+$$

Similar results are obtained with Mn^{+3} ions.

During the grafting reaction of acrylonitrile on starch, one obtains 95% yield of the grafted product and the molecular weight of grafted polyacrylonitrile chain is equal to 86,000. The content of the incidentally formed homopolymer is 4.2–4.9%. Replacing acrylonitrile by methyl methacrylate diminishes the grafting efficiency to 87% and the amount of incidentally formed poly(methyl methacrylate) is equal to 13.3%.

In order to characterize the grafted copolymers, one introduced the concept of grafting efficiency (G), total conversion (C_t), grafting efficiency (GE), and grafting frequency (GF), which are defined as follows:

$$G = \frac{\text{mass of grafted copolymer} - \text{mass of polymer}}{\text{mass of polymer}} \times 100\%$$

$$C_t = \frac{\text{mass of grafted polymer } - \text{ mass of homopolymer}}{\text{mass of polymer}} \times 100\%$$

$$GE = \frac{\text{mass of grafted polymer}}{\text{mass of copolymer } + \text{ mass of grafted polymer}} \times 100\%$$

$$GF = \frac{\text{mass of grafted polymer}}{\text{mass of polymer}} \cdot \frac{\overline{M}_{polymer}}{\overline{M}_{grafted\ polymer}} \times 100\%$$

In the case of grafting of acrylonitrile on starch by the mass of polymer, one understands the mass of starch and by mass of grafted polymer—the mass of polyacrylonitrile attached to starch.

Known are also ionic methods of grafting. Using the cationic method, it is possible to graft tetrahydrofuran on polyethylene or polypropylene.

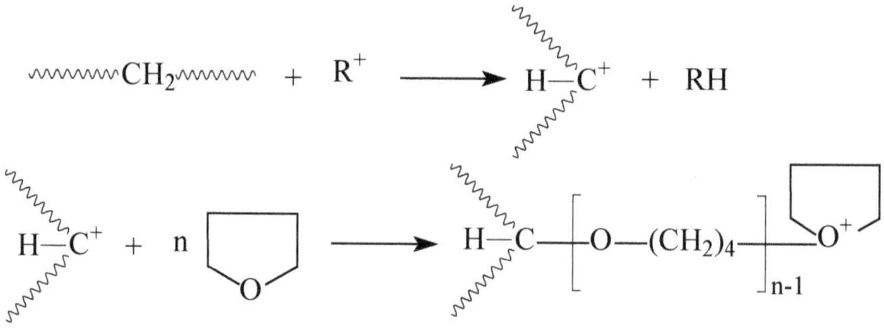

Polymers containing reactive side groups, can also transform ionic processes into grafted copolymers. For example, the chloromethylated polystyrene dissolved in carbon disulfide in the presence of aluminum tribromide reacts with 2-methylpropene, forming mainly polystyrene grafted with poly(2-methylpropene). Similarly, poly(vinyl chloride) undergoes a cationic grafting reaction by styrene in the presence of aluminum trichloride or by butadiene in the presence of the diethylaluminum chloride complex with cobalt.

The anionic grafting processes were also performed. An example of such a reaction is the grafting of polyacrylonitrile on poly(p-chlorostyrene) in tetrahydrofuran in the presence of sodium naphthalene:

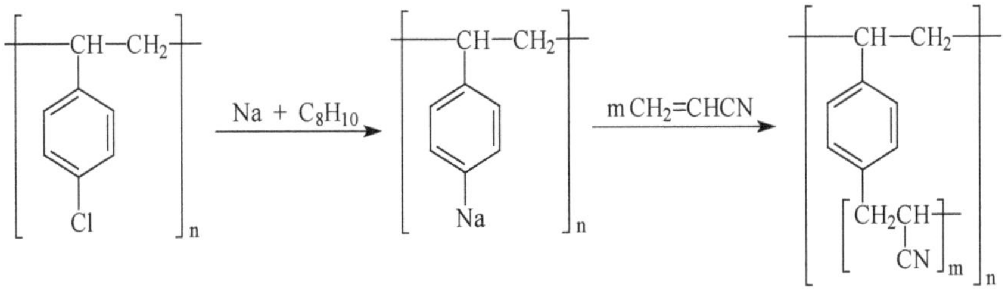

The unsaturated polymers can be grafted with vinyl monomers using the Ziegler–Natta-type catalysts. In the first stage, they are subjected to react with diethylaluminum hydride. One obtains the alkylaluminum polymeric compounds in this way. Then, by the reaction of halogens with metals such as titanium trichloride, one gets active centers initiating the grafting reaction. When using propylene, the method leads to the formation of side chains being an isotactic polypropylene:

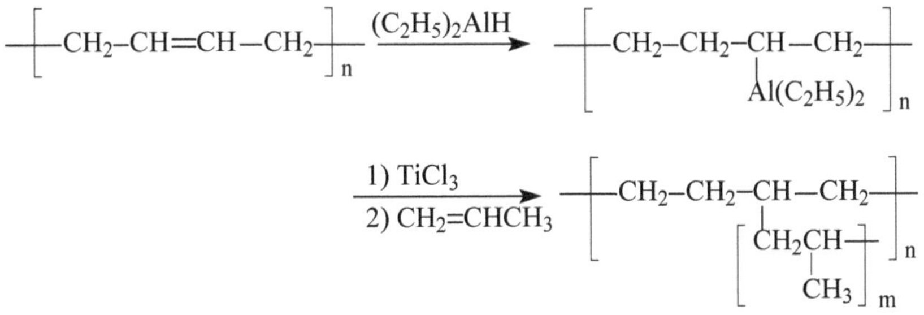

The above described polymer grafting examples do not cover all possibilities for conducting this important process.

7.2.7 CROSSLINKING OF SATURATED POLYMERS

Crosslinking is an important reaction of chemical modification of polymers making obtaining fusible and insoluble products possible, often with better properties than linear polymers. A small degree of crosslinking reduces the mobility of polymer chains and prevents their relative motion to each other, providing them properties of an elastomer.

During the tension or compression of partly cross-linked polymers, the extended lateral bonds cause return to the previous shape after stopping the deforming stress. The increase of crosslinking density of polymers reduces their flexibility and too dense crosslinking causes fragility of the product. The cross-linked polymers are often called duroplasts.

Depending on chemical structure of polymers, there are several ways of their crosslinking. In particular, one distinguishes:

- crosslinking of polymers containing no functional groups,

- crosslinking of unsaturated polymers,

- crosslinking of polymers containing reactive functional groups reacting with each other,

- crosslinking of polymers containing functional groups, able to react with other small or large molecule compounds, known as hardeners.

7.2.8 CROSSLINKING OF POLYMERS WITHOUT FUNCTIONAL GROUPS

Polymers without functional groups can cross-link using the free-radical reactions, initiated by the decay of organic peroxides as well as the UV radiation. The crosslinking reaction of such compounds can be also induced by radiation with streams of photons, electrons, neutrons, or a protons. Sometimes the reaction may take place under the influence of ionic catalysts of Friedel–Crafts type.

In order to cross-link a polymer with organic peroxides one prepares the so-called "preblends." The polymer is mixed with an organic peroxide and fillers (addition of inorganic or organic substances modifying material properties). The mixture is then annealed in forms at temperature allowing a decomposition of peroxide to free radicals. This method has found practical application in crosslinking of ethylene–propylene, polyester, and silicone rubbers.

The radiolysis of polymers under the influence of ionizing radiation can lead to both the crosslinking reaction and the degradation. In the case of high doses of radiation used dominates the degradation process. In other case (low dose), the reaction course depends on the polymer structure.

Geminallybisubstituted polymers, that is containing two identical substituents at the same carbon atom and the chlorinated polymers tend to split the chain, thus its degradation. The other types of vinyl polymers are easier to cure.

The mechanism of radiation-induced crosslinking is a free-radical process and proceeds through the stage of the extraction of hydrogen atom with the formation of a macroradical.

The reaction proceeds according to the scheme:

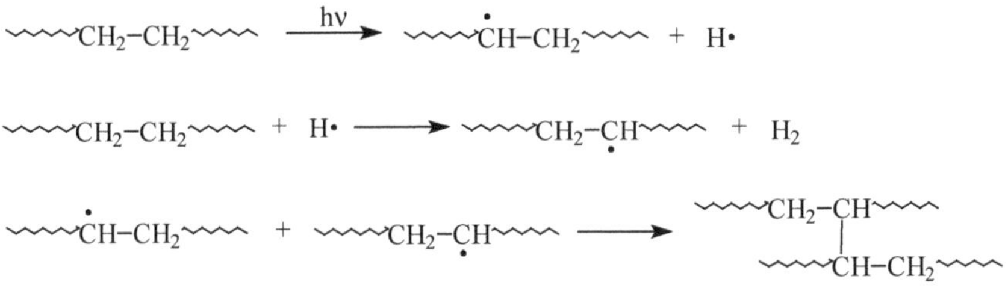

$$\text{CH}_2\text{-CH}_2 \xrightarrow{h\nu} \dot{\text{C}}\text{H-CH}_2 + \text{H}\cdot$$

$$\text{CH}_2\text{-CH}_2 + \text{H}\cdot \longrightarrow \text{CH}_2\text{-}\dot{\text{C}}\text{H} + \text{H}_2$$

$$\dot{\text{C}}\text{H-CH}_2 + \text{CH}_2\text{-}\dot{\text{C}}\text{H} \longrightarrow \begin{array}{c} \text{CH}_2\text{-CH} \\ | \\ \text{CH-CH}_2 \end{array}$$

In a similar way, runs the photolysis reaction of polymers under the influence of ultraviolet or visible light. But the process is more specific than the radiolysis. In this case to proceed, well-defined polymer structure elements or additives are necessary, enabling absorption of light radiation in an appropriate wavelength range.

If one adds some sensitizersto polymer, such as benzophenone, the absorption in ultraviolet causes excitation of sensitizer molecule and then splitting off the hydrogen atom from the polymer molecule. It leads to the formation of a macroradical capable of crosslinking by recombination with other macromolecules.

If the chromophore group is present in polymer molecule then the absorption of an energy quantum can lead to both the crosslinking as well as the degradation of polymer.

An example of such a reaction is the process taking place during the irradiation of vinyl polymers containing ester groups as side groups:

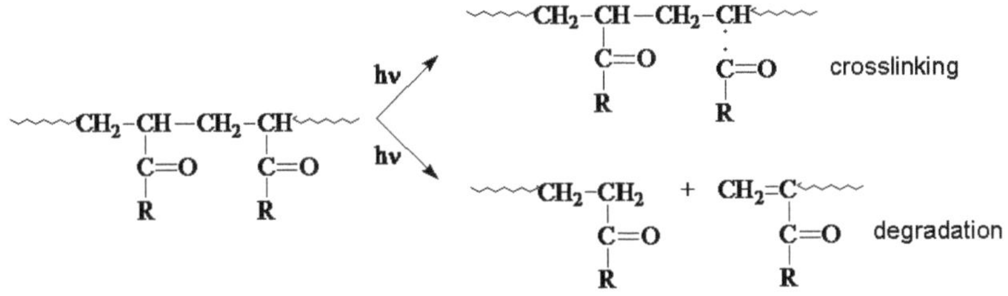

Copolymers of vinyl esters and fluorinated monomers, cross-linked by the action of ultraviolet radiation, are used as weather-resistant coatings for wood.

7.2.9 CROSS-LINKING OF UNSATURATED POLYMERS

The crosslinking process of unsaturated polymers is often called vulcanization. The vulcanization may be carried out under the influence of organic peroxides, sulfur, disulfur dichloride (S2Cl2), and other crosslinking agents such as the 1,3-phenylene disulfone acid.

In the case of crosslinking with organic peroxides, the split off of hydrogen occurs most frequently at the allyl carbon atom, that is, at the atom that is adjacent to the carbon atom forming a double bond. A macroradical is then formed and the crosslinking reaction takes place as a result of the recombination of radicals or through binding processes connected with a charge transfer:

$$\sim\sim CH_2\text{-}CH\text{=}CH\text{-}CH_2\sim\sim \ +\ RO\cdot \ \longrightarrow\ \sim\sim CH_2\text{-}CH\text{=}CH\text{-}\overset{\cdot}{CH}\sim\sim\ +$$

$$\begin{array}{c} \sim\sim CH_2\text{-}CH\text{=}CH\text{-}\overset{\cdot}{CH}\sim\sim \\[4pt] + \\[4pt] \sim\sim CH_2\text{-}CH\text{=}CH\text{-}\overset{\cdot}{CH}\sim\sim \end{array} \quad\longrightarrow\quad \begin{array}{c} \sim\sim CH_2\text{-}CH\text{=}CH\text{-}\underset{|}{CH}\sim\sim \\ \sim\sim CH_2\text{-}CH\text{=}CH\text{-}CH\sim\sim \end{array}$$

Or

$$\begin{array}{c} \sim\sim CH_2\text{-}CH\text{=}CH\text{-}\overset{\cdot}{CH}\sim\sim \\[4pt] + \\[4pt] \sim\sim CH_2\text{-}\overset{\cdot}{CH}\text{=}CH\text{-}CH_2\sim\sim \end{array} \quad\longrightarrow\quad \begin{array}{c} \sim\sim CH_2\text{-}CH\text{=}CH\text{-}\underset{|}{CH}\sim\sim \\ \sim\sim CH_2\text{-}CH\text{-}CH\text{-}CH_2\sim\sim \end{array}$$

The initiation of vulcanization using sulfur is carried out by attaching the active hydrogen atom to the sulfur molecules.

$$\sim\sim\sim CH_2\text{-}CH\text{=}CH\text{-}CH_2\sim\sim\sim \ +\ S_X \ \longrightarrow\ \sim\sim\sim CH_2\text{-}CH\text{=}CH\text{-}\overset{\cdot}{CH}\sim\sim\sim\ +\ S_X H$$

The formed radical reacts with the next sulfur molecule and the resulting addition product with radical on the sulfur atom joins the unsaturated carbon atom of another chain of polydiene.

$$S_X \ +\ \sim\sim\sim CH_2\text{-}CH\text{=}CH\text{-}\overset{\cdot}{CH}\sim\sim\sim \ \longrightarrow\ \sim\sim\sim CH_2\text{-}CH\text{=}CH\text{-}\underset{\overset{|}{S_X}}{CH}\sim\sim\sim$$

$$\begin{array}{c} \sim\sim\sim CH_2\text{-}CH\text{=}CH\text{-}\underset{\overset{|}{S_X}}{CH}\sim\sim\sim \\[4pt] + \\[4pt] \sim\sim\sim CH_2\text{-}CH\text{=}CH\text{-}CH_2\sim\sim\sim \end{array} \quad\longrightarrow\quad \begin{array}{c} \sim\sim\sim CH_2\text{-}CH\text{=}CH\text{-}\underset{\overset{|}{S_X}}{CH}\sim\sim\sim \\ \sim\sim\sim CH_2\text{-}\overset{\cdot}{CH}\text{-}CH\text{-}CH_2\sim\sim\sim \end{array}$$

Deactivationof the formedmacroradical takes place by transfering to the next polydiene molecule. The whole process runs on the principle of a chain reaction, up to a total exhaustion of the unbounded sulfur contained in the system.

The crosslinking of polydienes by using the sulfur is very slow. Therefore, one uses accelerators of this process for this purpose, such as tetramethylthiuram disulfide, dithiocarbamine acid derivatives, thiazole derivatives, diphenyl-guanidine, and the others.

In order to achieve the full effect of speeding up the process of vulcanization,additionally,one uses other activators such as zinc oxide and stearic acid.

Acceleration of the crosslinking process with sulfur reduces the number of cyclic monosulfide groups and increases the number of lateral mono- and disulfide bonds.

7.2.10 CROSSLINKING OF POLYMERS CONTAINING REACTIVE FUNCTIONAL GROUPS

If the polymer molecule contains reactive functional groups, capable of condensing with each other, than under appropriate conditions, intermolecular reactions combined with the crosslinking of polymer are possible.

Examples of this type of reactive polymers are phenol–formaldehyde resins of rezol type as well as urea or melamine–formaldehyde resins. A characteristic feature of such compounds is the presence of hydroxymethyl groups, also called methyl groups.

These groups react with each other with evolution of water or formaldehyde molecules. At the same time, a cross-linked polymer is formed, which contains methylene–ether or methylene bridges.

$$—CH_2OH \; + \; HOH_2C— \Big\langle \begin{matrix} —CH_2\text{--}O\text{--}CH_2 \; + \; H_2O \\ \\ —CH_2— \; + \; HCOH \end{matrix}$$

This reaction can be catalyzed by hydrogen or hydroxidel ions, or initiated at elevated temperatures. Since these reactions proceed readily at higher temperatures, therefore, such type of polymers are called thermosetting polymers.

Polymers containing reactive groups in the polymer chain can react with other suitable multifunctional compounds with the formation of cross-linked products. This type of polymers are called chemoreactive compounds. A striking example of this type of connections are polymers with epoxy groups, which can be cross-linked in a polymerization reaction, combined with opening of the oxirane ring as in the reactions with multifunctional amines or carboxylic acids.

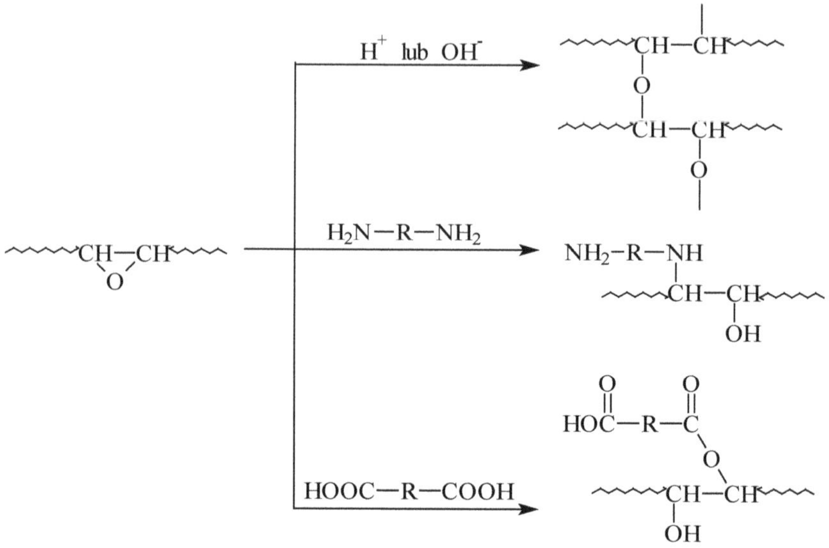

The others, after attachment to the epoxy group, free amino or carboxyl groups, react with epoxy groups of other polymer chains, forming cross-linked products.

Very good crosslinking agents, reacting with almost all reactive groups containing mobile hydrogen atoms, are multifunctional isocyanates, resulting in the formation of urethane bridges.

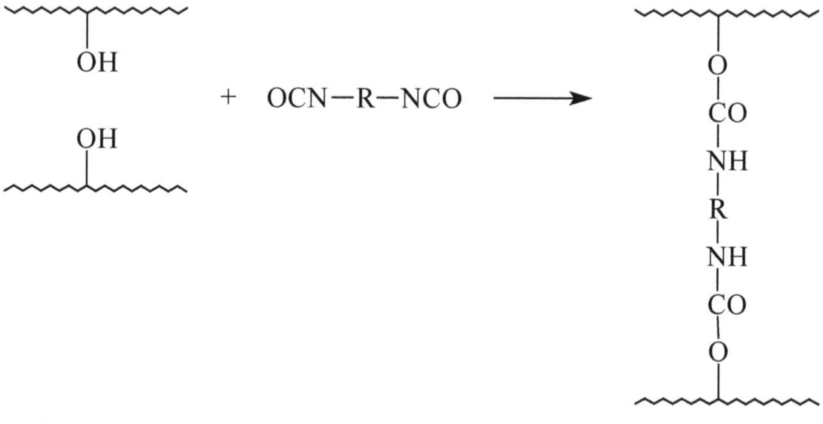

Highly reactive groups, enabling crosslinking, are chlorosulfonated groups or other groups containing mobile chlorine atom.

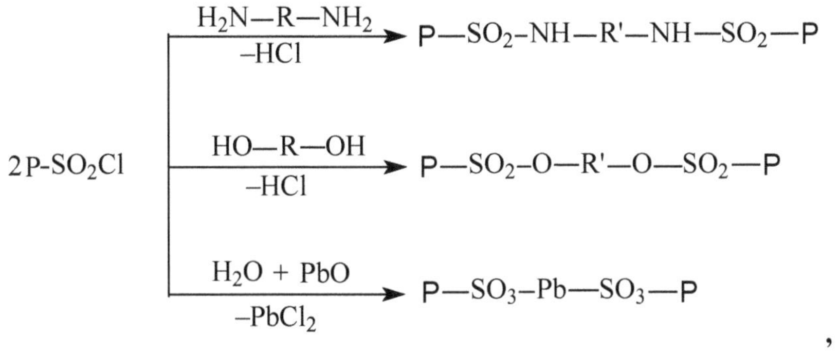

$$2\,P\text{-}SO_2Cl \begin{cases} \xrightarrow[-HCl]{H_2N\text{-}R\text{-}NH_2} & P\text{-}SO_2\text{-}NH\text{-}R'\text{-}NH\text{-}SO_2\text{-}P \\[2ex] \xrightarrow[-HCl]{HO\text{-}R\text{-}OH} & P\text{-}SO_2\text{-}O\text{-}R'\text{-}O\text{-}SO_2\text{-}P \\[2ex] \xrightarrow[-PbCl_2]{H_2O + PbO} & P\text{-}SO_3\text{-}Pb\text{-}SO_3\text{-}P \end{cases}$$

,

where P—polymer chain

Known are also possibilities of crosslinking the linear polymers through the formation of coordination bonds.

The Friedel–Crafts reaction enables a catalytic crosslinking of polymers by ionic mechanism. Itstypical example is the crosslinking of polystyrene with p-bis (chloromethyl)benzene in the presence of SnCl4.

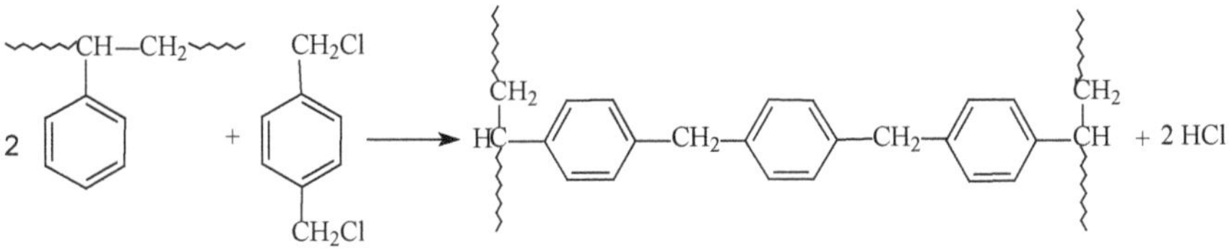

7.3 DEGRADATION OF POLYMERS

The degradation of polymers is a process in which the macromolecules decompose into smaller fragments. The latter may be molecules with smaller molecular weight or the products of partial decomposition with changed chemical composition as a result of split off or transformation of certain substituents. A special case of the degradation of polymers is the reaction of depolymerization leading to the monomer formation.

The process of degradation of polymers, run in a controlled way, is of a great practical importance. It allows for a reduction in molecular weight of polymer molecules in order to standardize the length of chains (a process called linearization) and to facilitate processing operations. Examples of practical use of controlled degradation is the mercerization of cellulose in the manufacturing of cellulose fibers, mastication of rubber before vulcanization obtaining peptides and amino acids by hydrolysis of proteins and the recovery of methyl methacrylate by depolymerization of poly(methyl methacrylate) waste and also the linearization of polyorganosiloxanes.

The uncontrolled polymer degradation process is often harmful, limiting their practical applications.

Factors affecting the degradation of polymers can be both physical and chemical.

7.3.1 DEGRADATION OF POLYMERS BY PHYSICAL AGENTS

Various physical factors affect natural polymers and are responsible for their degradation, such as heat, light illumination and radiation, ultrasounds and mechanical forces during grinding, rolling, and other mechanical processes.

The polymer degradation caused by the physical agents is often referred to as the polymer aging process.

The persistence of individual bonds in macromolecule depends on their dissociation energy, which causes disintegration of a particular bond. According to the Boltzmann distribution at any temperature T [K], there is a finite probability of "W" that a given bond has reached the critical level of binding and dissociation energy Ed.

$$W = e^{-\frac{E_d}{kT}},$$

where k is the Boltzmann constant.

As a result of dissociation of a specific bond, two macroradicals are formed, which undergo the dysproportionation reaction. A new macroradical and a polymer with shorter chain and with double bondarethen formed.

If the macroradical is located at the end of the chain, then a polymer rupture reaction can take place, during which a monomer molecule is released:

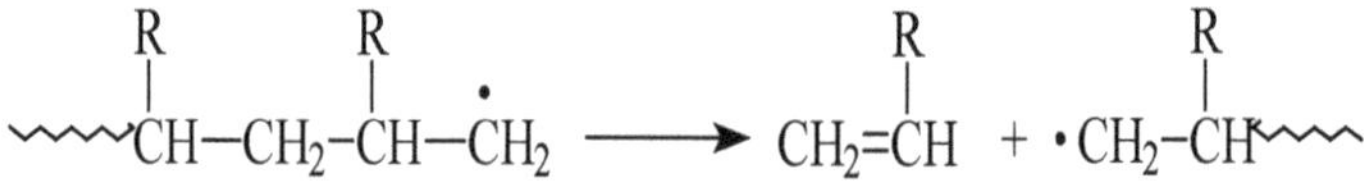

$$
\begin{array}{ccc}
R & R & \\
| & | & \bullet \\
\text{\textasciitilde\textasciitilde CH-CH}_2\text{-CH-CH}_2 & \longrightarrow & \text{CH}_2\text{=CH} \; + \; \bullet\text{CH}_2\text{-CH\textasciitilde\textasciitilde} \\
& & \begin{array}{cc} R & R \\ | & | \end{array}
\end{array}
$$

The polymer degradation process, running in the solid state, is accompanied by a parallelly progressing crosslinking process, which was discussed previously.

At sufficiently high temperatures, the polymer degradation occurs in the absence of oxygen.

The photodegradation of polymers occurs under the influence of ultraviolet radiation and sometimes visible.

The energy of photon radiation (quantum) for a given wavelength is determined by the equation:

$$E = h\upsilon = h\frac{c}{\lambda},$$

where:
h—Planck's constant, h = 6.623 ×10−27 erg s
v—frequency of radiation (1/s)
c—speed of light, c = 3 ×1010 cm/s
l—wavelength.
The photodegradation reaction takes place when the energy absorbed by polymer molecule is at least equal to the dissociation energy Ed.

Sometimes, however, the absorbed energy can be converted into heat or emitted radiation in form of light quanta.

$$\Phi = \frac{\text{liczba makrocząsteczek ulegających fotodegradacji}}{\text{liczba zaabsorbowanych kwantów świata przez polimer}}$$

To halt this process, one applies usually appropriate photostabilizers. However, in the case when a change of structure of hotodegradble polymers is needed, appropriate photosensitizers are then used, which facilitate this process.

Similarto the photolysis, but much easier, the radiation degradation of polymers proceeds. A characteristic feature of this process is the nearly complete lack of depolymerization, leading to the formation of monomers.

Different groups present in polymer chains exhibit different sensitivities to radiation. If the presence of certain groups, such as phenyl radical or double bonds increases the resistance of polymer to the ionizing radiation, then one speaks about the internal protective effect. Its action is not limited to the nearest mer only, but extends to the whole macromolecule.

7.3.2 CHEMICAL DEGRADATION

The degradation taking place under the influence of chemical agents causes ruptures of chains as a result of chemical reaction. This concerns mainly the hydrolytic reactions, running principally in the case of condensation polymers and containing easy to break chain elements, such as ester or amide groups. Very important chemical degradation reactions are the oxidation reactions and attachment of ozone.

The molecular oxygen present in air facilitates the process of degradation of polymers and makes it possible at the relatively low temperatures, even at room temperature after a sufficiently long period of time. This process takes place more easily when performed in the presence of singlet oxygen or ozone.

The reaction of degradation in the presence of oxygen proceeds, with its participation, through the stage of formation of hydrogen peroxides. The macroradical, formed by thermal dissociation or other radical reactions, reacts with oxygen, and then with another polymer molecule in transfer reaction associated with the formation of a hydrogen peroxide:

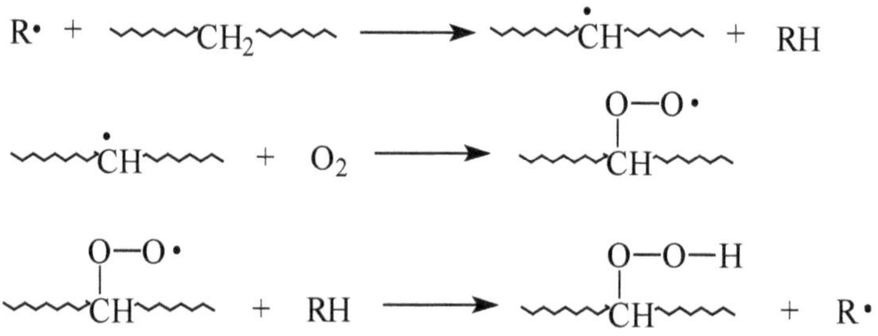

$$R\cdot + \sim\sim CH_2\sim\sim \longrightarrow \sim\sim\overset{\bullet}{C}H\sim\sim + RH$$

$$\sim\sim\overset{\bullet}{C}H\sim\sim + O_2 \longrightarrow \sim\sim\underset{\underset{CH}{|}}{\overset{O-O\cdot}{}}\sim\sim$$

$$\underset{\underset{CH}{|}}{\overset{O-O\cdot}{}}\sim\sim + RH \longrightarrow \sim\sim\underset{\underset{CH}{|}}{\overset{O-O-H}{}}\sim\sim + R\cdot$$

The formed hydrogen peroxides decompose easily by heating or by light illumination according to the scheme:

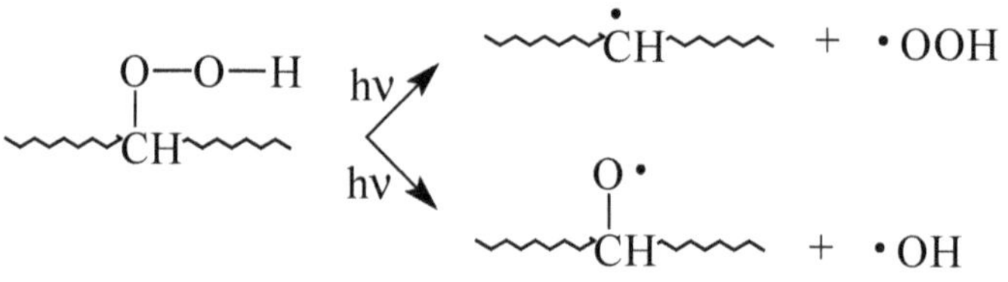

$$\underset{\underset{CH}{|}}{\overset{O-O-H}{}}\sim\sim \quad \overset{h\nu}{\underset{h\nu}{\Big<}} \quad \begin{array}{l} \sim\sim\overset{\bullet}{C}H\sim\sim + \cdot OOH \\[2ex] \sim\sim\underset{\underset{CH}{|}}{\overset{O\cdot}{}}\sim\sim + \cdot OH \end{array}$$

This causes a series of follow-up free-radical reactions. The oxy–macroradical is easily disrupted in b position. During the disrupture, an aldehyde group and a newmacroradical are formed at the chain end:

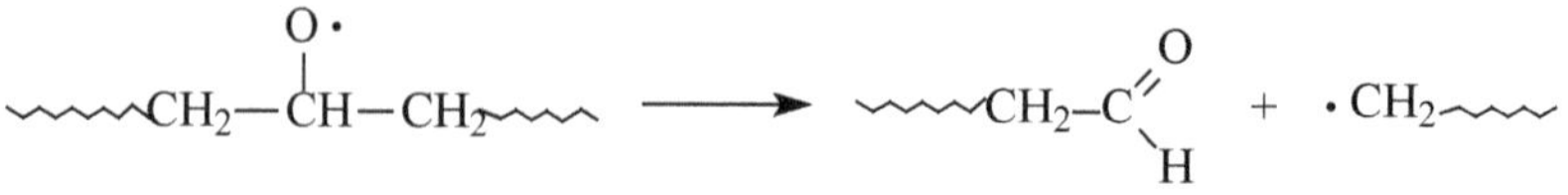

$$\sim\sim CH_2-\underset{\underset{H}{|}}{\overset{O\cdot}{C}}-CH_2\sim\sim \longrightarrow \sim\sim CH_2-C\overset{\displaystyle O}{\underset{\displaystyle H}{\Big\backslash}} + \cdot CH_2\sim\sim$$

Even easier, the ozone moleculecan be attached to the polymer. As a result of this reaction, a peroxide macroradical and a hydroxyl radical are formed.

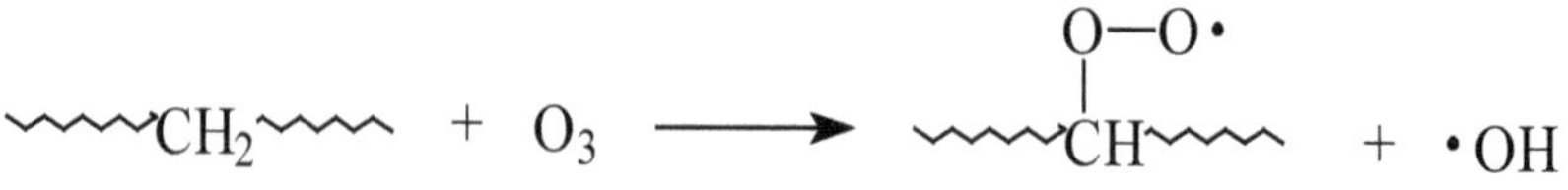

$$\sim\sim CH_2\sim\sim + O_3 \longrightarrow \sim\sim\underset{\underset{CH}{|}}{\overset{O-O\cdot}{}}\sim\sim + \cdot OH$$

If the polymer molecule contains double bonds, then the ozonides are formed, which subsequently decompose into aldehydes and acids.

$$\text{~~HC}=\text{CH~~} + O_3 \longrightarrow \text{~~HC} \underset{O}{\overset{O-O}{\diagup \diagdown}} \text{CH~~} \longrightarrow \begin{array}{c} \text{~~} C \overset{O}{\underset{H}{\diagup}} \\ + \\ \text{~~} C \overset{O}{\underset{H}{\diagup}} \end{array}$$

This phenomenon has been used recently for the fabrication of degradable polymers. The plasticsderived from them decompose under natural conditions and after use may be disposed of in waste dumps.

The currently used degradable polymers can be divided into:
- photodegradable,
- biodegradable,
- composites obtained from biodegradable polymers.

The photodegradable polymers, containing sensitizers, undergo the degradation process by exposure to ultraviolet radiation in the presence of oxygen.

The photodegradable polymers, among which the most important are the polyolefins, are intended mainly for the manufacture of disposable packagings.

An interesting application of the photodegradable polyethylene film is the protection of young plants growth in countries with dry climates. Crops are covered with foil containing a precise amount of sensitizer. The foil prevents the heat and the moisture loss. It also prevents the access of pests and weed growth and ensure a proper growth and development of crop root. After reaching a certain growth, the intense sunlight causes degradation of the film and its disintegration into small, degraded fragments. Breakthrough time of the films depends on the amount of added sensitizer, assuming a cloudless sky. The crushed film residue does not need to be collected and can be mixed with the soil during the plowing.

The biodegradable polymers are natural polymers, and polymers obtained by the modification of natural as well as those manufactured by various methods of chemical synthesis and biotechnological processes. The world's first biodegradable polymer is produced using the biotechnology methodcalled Bipol, produced by British company Zeneca. Bipol is a polyester derived from 3-hydroxybutyric and 3-hydroxypentanoic acids. The production process involves the fermentation of sugars with the use ofAlcaligenesentropus bacteria.

During the last 20 years,the properties of many other similar polyesters were explored. As result, the fatty polyhydroxy acids are now one of the larger group of thermoplastic polymers (more than 100 different types). In addition to the biodegradability, an important feature of these products is that they can be processed by the same methods as other thermoplastics

The microbiological processes are also used for the production of poly(lactic acid) (PLA), which is produced by the Lactobacillus bacteria in the process of fermentation of sugars. It is used for the manufacture of packaging materials and fibers for surgical applications as well as, among the others, for carpet production.

The poly(aspartic acid) is a completely biodegradable polyamide. It is manufactured by Bayer AG company since 1997 by reacting the maleic anhydride with ammonia. The polymer is used to soften water and to prevent the formation of deposits in pipes and in reservoirs.

An enormous advantage of biodegradable polymers is the possibility of recycling of used products and other waste by composting.

REFERENCES

Abusleme, J.; Giannetti, E. Emulsion Polymerization. Macromolecules 1991,24, 67–83.

Adesamya, I.; Classification of Ionic Polymers. Polym. Eng. Sci. 1988, 28, 1473.

Akbulut, U.; Toppare, L.; Usanmaz, A.; Onal, A. Electroinitiated Cationic Polymeryzation of Some Epoksides. Makromol. Chem. Rapid. Commun. 1983, 4, 259.

Andreas, F.; Grobe, K. Chemiapropylene;, WNT: Warszawa, 1974.

Barg, E. I. Technologiatworzywsztucznych; PWT: Warszawa, 1957.

Billmeyer, F. W. Textbook of Polymer Science; JohnWiley and Sons: New York, 1984.

Borsig, E.; Lazar, M. Chemical Modification of Polyalkanes. Chem. Listy1991, 85, 30.

Braun, D.; Czerwiński, W.; Disselhoff, G.; Tudos, F. Analysis of the Linear Methods for Determining Copolymerization Reactivity Ratios. Angew. Macromol. Chem. 1984, 125, 161.

Burnett, G. M.; Cameron, G. G.; Gordon, G.; Joiner, S. N. Solvent Effects on the Free Radical Polymerization of Styrene. J. Chem. Soc. Faraday Trans. , 11973, 69, 322.

Challa, R. R.; Drew, J. H.; Stannet, V. T.; Stahel, E. P. Radiation-Induced EmulsionPolymeryzation of Vinyl Acetate. J. Appl. Polym. Sci. 1985, 30, 4261.

Candau, F. Mechanism and Kinetics of the Radical Polymerization of Acrylamide in Inverse Micelles. Macromol. Chem. , MacromolSymp. 1990, 31, 27.

Ceresa, R. J. Kopolimeryblokoweiszczepione; WNT: Warszawa, 1965.

Charlesby, A. Chemiaradiacyjnapolimerów; WNT: Warszawa, 1962.

Chern, C. S.; Poehlein, G. W. Kinetics of Crosslinking Vinyl Polymerization. Polym. Plast. Technol. Eng. 1990, 29, 577.

Chiang, W. Y.; Hu, C. M. Studies of Reactions with Polymers. J. Appl. Polym. Sci. 1985, 30, 3895–4045.

Chien, J. C.; Salajka, Z. Syndiospecific Polymerization of Styrene. J. Polym. Sci. A1991, 29, 1243.

Chojnowski, J.; Stańczyk, W. E. Procesychlorowaniapolietylenu. Polimery1972, 17, 597.

Collomb, J.; Morin, B.; Gandini, A.; Cheradame, H. Cationic Polymerization Induced by Metal Salts. Europ. Polym. J. 1980, 16, 1135.

Czerwiński, W. K. Solvent Effects on Free-Radical Polymerization. Macromol. Chem. 1991, 192, 1285–1297.

Dainton, F. S. Reakcjełańcuchowe; PWN:Warszawa, 1963.

Ding, J.; Price, C.; Booth C. Use of Crown Ether in the Anionic Polymerization of Propylene Oxide. Eur. Polym. J. 1991, 27, 895–901.

Dogatkin, B. A. Chemiaelastomerów; WNT:Warszawa, 1976.

Duda, A.; Penczek, S. Polimeryzacjajonowaipseudojonowa. Polimery1989, 34, 429.

Dumitrin, S.; Shaikk, A. S.; Comanita, E.; Simionescu, C. Bifunetional Initiators. Eur. Polym. J. 1983, 19, 263.

Dunn, A. S. Problems of Emulsion Polymerization. Surf. Coat. Int. 1991, 74, 50.

Eliseeva, W. I.; Asłamazova, T. R. Emulsionnajapolimeryzacja v otsustivieemulgatora. Uspiechi-Chimii1991, 60, 398.

Erusalimski, B. L. Mechanism dezaktivacirastuszczichciepiej v procesachionnojpolimeryzacjivi-nylovychpolimerov. Vysokomol. Soed. A. 1985, 27, 1571.

Fedtke, M. Chimiczieskijereakcjipolimerrov;Khimia:Moskwa, 1990.

Fernandez-Moreno, D.; Fernandez-Sanchez, C.; Rodriguez, J. G.; Alberola, A. Polymerization of Styrene and 2-vinylpyridyne using AlEt-VClHeteroaromatic Bases. Eur. Polym. J. 1984, 20, 109.

Fieser, L. F.; Fieser, M. Chemiaorganiczna; PWN: Warszawa, 1962.

Flory, P. J. Nauka o makrocząsteczkach. Polimery1987, 32, 346.

Frejdlina, R. Ch.; Karapetian, A. S. Telomeryzacja; WNT: Warszawa, 1962.

Funt, B. L.; Tan, S. R. The Photoelectrochemical Initiation of Polymerization of Styrene. J. Polym. Sci. A1984, 22, 605.

Gerner, F. J.; Hocker, H.; Muller, A. H.; Shulz, G. V. On the Termination Reaction in the Anionic Polymerization of Methyl Methacrylate in Polar Solvents. Eur. Polym. J. 1984, 20, 349.

Ghose, A.; Bhadani, S. N. Electrochemical Polymerization of Trioxane in Chlorinated Solvents. Indian J. Technol. 1974, 12, 443,

Giannetti, E.; Nicoletti, G. M.; Mazzocchi, R. Homogenous Ziegler–Natta Catalysis. J. Polym. Sci. A1985, 23, 2117.

Goplan, A.; Paulrajan, S.; Venkatarao, K.; Subbaratnam, N. R. Polymerization of N,N'- methylene Bisacrylamide Initiated by Two New Redox Systems Involving Acidic PermanganateEur. Polym. J. 1983, 19, 817.

Gózdz, A. S.; Rapak, A. Phase Transfer Catalyzed Reactions on Polymers. Macromol. Chem. Rapid Commun. 1981, 2, 359.

Greenley, R. Z. An Expanded Listing of Revized Q and e Values. J. Macromol. Sci. Chem. A1980, 14, 427.

Greenley, R. Z. Recalculation of Some Reactivity Ratios. J. Macromol. Sci. Chem. A1980, 14, 445.

Guhaniyogi, S. C.; Sharma, Y. N. Free-Radical Modification of Polipropylene. J. Macromol. Sci. Chem. A1985, 22, 1601.

Guo, J. S.; El Aasser, M. S.; Vanderhoff, J. W. MicroemulsionPolymerization of Styrene. J. Polym. Sci. A1989, 27, 691.

Gurruchaga, M.; Goni, I.; Vasguez, M. B.; Valero, M.; Guzman, G. M. An Approach to the Knowledge of the Graft Polymerization of Acrylic Monomers onto Polysaccharides Using Ce (IV) as Initiator. J. Polym. Sci. Part C1989, 27, 149.

Hasenbein, N.; Bandermann, F. Polymerization of Isobutene with VOCl3 in Heptane. Macromol. Chem. 1987, 188, 83.

Hatakeyama, H. et al: Biodegradable Polyurethanes from Plant Components. J. Macromol. Sci. A. 1995, 32, 743.

Higashimura, T.; Law, Y. M.; Sawamoto, M. Living Cationic Polymerization. Polym. J. 1984, 16, 401.

Hirota, M.; Fukuda, H. Polymerization of Tetrahydrofuran. Nagoya-ski Kogyo KenkynshoKentynHokoku1984, 51.

Hodge, P.; Sherrington, D. C. Polymer-Supported Reactions in Organic Syntheses;John Wiley and Sons: New York, 1980;tłum. narosyjski Mir:Moskwa, 1983.

Inoue, S. Novel Zine Carboxylates as Catalysts for the Copolymerization of CO2with Epoxides. Makromol. Chem. Rapid Commun,1980, 1, 775.

Janović, Z. Polimerizacije I polimery;IzdanjaKemije Ind. :Zagrzeb, 1997.

Jedliński, Z. J. Współczesnekierunki w dziedziniesyntezypolimerów. Wiadomości Chem. 1977, 31, 607.

Jenkins, A. D.; Ledwith, A. Reactivity, Mechanism and Structure in Polymer Chemistry; JohnWiley and Sons: London, 1974; tłum. na rosyjski, Mir:Moskwa, 1977.

Joshi, S. G.; Natu, A. A. Chlorocarboxylation of Polyethylene. Angew. Makromol. Chem. 1986, 140, 99–115

Kaczmarek, H. polimery, a środowisko. Polimery1997, 42, 521.

Kang, E. T.; Neoh, K. G. Halogen Induced Polymerization of Furan. Eur. Polym. J. 1987, 23, 719.

Kang, E. T.; Neoh, K. G.; Tan, T. C.; Ti, H. C. Iodine Induced Polymerization and Oxidation of Pryridazine. Mol. Cryst. Lig. Cryst. 1987, 147, 199.

Karnojitzki, V. Organiczieskijepierekisy; I. I. L. :Moskwa, 1961.

Karpiński, K.; Prot, T. Inicjatoryfotopolimeryzacjiifotosieciowania. Polimery1987, 32, 129.

Keii, T. Propene Polymerization with MgCl2-Supported Ziegler Catalysts. Macromol. Chem. 1984, 185, 1537.

Khanna, S. N.; Levy, M.; Szwarc, M. Complexes Formed by Anthracene with 'Living' Polystyrene. Dormant Polym. Trans. Faraday Soc. 1962,58, 747–761.

Koinuma, H.; Naito, K.; Hirai, H. Anionic Polymerization of Oxiranes. Macromol. Chem. 1982, 183, 1383.

Korniejev, N. N.; Popov, A. F.; Krencel, B. A. Komplieksnyjemetalloorganiczeskijekatalizatory; Khimia: Leningrad, 1969.

Korszak, W. W. Niekotoryjeproblemypolikondensacji. Vysokomol. Soed. A1979, 21, 3.

Korszak, W. W. Technologiatworzywsztucznych; WNT: Warszawa, 1981.

Kowalska, E.; Pełka, J. Modyfikacjatworzywtermoplastycznychwłóknamicelulozowymi. Polimery2001, 46, 268.

Kozakiewicz, J.; Penczek, P. Polimeryjonowe. Wiadomości Chem. 1976, 30, 477.

Kubisa, P. Polimeryzacjażyjąca. Polimery1990, 35, 61.

Kubisa, P.; Neeld, K.; Starr, J.; Vogl, O. Polymerization of Higher Aldehydes. Polymer1980, 21, 1433.

Kucharski, M. Nitroweiaminowepochodnepolistyrenu. Polimery1966, 11, 253.

Kunicka, M.; Choć, H. J. HydroliticDegradation and Mechanical Properties of Hydrogels. Polym. Degrad. Stab. 1998, 59, 33.

Lebduska, J.; Dlaskova, M. Roztokovakopolymerace. ChemickyPrumysl1990, 40, 419–480.

Leclerc, M.; Guay, J.; Dao, L. W. Synthesis and Characterization of Poly(alkylanilines). Macromolecules1989, 22, 649.

Lee, D. P.; Dreyfuss, P. Triphenylphesphine Termination ofTetrahydrofuran Polymerization. J. Polym. Sci. ,Polym. Chem. Ed. 1980, 18, 1627.

Leza, M. L.; Casinos, I.; Guzman, G. M.; Bello, A. Graft Polymerization of 4-vinylpyridineonto Cellulosic Fibres by the Ceric on Method. Angew. Macromol. Chem. 1990, 178, 109–119.

Lopez, F.; Calgano, M. P.; Contrieras, J. M.; Torrellas, Z.; Felisola, K. Polymerization of Some Oxiranes. Polym. Int. 1991, 24, 105.

Lopyrev, W. A.; Miaczina, G. F.; Szevaleevskii, O. I.; Hidekel, M. L. Poliacetylen. Vysokomol. Soed. 1988, 30, 2019.

Mano, E. B.; Calafate, B. A. L. Electrolytically Initiated Polymerization of N-vinylcarbazole. J. Polym. Sci. ,Polym. Chem. Ed. 1983, 21, 829.

Mark, H. F. Encyclopedia of Polymer Science and Technology; Concise 3rd ed.; Wiley-Interscience: Hoboken, New York, 2007.

Mark, H.; Tobolsky, A. V. Chemiafizycznapolimerów; PWN: Warszawa, 1957.

Matyjaszewski, K.; Davis, T. P. Handbook of Radical Polymerization; Wiley-Interscience: New York, 2002.

Matyjaszewski,K.; Mülle, A. H. E. *50 Years of Living Polymerization*;Prog. Polym. Sci. 2006, 31, 1039–1040.

Mehrotra, R.; Ranby, B. Graft Copolymerization onto Starch. J. Appl. Polym. Sci. 1977, 21, 1647.

Morrison, R. T.; Boyd, N. Chemiaorganiczna; PWN: Warszawa, 1985.

Morton, M. Anionic Polymerization; Academic Press: New York, 1983.

Munk, P. Introduction to Macromolecular Science; JohnWiley &Sons: New York, 1989.

Naraniecki, B. Cieczemikroemulsyjne. Przem. Chem. 1999, 78(4), 127.

Neoh, K. G.; Kang, E. T.; Tan, K. L. Chemical Copolymerization of Aniline with Halogen-Substituted Anilines. Eur. Polym. J. 1990, 26, 403.

Nowakowska, M.; Pacha, J. Polimeryzacja olefin wobeckompleksówmetaloorganicznych. Polimery1978, 23, 290.

Ogata, N.; Sanui, K.; Tan, S. Synthesis of Alifatic Polyamides by Direct Polycondensation withTriphenylphosphine. Polym. J. 1984, 16, 569.

Onen, A.; Yagci, Y. Bifunctional Initiators. J. Macromol. Sci. Chem. A1990, 27, 743, Angew. Makromol. Chem. 1990, 181, 191.

Osada, Y. Plazmiennajapolimerizaciaiplazmiennajaobrabotkapolimerov. Vysokomol. Soed. A1988, 30, 1815.

Pasika, W. M. Cationic Copolymerization of Epichlorohydrinwith Styrene Oxide and Cyclo hexene Oxide. J. Macromol. Sci. Chem. A1991, 28, 545.

Pasika, W. M. Copolymerization of Styrene Oxide and Cyclohexene Oxide. J. Polym. Sci. Part A1991, 29, 1475.

Pielichowski, J.; Puszyński, A. Preparatykamonomerów; W. P. Kr. :Kraków, 2007, pp 67–83.

Pielichowski, J.; Puszyński, A. Preparatykapolimerów; TEZA WNT:Kraków, 2005.

Pielichowski, J.; Puszyński, A. Technologiatworzywsztucznych; WNT: Warszawa, 2003.

Pielichowski, J.; Puszyński, A. Wybranedziały z technologiichemicznejorganicznej. W. P. Kr. : Kraków.

Pistola, G.; Bagnarelli, O. Electrochemical Bulk Polymeryzation of Methyl Methacrylate in the Presence of Nitric Acid. J. Polym. Sci. , Polym. Chem. Ed. 1979, 17, 1002.

Pistola, G.; Bagnarelli, O.; Maiocco, M. Evaluation of Factors Affecting the Radical Electropolymerization of Methylmethacrylate in the Presence of HNO3. J. Appl. Electrochem. 1979, 9, 343.

Połowoński, S. Technikipomiarowo-badawcze w chemiifizycznejpolimerów; W. P. L. :Łódź, 1975.

Porejko, S.; Fejgin, J.; Zakrzewski, L. Chemiazwiązkówwielkocząsteczkowych; WNT: Warszawa, 1974.

Pracazbiorowa: Chemiaplazmyniskotemperaturowej; WNT: Warszawa, 1983.

Puszyński, A. Chlorination of Polyethylene in Suspension. Pol. J. Chem. 1981, 55, 2143

Puszyński, A.; Godniak, E. Chlorination of Polyethylene in Suspension in the Presence of Heavy Metal Salts. Macromol. Chem. RapidCommun1980, 1, 617.

Puszyński, A.; Dwornicka, J. Chlorination of Polypropylene in Suspension. Angew. Macromol. Chem. 1986, 139, 123.

Puszyński, J. A.; Miao, S. Kinetic Study of Synthesis of SiC Powders and Whiskers in the Presence KClO3 and Teflon. Int. J. SHS1999, 8(8), 265.

Rabagliati, F. M.; Bradley, C. A. Epoxide Polymerization. Eur. Polym. J. 1984, 20, 571.

Rabek, J. F. Podstawyfizykochemiipolimerów; W. P. Wr. :Wrocław, 1977.

Rabek, T. I. Teoretycznepodstawysyntezypolielektrolitówiwymieniaczyjonowych;PWN: Warszawa, 1962.

Regas, F. P.; Papadoyannis, C. J. Suspension Cross-linking of Polystyrene With Friedel-Crafts Catalysts. Polym. Bull. 1980, 3, 279.

Rodriguez, M.; Figueruelo, J. E. On the Anionic Polymerization Mechanism of Epoxides. Macromol. Chem. 1975, 176, 3107.

Roudet, J.; Gandini, A. Cationic Polymerization Induced by Arydiazonium Salts. Macromol. Chem. Rapid Commun. 1989, 10, 277.

Sahu, U. S.; Bhadam, S. N. Triphenylphosphine Catalyzed Polymerization of Maleic Anhydride. Macromol. Chem. 1982, 183, 1653.

Schidknecht, C. E. Polimerywinylowe; PWT: Warszawa, 1956.

Szwarc, M.; Levy, M.; Milkovich, R. Polymerization Initiated by Electron Transfer to Monomer. A New Method of Formation of Block Copolymers. J. Am. Chem. Soc. 1956,78, 2656–2657.

Szwarc, M. 'Living' Polymers. Nature1956, 176, 1168–1169.

Sen, S.; Kisakurek, D.; Turker, L.; Toppare, L.; Akbulut, U. Electroinitiated Polymerization of 4-Chloro-2,6-Dibromofenol. New Polym. Mater. 1989, 1, 177.

Sikorski, R. T. Chemicznamodyfikacjapolimerów. PraceNauk. Inst. Technol. Org. Tw. Szt. P. Wr. 1974, 16, 33.

Sikorski, R. T. Podstawychemiiitechnologiipolimerów; PWN: Warszawa, 1984.

Sikorski, R. T.; Rykowski, Z.; Puszyński, A. Elektronenempfindliche Derivate von Poly-methylmethacrylat. Plaste und Kautschuk1984, 31, 250.

Simionescu, C. I.; Geta, D.; Grigoras, M. Ring-Opening Isomerization Polymeryzation of 2-Methyl-2-Oxazoline Initiated by Charge Transfer Complexes. Eur. Polym. J. 1987, 23, 689.

Sheinina, L. S.; Vengerovskaya, Sh. G.; Khramova, T. S.; Filipowich, A. Y. Reakcjipiridinoviichczietvierticznychsoliej s epoksidnymisoedineniami. Ukr. Khim. Zh. 1990, 56, 642.

Soga, K.; Toshida, Y.; Hosoda, S.; Ikeda, S. A Convenient Synthesis of a Polycarbonate. Makromol. Chem. 1977, 178, 2747.

Soga, K.; Hosoda, S.; Ikeda, S. A New Synthetic Route to Polycarbonate. J. Polym. Sci. ,Polym. Lett. Ed. 1977, 15, 611.

Soga, K.; Uozumi, T.; Yanagihara, H.; Siono, T. Polymerization of Styrene with Heterogeneous Ziegler-Natta Catalysts. Macromol. Chem. Rapid. Commun. 1990, 11, 229.

Soga, K.; Shiono, T.; Wu, Y.; Ishii, K.; Nogami, A.; Doi, Y. Preparation of a Stable Supported Catalyst for Propene Polymerization. Makromol. Chem. Rapid Commun. 1985, 6, 537.

Sokołov, L. B.; Logunova, W. I. Otnositielnajareakcionnosposobnostmonomeroviprognozirovanicpolikondensacji v emulsionnychsistemach. Vysokomol. Soed. A. 1979, 21, 1075.

Soler, H.; Cadiz, V.; Serra, A. Oxirane Ring Opening with Imide Acids. Angew. Makromol. Chem. 1987, 152, 55.

Spasówka, E.; Rudnik, E. Możliwościwykorzystaniawęglowodanów w produkcjibiodegradowalnychtworzywsztucznych. Przemysł Chem. 1999, 78, 243.

Stevens, M. P. Wprowadzenie do chemiipolimerów; PWN: Warszawa, 1983.

Stępniak, I.; Lewandowski, A. Elektrolitypolimerowe. Przemysł Chem. 2001, 80(9), 395.

Strohriegl, P.; Heitz, W.; Weber, G. Polycondensation using Silicon Tetrachloride. Macromol. Chem. Rapid Commun. 1985, 6, 111.

Szur, A. M. Vysokomolekularnyjesoedinenia;VysszajaSzkoła:Moskwa, 1966.

Tagle, L. H.; Diaz, F. R.; Riveros, P. E. Polymeryzation by Phase Transfer Catalysis. Polym. J. 1986, 18, 501.

Takagi, A.; Ikada, E.; Watanabe, T. Dielectric Properties of Chlorinated Polyethylene and Reduced Polyvinylchloride. Memoirs Faculty Eng. ,Kobe Univ. 1979, 25, 239.

Tani, H. Stereospecific Polymerization of Aldehydes and Epoxides. Adv. Polym. Sci. 1973, 11, 57.

Vandenberg, E. J. CoordynationCopolymeryzation of Tetrahydrofuran and Oxepane with Oxetanes and Epoxides. J. Polym. Sci. Part A1991, 29, 1421.

Veruovic, B. Navody pro laboratornicviceni z chemiepolymer; SNTL: Praha, 1977.

Vogdanis, L.; Heitz, W. Carbon Dioxide as a Mmonomer. Macromol. Chem. Rapid Commun. 1986, 7, 543.

Vollmert, B. Grundriss der Makromolekularen Chemic; Springer-Verlag: Berlin, 1962.

Wei, Y.; Tang, X.; Sun, Y. A study of the Mechanism of Aniline Polymeryzation. J. Polym. Sci. Part A1989, 27, 2385.

Wei, Y.; Sun, Y.; Jang, G. W.; Tang, X. Effects P-Aminodiphenylamine on Electrochemical Polymerization of Aniline. J. Polym. Sci. ,Part C1990, 28, 81.

Wei, Y.; Jang, G. W. Polymerization of Aniline and Alkyl Ringsubstituted Anilines in the Presence of Aromatic Additives. J. Phys. Chem. 1990, 94, 7716.

Wiles, D. M. Photooxidative Reactions of Polymers. J. Appl. Polym. Sci. , Appl. Polym. Symp. 1979, 35, 235.

Wilk, K. A.; Burczyk, B.; Wilk, T. Oil in Water Microemulsions as Reaction Media. 7th International Conference Surface and Colloid Science, Compiegne, 1991.

Winogradova, C. B. Novoe v obłastipolikondensacji. Vysokomol. Soed. A1985, 27, 2243.

Wirpsza, Z. Poliuretany; WNT: Warszawa, 1991.

Wojtala, A. Wpływwłaściwościorazotoczeniapoliolefinnaprzebiegichfotodegradacji. Polimery 2001, 46, 120.

Xue, G. Polymerization of Styrene Oxide with Pyridine. Macromol. Chem. Rapid Commun. 1986, 7, 37.

Yamazaki, N.; Imai, Y. Phase Transfer Catalyzed Polycondensation of Bishalomethyl Aromatic compounds. Polym. J. 1983, 15, 905.

Yasuda, H. Plasma Polymerization; Academic Press, Inc. : Orlando, 1985.

Yuan, H. G.; Kalfas, G.; Ray, W. H. Suspension Polymerization. J. Macromol. Sci. Rev. Macromol. Chem. Phys. 1991, C 31, 215.

Zubakowa, L. B.; Tevlina, A. C.; Davankov, A. B. Sinteticzeskijeionoobmiennyjemateriały;Chim ia:Moskwa, 1978.

Zubanov, B. A. Viedienie v chimiupolikondensacionnychprocesov;Nauka:Ałma Ata, 1974.

Żuchowska, D. Polimerykonstrukcyjne; WNT: Warszawa, 2000.

Żuchowska, D. Strukturaiwłaściwościpolimerówjakomateriałówkonstrukcyjnych; W. P. Wr. :Wrocław, 1986.

Żuchowska, D.; Steller, R.; Meisser, W. Structure and Properties of Degradable Polyolefin–Starch Blends. Polym. Degrad. Stab. 1998, 60, 471.

CHAPTER 8

EFFICIENCY AND APPLICATION OF POLYSTYRENE PLASTICS BLENDS IN ENVIRONMENT PROTECTION

R.R. USMANOVA[1] and G.E. ZAIKOV[2]

[1]Ufa State Technical University of Aviation, 12 Karl Marks Str., Ufa 450000, Bashkortostan, Russia, E-mail: usmanovarr@mail.ru
[2]N.M. Emanuel Institute of Biochemical Physics, Russian Academy of Sciences, [4]Kosygin Str., Moscow 119334, Russia, E-mail: chembio@sky.chph.ras.ru

CONTENTS

ABSTRACT

An assessment of the cost-effectiveness of measures to protect the environment was carried out. Relations are obtained to calculate the damage caused by atmospheric emissions in the production of polystyrene. Criteria for technical and economic evaluation of the effectiveness of gas purification system were developed. A method was developed for estimating gas treatment facilities, allowing the design stage to make a comparative analysis of competing systems, taking into account the costs of implementation of environmental measures.

8.1   CURRENT STATE OF A PROBLEM

Continuously growing production capacity of plastics used in the technology of new materials leads to continuous improvement of gas cleaning used in industrial processes. In the coming years, this improvement will be in line with the trend of the gas cleaning in general. In the future, preference will be given to methods and apparatus providing fine cleaning of exhaust gas streams from harmful substances. Optimization problem is close, and is connected with various alternatives of designs, characteristic for designing a gas-cleaning installation of constructions. The survey of the domestic and foreign literature[1,2,3,4] has revealed a total absence of attempts of the solution of a problem of optimization in the stated aspect. Moreover, it is not revealed that attempts to optimize are constructive technological parameters of any unique apparatus within the limits of its use in concrete conditions.

Preliminary studying of a problem has shown its complexity. The way to its solution as it is installed, passes through construction in the mathematical form of technical and economic models of optimized installations, that is, the equations in which constructive technological, technical, and economic parameters would be connected together. Now such model exists only with reference to cyclonic installations.[2,5]

From the experiment, it is observed that a special attention is given to preparation and the analysis of technical projects on designing, bases of quality of the design are put at these early stages of work.

It is possible to note the momentous defects to overcome what is necessary in the proximal years.

1. Insufficiency of the nomenclature, a gas-cleaning installation, and its lag from growing powers of the industry.
2. Weakness of computational baseline in which predominates empiric.
3. Absence of strict scientific criteria for designing a gas-cleaning installation of constructions with number of steps of clearing two and more.

For the specified reason at designing of such constructions, the big role is played by purely heuristic factor.

4. The weakest and improbable working out of the questions connected with drawing by the flying emissions of a damage to a circumambient and, accordingly, with definition of economic benefit of liquidation of this damage.

These and other unresolved problems should be solved by the development engineers, initiating to master designing a gas-cleaning installation of constructions. Central tasks that were put by working out of a new build of a rotoklon with inner circulation of a liquid:

- to produce criteria of a technical and economic estimation of a system's effectiveness of environment protection against pollution;
- to create the apparatus with a wide range of change in operating conditions and a wide scope, including for clearing of gases of the basic industrial assemblies of a finely divided dust.

8.2 CALCULATION OF THE PREVENTED DAMAGE FROM ATMOSPHERIC AIR POLLUTION

The prevented ecological damage from pollutant emissions in an aerosphere—an estimation in the monetary form of possibly negative aftereffects from pollutant emissions, which in an observed time span, was possible to avoid as a result of activity of supervising authorities in metrology, conducting a complex of provisions, implementation of nature protection programs. At implementation, a gas-cleaning installation was momentous to define magnitude of an economic damage in the set region, prevented as a result of conducting of nature protection provisions on air protection from pollutant emissions.[6,17,7–18]

The integrated estimation of magnitude of the prevented damage from pollutant emissions can be spent on an aerosphere as one large source or group of sources, and for region as a whole. In the capacity of sized-up group of sources, all sources in the given city and region can be observed as a uniform source.

Prevention of economic damage from the emission of pollutants into the air, where thousand rubles, Y_p, is given by:

$$Y_p = Y_y \cdot M_{pr} \cdot K_e \cdot I_\partial$$

where Y_p: a parameter of a specific damage from pollutant emissions in a free air in observed economic region of Russian Federation, Russian rouble/ton;

K_e: factor of an ecological situation and the ecological significance of a condition of a free air of territories of economic region of Russia;

I_0: an index on the industries, installed by Ministry of Economics of Russia. We accept equal 1;

M_{pr}: a pollutant emission reduced mass in the region, scaled down as a result of conducting matching nature protection provisions, thousand tons/year;

K_{ei}: factor of relative ecological–economic hazard to i th pollutant;

N: quantity of considered pollutants.

$$M_{pr} = \Delta M - M_c$$
$$\Delta M = M_1 - M_2 + M_{new},$$

where ΔM: total volume of a reduced mass of the scaled down waste interception, thousand conditional, tons/year;

M_c: a reduced mass of the waste interception, which has been scaled down as a result of slump in production in region, thousand conditional, tons/year;

M_1 and M_2: accordingly, a waste interception reduced mass on the beginning and the end of the design period, thousand conditional, tons/year;

M_{new}: a reduced mass of waste interception of the new factories and manufactures, thousand conditional, tons/year.

For calculation of a reduced mass of pollution, the confirmed values of maximum permissible concentration (maximum concentration limit) of pollutants in water of ponds are used. By means of maximum concentration limit factors of ecological and economic hazard of pollutants as magnitude, return maximum concentration limit is defined.

The reduced mass of pollutants pays off by formula:

$$M = \sum_{i=1}^{N} m_i \cdot K_{ei},$$

where m_i: mass of actual waste interception of pollutant in water installations of observed region, tons/year;

K_{ei}: factor of relative ecological–economic hazard to pollutant;

N: quantity of considered pollutants.

i: Substance number in the table.

8.3 TECHNICAL AND ECOLOGICAL ASSESSMENT GAS-CLEANING PLANT SAMPLING

In ecology and harmonious exploitation bases, estimations of economic efficiency of nature protection provisions are resulted. The problem is put to inject into calculation of a damage to a circumambient Y operational parameters of the given clearing installation to pass to relative magnitudes. It will allow to scale down number of the factors that are not influencing functioning of system to devise methods of calculation of relative efficiency of a gas-cleaning installation of the constructions, giving the chance to choose the most rational approaches and the equipment of systems of trapping of harmful making atmospheric emissions. In the most general event, the damage A_t caused by atmospheric emissions can be computed as $Y = B \cdot M$. The reduced mass of emission incorporating N components, will be computed in an aspect:

$$M = \sum_{i=1}^{N} A_i \cdot m_i$$

The emission mass m_i is proportional to an overshoot through system

$$m_i = (1 - \eta_i) \cdot m_{oi}$$

In an industrial practice, it is normally set or the share of concrete pollution in departing gas is known, C_{oi}. We will consider gas that is diluted enough so that its density ρ does not depend on the presence of admixtures

$$m_{oi} = C_{oi} \cdot \rho \cdot Q .$$

Let's compute a damage caused to an aerosphere, per unit masses of trapped pollution Y_m

$$Y_m = \frac{B \cdot \sum_{i=1}^{N} A_i \cdot (1 - \eta_i) \cdot C_{oi}}{\sum_{i=1}^{N} \eta_i \cdot C_{oi}} . \tag{8.1}$$

If to size up the gas-cleaning plant on average parameters, $\eta_i = \eta$

$$A = \frac{1}{N} \cdot \sum_{i=1}^{N} A_i \cdot C_{oi} \qquad Y_m = \frac{B.A.(1 - \eta)}{\eta}$$

Let's formulate a principle of ecological efficiency of nature protection provisions for at least a damage put to a circumambient. Purpose function in this case will appear in the form of $Y_m \to min$. Magnitude Y_m diminishes with value growth

$$E = \frac{\sum\limits_{i=1}^{N} \eta_i \cdot C_{oi}}{B.\sum\limits_{i=1}^{N} A_i \cdot (1 - \eta_i) \cdot C_{oi}} \qquad (8.2)$$

We will consider magnitude E as criterion of ecological efficiency of nature protection provisions. The criterion of relative ecological efficiency Θ is representable in the form of the relation of values E computed for compared alternative E_1 and the base accepted in the capacity of E_0

$$\Theta = E_1/E_0.$$

In case of the single-component pollution of criterion of relative ecological efficiency, we will find Θ as

$$\Theta = \frac{\eta_1}{\eta_0} \cdot \frac{1-\eta_0}{1-\eta_1}. \qquad (8.3)$$

Thus, for two gas-cleaning plants of the concrete manufacture, different in separation extent, $\eta_1 \neq \eta$, relative ecological efficiency of system is sized up by technological parameter $\Theta \to max$. The prevented damage Y_p compute as a difference between economic losses of two competing alternatives as $Y_p = Y_0 - Y_1$

We will be restricted to an event of comparison of two alternatives of the clearing of gas emissions intended for the same manufacture with fixed level of technological perfection. In the capacity of base alternative Y_0, we will accept the greatest possible damage such as atmospheric emissions of the manufacture, which technological circuit design does not provide a clearing stage, $\eta_{i0} = 0$. For the fixed technological circuit design of manufacture efficiency of a stage of clearing, we will size up in shares from the maximum damage $E_n = Y_p/Y_0$

$$E_n = \frac{y_p}{y_o} = \frac{\sum\limits_{i=1}^{N} A_i \cdot C_i \cdot \eta_i}{\sum\limits_{i=1}^{N} A_i \cdot C_{oi}} \qquad (8.4)$$

If to consider that all components of harmful emission with aggression average indexes are trapped in equal extents ($A_i = A$, $\eta_i = \eta$), we come to $E_n = \eta$. Thus, the eurysynusic extent of trapping η is a special case of criterion of ecological efficiency E_n computed for one-parametric pollution or for emission with average characteristics. At sampling of the gas-cleaning plant, it is necessary to give preference to the installation securing higher values of criterion E_n.

The gas-cleaning installation demands expenses Z (rbl/c) on the creation and functioning. These charges can essentially differ depending on the accepted method of clearing of gas emissions and should be taken into consideration at an estimation of the general damage. For example, air purification from a dust cyclone separators will have even more low cost in the "dry" and "wet" condition at which it is necessary to provide additional charges on water, pumping devices, neutralization of runoffs, etc. At the same time, centrifugal separators are not suitable for clearing of gaseous impurities. We will use relative parameters, that is, to consider a gain of the prevented damage $\Delta Y_p = Y_1 - Y_2$ on rouble of expenses ΔZ.

Function of the purpose $Y_p \rightarrow max$ will register in an aspect

$$E_n = (\Delta Y_n / \Delta Z) \rightarrow max.$$

We will be restricted to consider a method of calculation of magnitude Yp for centrifugal dust traps widely used in practice. We will define in maintenance costs, a variable component of the power inputs connected with a water resistance of the apparatus ΔP (Pascal). The head loss $\Delta H = \Delta P / \rho$ (J/kg) is definable from the Bernoulli's theorem, which has been written down for entrance and target cross-sections of the gas passage. The energy consumption will be computed as $I = \Delta P \cdot Q$ (J/s).

Power inputs Z_e in terms of energy costs C_e (Russian rouble/J) is definable from a relationship

$$Z_e = C_e \cdot Q \cdot \Delta P$$

The prevented damage Y_p computed on rouble of expenses will be

$$E_n = \frac{B \cdot \rho \cdot \sum_{i=1}^{N} A_i \cdot C_{oi} \cdot \eta_i}{C_e \cdot \Delta P} \tag{8.5}$$

The criterion of relative ecological efficiency of the whirlwind apparatus, $\Theta_n = E_{n1} / E_{n0}$, is computed on values E_n for two installations E_{n1} and E_{n0}, from which one is accepted for base E_{n0}. At transition to average magnitudes

$$\Theta_i = \frac{\Delta P_0 \cdot \eta_1}{\Delta P_1 \cdot \eta_0} . \tag{8.6}$$

Let's apply the results received earlier according to efficiency of clearing of gas emissions to the comparative analysis of dust extractors of centrifugal action by criterion Θ. As the base two-stage installation of type «cyclone C-6» is accepted. Results of the comparative analysis are introduced in Table 8.1. As the introduced data show, relational, technical, and ecological criterion to efficiency Θ reflects logic of process of dust separation—the above a device purification efficiency η, the magnitude Θ is more. In this case, instead of qualitative ascertaining of the fact the quantitative assessment of efficiency of the clearing of gas emissions is offered, allow to define in what degree competing systems differ from each other.

TABLE 8.1 Comparative Technical and Economic Indicators of Offered and Base Dust-Collecting Plants.

No p/p	The indicator name	Unit measurements	Basic hardware	New installation
1	Productivity	m^3/s	6	6
2	Hydraulic resistance	Pascal	2100	1350
3	Factor of hydraulic resistance		10.8	6.9
4	Concentration of emissions after installation	Mg/m^3	127.5	39
5	The occupied space in the plan	m^2	17	5.3
6	Metal consumption	m^2	6.8	4.9
7	The general power consumption	kWh	30	31
8	The specific expense of the electric power on clearing 1000 m^3 emissions	kWh	0.6	0.475
9	Intensity of flow	kg/h	–	0.5
10	Criterion of efficiency η		1.0	1.85

In Figure 8.1, results of researches of separating ability of a rotoklon depending on conditional speed of gas υ, calculated on full section of the device are introduced.

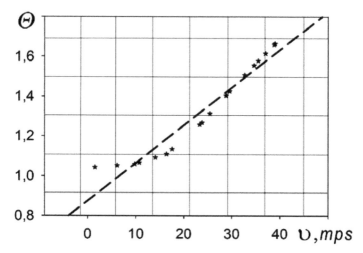

FIGURE 8.1   Relative technical and economic efficiency of a rotoklon.

Comparison of considered devices by means of criterion relative to technical and economic efficiency Θ is spent. The cited data have illustrative character and show possibilities of application of criteria of technical and economic efficiency Θ for comparison purposes of gas-cleaning plant devices.

8.4   RESULTS OF INDUSTRIAL TESTS OF THE GAS-CLEANING PLANT ON THE BASIS OF DEVICE "ROTOKLON"

Figure 8.2 shows a schematic diagram of the cleaning process. In this scheme, a combined gas stream from one of the fan apparatus 2 is supplied to the furnace heater 3, wherein the heat of combustion of the natural gas is heated to temperature from 350 to 400 °C. Heated gases at a flow rate of up to 200 s^{-1} enter RotoClone 4. The outgoing gases from the apparatus, heated to 350–400 °C, are vented to the atmosphere through the heat exchanger 5.

The slope breeching 2 is actually the hollow scrubber, 400 mm in diameter, working in the evaporation cooling regime. At work, gases arrive from it in a Venturi scrubber 3, consisting of two cylindrical columns in diameter of 1000 mm with the general bunker. In each column of a scrubber, it is established on three atomizers. The Venturi scrubber of the first step of clearing has a mouth which is 100 mm in diameter and is irrigated with water from an atomizer established in front of the confusor.

Gases after a slope breeching, first passes through the bunker, the drop catcher 4, and then in a rotoklon of the another step, which consists from the heat exchanger 7. The exhaust of gases from the furnace is carried out by vacuum pump VVN-50 established behind devices of clearing of gas emissions. Purified gases are deduced in atmosphere.

Regulation of pressure of gases under a furnace roof and the expense of gases is carried out by a throttle in front of the vacuum pump. Slurry water from devices of clearing of gas emissions flows off by gravity in a tank of a hydro shutter 9, whence also by gravity arrives in a slurry tank. From a slurry tank water on two slurry clarifier is taken away on water purification. After clarification, chemical processing, and cooling, water is fed again by the pump on irrigation of gas-cleaning installations.

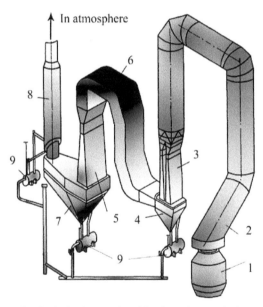

FIGURE 8.2   The technological scheme of purification of gas emissions plastics production. (1) mixer flows, (2) gas exit branch, (3) Venturi scrubber, (4) bunker—the drop catcher, (5) a rotoklon, (6) gas pipeline, (7) the heat exchanger, (8) exhaust pipe, and (9) tank—a hydraulic hitch.

The dust in gases, differs high dispersion (to 80 wt.% of particles less than 5/6 microns). In Table 8.2, the compound of a dust of exhaust gases is resulted.

In tests for furnaces, almost constant electric regime that secured with identity of conditions at which parameters of systems of dust separation characterize was supported. The furnace worked on the fifth–seventh steps of pressure at fluctuations of capacity 14.5/17.5 megawatt.

TABLE 8.2   Results of Post-Test Examination.

Compound	Requisite concentration, g/m^3	Concentration after clearing, g/m^3
Dust	0.02	0.00355
NO_2	0.10	0.024
SO_2	0.03	0.0005
CO	0.01	0.0019

The quantity of dry gases departing from the furnace made 1500/2000 m^3/h. The temperature of gases before clearing of gas emissions equaled 750/850 °C and humidity did not exceed 4/5% (on volume).

TABLE 8.3   Results of Calculation of a Payment for Pollutant Emission.

The list of pollutants (the substance name)	It is thrown out for the accounting period, t/year			The base specification of a payment within admissible specifications, a Russian rouble/t	The size of a payment for a maximum permissible emission, Russian rouble/year	The base specification of a payment within the established limits, a Russian rouble/t	Total a payment on the enterprise, a Russian rouble/year
	In total	Including					
		VPE	MPE				
1	2	3	4	5	6	7	8
The inorganic Dust	19.710		–	21	228.65	105	228.65
Nitrogen dioxide	105.1		–	52	379.19	260	379.19
Carbon monoxide	288.2		–	0.6	2.99	3	2.99
Sulfur dioxide	197.1		–	40	539.14	200	539.14
Total					1149.97		1149.97

In Table 8.3, results of calculation of a payment for pollutant emission of system of dust separation are shown.

Thus, we have chosen the scheme of clearing of gases, which allows the lower concentration of pollutants to preset values, and consequently, to lower payments by the enterprise for emissions.

8.5   CONCLUSIONS

1. On the basis of a method of an estimation of economic efficiency of the carried out nature protection actions parities for calculation of the damage put to environment by atmospheric emissions of manufacture are received. Transition to relative indicators has allowed to scale down number of the factors that are not influencing process of clearing of gas emissions

2. Methods of an estimation of a gas-cleaning installation of the constructions are devised, allowing on a design stage to make the comparative analysis of competing systems in terms of expenses for realization of nature protection actions

3. Criteria of technical and ecological estimation of the gas-cleaning plants incorporating both economic and technology factors are developed.

KEYWORDS

- ecological efficiency
- environmental damage
- atmospheric emissions
- rotoklon, efficiency criterion
- the maximum permissible concentration

REFERENCES

1. Belevitsky, A. M. Economy and Technical and Economic Optimisation of Dust-Collecting Plants (on an Example of Installations of Cyclonic Dust Separation): Help Supervising Material. Clearing of Gas Emissions. *Ind. Sanit. Gas Clean.* **1982**, *6*, 17–32.

2. Brodsky, Y. N.; Melnikova, L. N. Sanitary Cleaning Styrenated Gases. *Ind. Sanit. Gas Clean.* **1985**, *5*, 22–31.

3. Kouzov, I. A.; Malgin, A. D.; Skryabin, M. *Dust Element of Gases and Air in a Chemical Industry*; Khimia Publication: Leningrad, 1982; p 256.

4. Timonin, A. S. *Fundamentals of Design and Calculation of Chemical Technology and Environmental Protection Equipment*; K: Kaluga 2007; Vol. 2, pp 45–68.

5. Nikanorov, A. A. Sources and Methods of Clearing of Gas Bursts от Ozone. M: Nonferrous Metals the Information, Nova Science Publisher: New York, 1985.
6. *Enterprise Economy*: *The Textbook for High Schools*; Tumina, V. M., Ed.; Nova Science Publisher: New York , 2006, pp 104–123.
7. Eldyshev, Y. N. World Against Plastic Bags. *Ecol. Life* **2011**, *7*, 32–39.
8. Ivanov, I. P.; Kogan, B. I.; Bulls, A. P. *Engineering Ecology*; Kogana, B. I., Ed.; NSGTU: Novosibirsk, 1995; p 321.
9. Weinstein, M.; Krasnik, I. M.; Durable, C. D. Industrial and Sanitary Clearing of Gases. *Ind. Sanit. Gas Clean.* **1977**, *5*, 3–14.
10. *The Equipment for Clearing of Gas Emissions. The Catalogue. Environmental Control;* Khimia Publication: Moscow, 1992; p 245.
11. *Clearing and a Regeneration of Plant Emissions*; Maksimova, F., Ed.; The Wood Industry: Moscow, 1981, pp 39–45.
12. *Environmental Control*; Belova, S. V., Ed.; The Higher School: Moscow, 1991, pp 79–90.
13. Rodions, A. I.; Klushin, C. H.; Sister, C. *Technological Processes of Environmental Safety*; N. Bochkarevoj's Publishing House: Kaluga, 2000, pp 99–112.
14. Straus, V. *Industrial Clearing of Gases. The Lane with English;* Chemistry: Moscow, 1981, pp 102–110.
15. Skryabin, J. I. *The Industrial Dust Atlas*; Tsintihim-neftemash: Moscow, 1982, pp 54–67.
16. *Technics of Environment Protection*; Radionov, A. I., Klushin, V. N, Gorshechnikov, N. S., Ed.; Metallurgy: Moscow, 1989, pp 47–60.
17. *A Time Typical Technique of Definition of Economic Efficiency of Realisation of Nature Protection Actions and an Estimation of the Damage Caused to a National Economy by Environmental Pollution*; Gidro-Meteo-Publishing House: Leningrad, 1986; p 125.
18. *Enterprise Economy: The textbook for high schools*; Gorfinkelja, V. J., Shvandara, V. A., Eds.; JUNITI-IS: Moscow, 2007, pp 56–73.
19. Denisenko, L. I., et al. *Quality Monitoring and Clearings of Plant Emissions of Various Manufactures of Fluoric Connections*; Tsintihim-neftemash: Moscow, 1982; p 334.

CHAPTER 9

MATHEMATICAL MODEL OF KINETIC OF RADICAL POLYMERIZATION OF METHYL METHACRYLATE

S. V. KOLESOV[1], V. M. YANBORISOV[2], D. A. SHIYAN[3], A. O. BURAKOVA[3], K. A. TERESHCHENCO[3], N. V. ULITIN[3], and G. E. ZAIKOV[4]

[1]Institute of Organic Chemistry, Ufa Scientific Center, Russian Academy of Sciences, 71 October Prosp., Ufa 450054, Republic of Bashkortostan, Russian Federation, Tel.: +7 (342) 235-61-66; E-mail: kolesov@anrb.ru
[2]Ufa State University of Economics and Service, 24 Aknazarova Str.,Ufa 450022, Republic of Bashkortostan, Russian Federation, Tel.: +7 (347) 252-49-22; E-mail: YanborisovVM@mail.ru
[3]Kazan National Research Technological University, 68 Karl Marx Str., Kazan 420015, Republic of Tatarstan, Russian Federation, Tel.: +7 (843) 231-95-46; E-mail: n.v.ulitin@mail.ru
[4]Emanuel Institute of Biochemical Physics, Russian Academy of Sciences, 4 Kosygina Str., Moscow 119334, Russian Federation, Tel.: +7(499)137-41-01; E-mail: chembio@sky.chph.ras.ru

CONTENTS

ABSTRACT

The mathematical model of kinetic of radical polymerization of methyl methacrylate with consideration of autoacceleration has been developed using the model of the reaction rate constant of chain termination (the authors of the model being Hui and Hamielec). The parameters of the model have been determined by solving the inverse kinetic problem. The adequacy of the developed kinetic description has been proved by comparing of the theoretically calculated and experimental kinetic curves, and also the number average and weight average molecular weight of polymethyl methacrylate.

9.1   INTRODUCTION

Nowadays, the synthesis of polymethyl methacrylate by reaction injection molding allows to produce air glass having high values of haze, and strength properties. During reaction injection molding, the viscosity increases in radically polymerizable mass, which leads to abnormal growth rate (autoacceleration) of methyl methacrylate polymerization. This feature of polymethyl methacrylate synthesis having a strong influence on its molecular weight characteristics, optical, and strength properties.

According to references,[1–6] to find an answer to this question is possible using the mathematical model of the kinetics of radical polymerization of methyl methacrylate.

The development of the kinetics model was the aim of this scientific research.

9.2   EXPERIMENTAL

9.2.1   INITIAL SUBSTANCES

Methyl methacrylate is twice distilled in vacuum. The purity of the methyl methacrylate was monitored by ^{1}H NMR.

Benzoyl peroxide (BP) and azoisobutyronitrile (AIBN) was twice recrystallized from methanol and dried at 293 K under vacuum.

Solvents were purified by standard methods.[7]

9.2.2   EXPERIMENT METHODS

For the purpose of synthesis of polymethyl methacrylate, samples of the reaction mixtures (initiator + methyl methacrylate) were preliminary degassed in the "freeze–thaw" conditions to a residual pressure of 1 Pa. The kinetics of radical polymerization of methyl methacrylate were studied by dilatometric method (sealing liquid glycerin).[8] The conversion of methyl methacrylate was calculated by the formula

$$U = (\Delta V / (V_0 k))100\%$$

where U is a conversion of methyl methacrylate; V_0 is an initial volume of the reaction mixture; ΔV is a change of volume of the reaction mixture during the process; k is a coefficient of contraction.[9,10]

The error in determining the kinetic curves (plot curves dependences of methyl methacrylate conversion versus time) should not exceed 3–5% (each experiment was repeated three times).

The mass obtained in each test was dissolved in acetone and polymethyl methacrylate was precipitated 10-fold quantity of ethanol. The polymethyl methacrylate samples were dried in vacuum at 313 K.

Gel permeation chromatography was used to determine the number average and weight average molecular weights of the polymethyl methacrylate samples. Analyses were performed on a chromatograph Waters GPC 2000 System (eluent chloroform, flow rate 0.5 mL/min). Chromatograph column system was calibrated per polystyrene standards. Measuring fault of molecular weight characteristics polymethyl methacrylate samples, according to Quanyun, does not exceed 10%.

9.2.3 DEVELOPMENT OF KINETIC DESCRIPTION OF THE PROCESS

The method of bulk polymerization was considered. The kinetics of radical polymerization of methyl methacrylate according to Matyjaszewski included the following reactions: dissociation of initiator (d), chain propagation (p), chain transfer to monomer (tr), chain termination by recombination (rec), and chain termination by disproportionation (disp) values of the rate constants of corresponding elementary stages.

According to the law of mass action, the kinetics of the process has been formalized by the system of differential equations describing the instantaneous rate of components' changing concentrations:

$$\frac{d[C_i]}{dt} = f\left([C_1],\ [C_2],...,[C_i],\ k_{\mathrm{d}},\ k_{\mathrm{p}},...,\ k_{\mathrm{disp}}\right), \tag{9.1}$$

where $[\tilde{N}_i]$ is a concentration of components of reaction system, mol/L; k_d, k_p,..., k_{disp} are the rate constants of respective reactions.

The resulting system of eq 9.1 containing the unknown kinetic constants k_p, k_{tr}, k_{rec}, and k_{disp}, which have been found in course of solving the inverse kinetic problem in two stages.

9.2.4 SOLUTION OF THE INVERSE KINETIC PROBLEM

To solve the system of eq 9.1 at any temperature, it is necessary to know the temperature dependence on the rate constants of the elementary reactions. At the time of task description, temperature dependences that are recovered directly from the polymerization experiment data, which can be absolutely trusted, have been obtained only for k_d constant (Table 9.1). Temperature dependence of k_{disp} and k_{rec} constants, which take into account the autoacceleration, for this process are unknown. To k_p and k_{tr}, different sources,[1] show quite different temperature dependences, so they are also considered by us as unknown.

TABLE 9.1 The Kinetics Parameter of the Process of Methyl Methacrylate Radical Polymerization in Bulk.

Parameter	Reference
Initiation efficiency $f(\text{AIBN}) = 0.5$	11
Initiation efficiency $f(\text{BP}) = 1$	12
$k_d(\text{AIBN}) = 1.053 \times 10^{15} \cdot e^{-15440/T}$, s^{-1}	13
$k_d(\text{BP}) = 1.18 \times 10^{14} \cdot e^{-15097/T}$, s^{-1}	12
$k_p = 2.5 \times 10^6 \cdot e^{-2823/T}$, L/(mol·s)	*
$k_{tr} = 2.79 \times 10^5 \cdot e^{-5450/T}$, L/(mol·s)	*
$k_{disp} = 9.8 \times 10^7 \cdot e^{-353/T} \cdot e^{-2(A_1 \times U + A_2 \times U^2 + A_3 \times U^3)} / \left(3.956 \times 10^{-4} \cdot e^{2060/T} + 1\right)$, L/(mol·s),	
$k_{rec} = 3.956 \times 10^{-4} \cdot e^{2060/T} \cdot k_{disp}$, L/(mol·s),	*
$A_1 = 31.52 - 0.074 \cdot T$,	
$A_2 = 14.92 - 0.059 \cdot T$,	
$A_3 = -38.66 + 0.132 \cdot T$.	

* **The** results of this investi

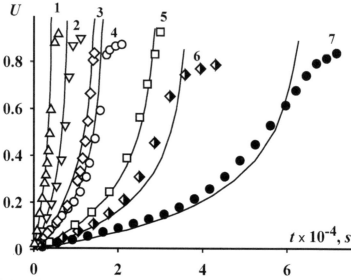

FIGURE 9.1 Conversion curves of methyl methacrylate radical polymerization in bulk, initiated by BP (dot line is an experiment data, line is a calculation): $[BP]_0 = 0.003$ (1, 2, 4, 6), 0.001 mol/L (3, 5, 7), $T = 353$ (1), 343 (2, 3), 333 (4, 5), 323 K (6, 7).

Temperature dependences of the k_p, k_{tr}, k_{disp}, and k_{rec} constants have been found through the sequential two-step solution of the inverse kinetic problem.

9.3   DETERMINATION OF K_p, K_{disp}, AND K_{rec} CONSTANTS BY EXPERIMENTAL CONVERSION CURVES OF METHYL METHACRYLATE

The model of reaction rate constant of chain termination[6] has been used to determine the temperature dependences of k_p, k_{disp}, and k_{rec} constants. Thus, that the difference between the theoretical and experimentally obtained conversion curves of methyl methacrylate has been a minimal. According to this model, the rate constant of chain termination reaction is represented as follows:

$$k_t = k_{t0}e^{-2(A_1U+A_2U^2+A_3U^3)}, \text{L/(mol·s)} \qquad (9.2)$$

where k_{t0} is the common reaction rate constant of chain termination without autoacceleration (for methyl methacrylate radical polymerization $k_{t0} = k_{disp0} + k_{rec0} = 9.8 \cdot 10^7 e^{-353/T}$, L/(mol·s),[11-17,21,22] T is the polymerization temperature, here and further on in K); A_1, A_2, and A_3 are the parameters of the model.

It was accepted that $k_{rec} / k_{disp} \approx k_{rec_0} / k_{disp_0}$

For the most of monomers in vinyl range, $k_{tr} \ll k_p$ (at least 3–4 orders),[12] that is, consumption of monomer is mainly due to reaction of chain propagation. Therefore, at this stage of the k_{tr} constant was neglected (assumed that $k_{tr} = 0$).

The temperature dependences of k_p, k_{disp}, and k_{rec} constants, obtained by us are presented in Table 9.1. The results of the comparison of experimental and kinetic curves, calculated according to the description of the process, are shown in Figure 9.1.

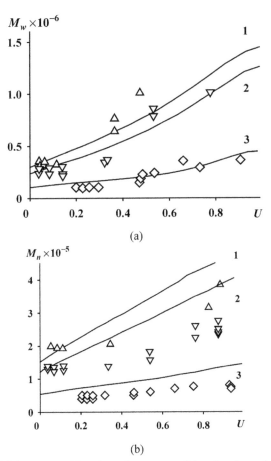

(a)

(b)

FIGURE 9.2   Weight average (a) and number average (b) molecular weight of polymethyl methacrylate obtained during AIBN initiated radical polymerization in bulk ([AIBN]$_0$ = 0.0155 (1), 0.0258 mol/L, (2, 3)) versus the conversion of methyl methacrylate (dot line is an experiment data[20], line is a calculation) at 343 (1, 2) and 363 K (3).

9.4 FINDING k_{tr} BY USE OF EXPERIMENTAL DATA OF POLYMETHYL METHACRYLATE MOLECULAR WEIGHT CHARACTERISTICS

Because the consumption of methyl methacrylate is mainly taking place due to the chain reaction, that is, reaction of chain transfer to methyl methacrylate does not contribute to the behavior of conversion curves of methyl methacrylate: the k_{tr} constant in the first stage of solving the inverse kinetic problem was assumed to be zero. However, the reaction of chain transfer to monomer contributes significantly to the molecular weight distribution and values of molecular weight characteristics of polymethyl methacrylate. Therefore, at this stage with other known constants, the k_{tr} constant has been found by comparing the theoretical and experimental values for number average and weight average molecular weight of polymethyl methacrylate. It was based on the developed kinetic description of the process of the radical polymerization of methyl methacrylate.

For this, first the system of eq 9.1 was rewritten with respect to the moments of the molecular weight distribution of polymethyl methacrylate using the method of generating functions.[18] Thereafter, the number average (M_n) and weight average (M_w) molecular weight of the polymethyl methacrylate have been obtained from the solution of system of equations by the correlations[18]: $M_n = (\sum \mu_1 / \sum \mu_0) \cdot M_o$, $M_w = (\sum \mu_2 / \sum \mu_1) \cdot M_o$, where $\sum \mu_0$, $\sum \mu_1$, and $\sum \mu_2$ are the sum of all zero, first, and second moments of the polymethyl methacrylate molecular weight distribution, mol/L; $M_m = 100$ is the molecular weight of methyl methacrylate.

Trying to achieve the minimum difference between the experimental[19] and calculated curves of molecular weight characteristics (Fig. 9.2, curves 2 and 3), the temperature dependence of the k_{tr} constant was found, which is shown in Table 9.1.

9.5 ADEQUACY VERIFICATION OF THE PROCESS MATHEMATICAL MODEL

The adequacy verification of mathematical model of the kinetic process was performed by the Fisher criteria.[20]

The average molecular weights of polymethyl methacrylate (Fig. 9.2, curve 1) and the conversion curves (Fig. 9.3), which had not been used in the solution of the inverse kinetic problem, were used instead as experimental data to verify the adequacy of the mathematical model of the process.

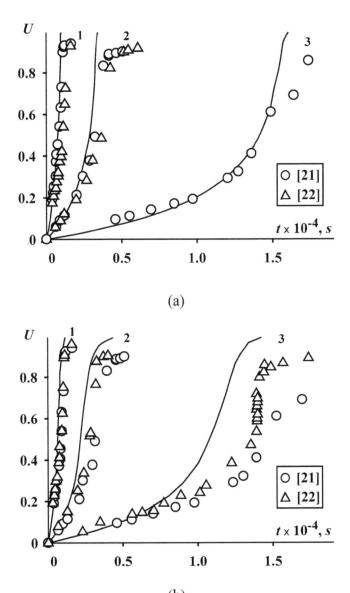

FIGURE 9.3   The conversion curves for the AIBN initiated radical polymerization of methyl methacrylate in bulk ([AIBN]$_0$= 0.0155 (a), 0.0258 mol/L (b)) (dot line is an experiment data, line is a calculation) at 363 (1), 343 (2), and 323 K (3).

It can be concluded that the mathematical model of radical polymerization process of methyl methacrylate developed by this research, considering the autoacceleration, is adequate to the experimental data, because calculated values of the Fisher criterion for each parameter have not exceeded normative values of the Fisher criterion.

9.6 CONCLUSION

Thus, this research shows the results of the kinetic mathematical modeling of the process of radical polymerization of methyl methacrylate. A distinctive feature of the developed mathematical model of the process was consideration of autoacceleration within the empirical model of the reaction rate constant of chain termination (the authors of the model are Hui and Hamielec). Only this was allowed to adequately describe the conversion curves of the process as well as number average and weight average molecular weights of polymethyl methacrylate in a wide range of temperatures and the concentrations of reagents.

KEYWORDS

- **autoacceleration**
- **methyl methacrylate**
- **polymethyl methacrylate**
- **radical polymerization**

REFERENCES

1. Matyjaszewski, K. *Handbook of radical polymerization*; John Wiley and Sons: Hoboken, Inc. Publication, 2002; p 936.
2. Achilias, D. S. *Macromole. Theor. Simul.* **2007**, *16* (*4*), 319.
3. Nayafi, M.; Roghani-Mamaqani, H.; Haddadi-Asl, V.; Salami-Kalajahi, M. *Adv. Polym. Technol.* **2011**, *30* (*4*), 257.
4. Ulitin, N. V.; Nasyrov, I. I.; Deberdeev, T. R.; Berlin, A. A. *Rus. J. Phys. Chem. B.* **2012**, *6* (*6*), 752.
5. Ulitin, N. V.; Deberdeev, T. R.; Deberdeev, R. Ya.; Berlin, A. A. *Doklady Phys. Chem.* **2012**, *446* (*2*), 180.
6. Hui, A. W.; Hamielec, A. E. *J. App. Polym. Sci.* **1972**, *16*, 749.
7. Svoronos, P.; Sarlo, E.; Kulawiec, R. *Organic Chemistry Laboratory Manual*; McGraw-Hill Education: Brooklyn, 1996; p 352.
8. Wang, X.; Ruckenstein, E. *J. App. Polym. Sci.* **1993**, *49* (*12*), 2179.
9. Denisov, E. T.; Afanas'ev, I. B. *Oxidation and Antioxidants in Organic Chemistry and Biology*. Boca Raton, FL: CRC Press, Taylor and Francis Group, 2005; p 987.
10. Quanyun, A. X. *Ultra-high Performance Liquid Chromatography and its Applications*; John Wiley & Sons, Inc.: New Jersey, 2013; p 317.
11. Kuzub, L. I.; Peregudov, N. I.; Irzhak, V. I. *Polym. Sci.* **2005**, *47*(A) *10*, 1063.

12. Bagdasaryan, Kh. S. *Theory of Free-Radical Polymerization*; Jerusalem: Israel Program for Scientific Translations, Hartford, Conn., D. Davey, 1968; p 328.
13. Tobolsky, A. V.; Baysal, B. *Polym. Sci.* **1953**, *11* (*5*), 471.
14. Baillagou, P. E.; Soong, D. S. *Chem. Eng. Sci.* **1985**, *40* (*1*), 87.
15. Mahabadi, H. K.; O'Driscoll, K. F. *J. Macromol. Sci. Chem.* **1977**, *11* (A) (*5*), 967.
16. Bevington, J. C.; Melville, H. W.; Taylor, R. P. *Polym. Sci.* **1954**, *14* (*2*), 463.
17. Brandrup, J.; Immergut, E. H.; Grulke, E. A. *Polymer Handbook*; Wiley Interscience: USA, 1999; p 2317.
18. Smithson, M. *Confidence Intervals*; Taylor & Fransis Group: New York, 2006; Vol. 140, p 93.
19. Balke, S. T.; Hamielec, A. E. *J. App. Polym. Sci.* **1973**, *17,* 905.
20. Kumar, A.; Gupta, R. K. *Fundamentals of Polymer Engineering, Revised and Expanded*; Marcel Dekker: New York, 2003; p 712.
21. Balke, S.; Garcia-Rubio, L.; Patel, R. *Polym. Eng. Sci.* **1982**, *22* (*12*), 777.
22. Marten, F. L.; Hamielec, A. E. *ACS Symp.* **1979**, *104*, 43.

CHAPTER 10

ISOPRENE POLYMERIZATION ON NEODYMIUM CATALYST SYSTEM

K. A. TERESHCHENCO[1], E. B. SHIROKIKH[1], V. P. ZAKHAROV[2],
D. A. SHIYAN[1], N. V. ULITIN[1], and G. E. ZAIKOV[3]

[1]Kazan National Research Technological University, 68 Karl Marx Str., Kazan
420015, Republic of Tatarstan, Russian Federation, Tel.: +7 (843) 231-95-46, E-mail:
n.v.ulitin@mail.ru
[2]Bashkir State University, 32 Validy Str., Ufa 450076, Republic of Bashkortostan,
Russian Federation, Tel.: +7(347)272-63-70, E-mail: zaharovvp@mail.ru
[3]Emanuel Institute of Biochemical Physics, Russian Academy of Sciences,
[4]Kosygina Str., Moscow 119334, Russian Federation, Tel.: +7(499)137-41-01, E-mail:
chembio@sky.chph.ras.ru

CONTENTS

ABSTRACT

A mathematical model of the kinetics ion-coordination polymerization of iso-prene ($NdCl_3 \cdot nROH-Al(i\text{-}C_4H_9)_3$-piperylene is catalyst system) has been developed. The individual rate constants of elementary reactions have been determined as a result of solving the inverse kinetic problem. The adequacy of the model is confirmed by comparing the calculated and experimental data. Correlation of constants with the composition of the catalyst system is demonstrated. By way of numerical experiment the effect of the process controlling parameters (catalyst system and monomer concentrations) has been determined on molecular weight characteristics of the polymer.

10.1 INTRODUCTION

Ziegler–Natta neodymium catalyst systems have been used for synthesis of polydiene over half a century,[1] but, nevertheless, they do not cease to be the subject of intense research. The reason of being higher interest is that unlike titanium catalyst systems, use of neodymium systems allows to carry out polymerization of dienes with high efficiency and stereospecificity, and also to regulate the molecular weight characteristics of polymer in a wide range.[1,2]

Polydiene molecular weight characteristics variation is achieved by varying type and amount of cocatalyst (organometallic compound), and polymerization conditions. Furthermore, not only the nature of the cocatalyst affects molecular characteristics of polydiene but electron donor compounds is a modifier as well, those are needed to significantly improve the catalyst activity.[2] As such, alcohol ROH and piperylene is used as modifier in commercial synthesis of polyisoprene on Ziegler–Natta neodymium catalyst system.[3] However, it should be noted that theoretical foundations of influencing the amount of ROH as part of a neodymium catalyst complex on kinetics and molecular weight characteristics of synthesized polyisoprene have not been studied to date. Meanwhile, the development of effective methods to control the molecular weight characteristics of polydienes is impossible with no detailed analysis of the process. Therefore, the aim of this work is identification and determination of the effect of process controlling parameters (concentration of components of polymerization mix and composition of catalyst) on kinetics of polymerization and molecular weight characteristics of polyisoprene. This polyisoprene has been synthesized making use of catalyst system $NdCl_3 \cdot nROH-Al(i\text{-}C_4H_9)_3$-piperylene by the methods of mathematical modeling (where n is a molecule number of ROH in the composition catalyst system, $n = 0.75 - 3$, according to this study[4]), and solvent used is isopentane.

The experimental data used here have been taken from Refs.3 and 4

10.2   MATHEMATICAL MODEL

This aim has been achieved in three stages: Mathematical formalism development, describing kinetics of the process (solution of the direct kinetic problem), finding the values of kinetic rate constants (solution of the inverse kinetic problem), and numerical experiments.

Step 1: Solution of the direct kinetic problem

According to Refs. 3 and 4, in the process of isoprene polymerization on the $NdCl_3 \cdot nROH\text{-}Al(i\text{-}C_4H_9)_3$-piperylene catalyst system (isopentane is a solvent, temperature is 25 °C), two types of active site are possible. It is well known[1] that the rate of reaction initiation is much higher than the chain propagation reaction rate, and the reaction initiation is not a limiting factor. Therefore, when compiling the kinetics, it is assumed that the initiation is of instantaneous nature.

In accordance with this, the kinetics is as follows:

1. propagation

$$R_j(i) + M \xrightarrow{k_{p,j}} R_j(i+1),$$

2. transfer on the organoaluminum compound

$$R_j(i) + A \xrightarrow{k_{ta,j}} P(i) + R_j(0),$$

3. transfer to the monomer

$$R_j(i) + M \xrightarrow{k_{tm,j}} P(i) + R_j(1),$$

4. transfer to the polymer

$$R_2(i) + P(k) \xrightarrow{k_{tp2}} R_2(i+k),$$

5. transition types of active sites into each other

$$R_1(i) \underset{k_{c21}}{\overset{k_{c12}}{\rightleftarrows}} R_2(i),$$

6. destruction of active sites

$$R_j(i) \xrightarrow{k_t} D,$$

where j is the type of active sites, $j = 1, 2$; $R_j(0)$ is the j type original active sites; "M" is monomer; "A" is triisobutyl aluminum; $R_j(i)$ is growing polymerization chain on the j type of active site, containing i units of isoprene; $P(\ )$ is nonactive polymerization chain, containing i units of isoprene; "D" is nonactive polymer, formed as a result of active site destruction (in further calculations is negligible due to its low concentration), $k_{p,j}$, $k_{ta,j}$, $k_{tm,j}$, $k_{tp2,j}$, $k_{c,xy}$, k_t are the rate constants of chain propagation, chain transfer to the organoaluminum compound, chain transfer to monomer, chain transfer to polymer, transition x type active site into the y type active site, and destruction of the active site, respectively.

The system of differential equations was recorded on the basis of this kinetics using the mass action law. This system might describe concentration changes of all components of the polymerization mass. The obtained system of equations containing infinite amounts has been converted to a system of differential equations using the method of generating functions for the statistical moments of the molecular-weight distribution of polyisoprene[5]:

$$\frac{d[M]}{dt} = f\left([M],[A],\mu_{10},...,\mu_{xy},k_{p,j},k_{ta,j},k_{tm,j},k_{c,xy},k_t\right),$$

$$\frac{d[A]}{dt} = f\left([M],[A],\mu_{10},...,\mu_{xy},k_{p,j},k_{ta,j},k_{tm,j},k_{c,xy},k_t\right),$$

$$\frac{d\mu_{xy}}{dt} = f\left([M],[A],\mu_{10},...,\mu_{xy},k_{p,j},k_{ta,j},k_{tm,j},k_{c,xy},k_t\right),$$

In this equation system:

[M] is monomer concentration (isoprene), mol/L,

[A] is triisobutyl aluminum concentration, mol/L,

$\mu_{1k} = \sum_{i=0}^{\infty} i^k [R_1(i)]$, $\mu_{2k} = \sum_{i=0}^{\infty} i^k [R_2(i)]$, $\mu_{3k} = \sum_{i=0}^{\infty} i^k [P(i)]$ are statistical moments of

molecular-weight distribution of polyisoprene (k is the order of the statistical moment), mol/L.

The main molecular weight characteristics of polyisoprene produced, that is, number-average (M_n) and weight-average molecular weight (M_w), polydispersity coefficient (K_{PD}) are expressed in terms of moments in a simple way (m_0 is molar mass of isoprene):

$$M_n = m_0 \frac{\sum_{k=1}^{3} \mu_{k1}}{\sum_{k=1}^{3} \mu_{k0}}, \quad M_w = m_0 \frac{\sum_{k=1}^{3} \mu_{k2}}{\sum_{k=1}^{3} \mu_{k1}}, \quad K_{PD} = \frac{M_w}{M_n} = \frac{\sum_{k=1}^{3} \mu_{k2} \times \sum_{k=1}^{3} \mu_{k0}}{\left(\sum_{k=1}^{3} \mu_{k1}\right)^2}.$$

Thus, the developed mathematical formalism allows carrying out numerical calculations, however, it requires to know all values of rate constants of elementary reactions of the process kinetics. The inverse kinetic problem has been solved to find values in question.

Step 2: Solution of the inverse kinetic problem

In general, the inverse kinetic problem is formulated as follows: to find a set of the constants that describes the experimental data most accurately.

Solution of the inverse kinetic problem was carried out in two stages. First of all, the rate constants of chain propagation, destruction, and transition of active site types in each other have been calculated by means of minimizing the sum of absolute residuals between experimental data (dynamics of active site and conversion curves) and calculated values. These values have been determined from the analytical solution of the system of differential equations for the monomer concentration, and total concentrations of the active site types 1 and 2:

$$\frac{d\mu_{10}}{dt} = -\left(k_{c12} + k_t\right)\mu_{10} + k_{c21}\mu_{20},$$

$$\frac{d\mu_{20}}{dt} = -\left(k_{c21} + k_t\right)\mu_{20} + k_{c12}\mu_{10},$$

$$\frac{d[M]}{dt} = -\left(k_{p1}\mu_{10} + k_{p2}\mu_{20}\right)[M].$$

The experimental conversion data fitting results are shown in Figure 10.1.

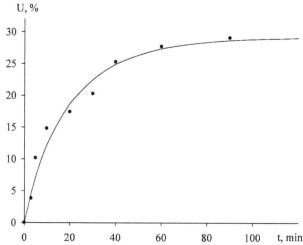

FIGURE 10.1 Conversion of isoprene (U) as a function of time (t) for the isoprene polymerization on the $NdCl_3 \cdot 0.75ROH$-$Al(i$-$C_4H_9)_3$-piperylene catalyst system (dotted line is an experiment data, line is a calculation; initial concentration of catalyst system is $[Nd]_0 = 0.0007$ mol/L, initial concentration of monomer is $[M]_0 = 1.5$ mol/L, isopentane is a solvent, temperature is 25 °C).

At the second stage, rate constants of chain transfer have been determined by minimization of functional $Z(k_{tm,j}, k_{ta,j})$. They were expressing the sum of absolute residuals between experimental data and the calculated values (Fig. 10.2) of polyisoprene number average molecular weight and weight average molecular weight:

$$Z(k_{tm1}, k_{tm2}, k_{ta1}, k_{ta2}) = \sum_i \left(M_{n,i}^{exp} - M_{n,i}^{calc} \right)^2 + \sum_i \left(M_{w,i}^{exp} - M_{w,i}^{calc} \right)^2.$$

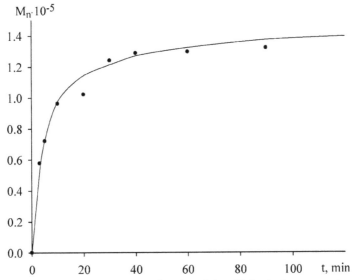

FIGURE 10.2 M_n of polyisoprene as a function of time (t) for the isoprene polymerization on the NdCl$_3\cdot$0.75ROH-Al(i-C$_4$H$_9$)$_3$-piperylene catalyst system (dotted line is an experiment data, line is a calculation; initial concentration of catalyst system is $[Nd]_0 = 0.0007$ mol/L, initial concentration of monomer is $[M]_0 = 1.5$ mol/L, isopentane is a solvent, temperature is 25 °C).

Calculation of optimal values of constants has been obtained using Hook–Jeeves optimization algorithm,[6, 7–8] which is characterized by fast convergence and ability to perform specific search.

The results of calculations are shown in Table 10.1, where as a unit, rate constants of elementary kinetic reactions were adopted for the case of $n = 0.75$, and the other values were in reference to the first ones.

During the analysis of the obtained constants by solving the inverse kinetic problem, the dependence of their magnitudes on the content of ROH in catalytic system has been found. Figure 10.3 demonstrates that during increase of n-number, it happens that actual linear increase of rate constants of chain propagation reaction occurs (Fig. 10.3a), and a linear decrease of the destruction reaction rate constant of active site as well (Fig. 10.3b).

TABLE 10.1   The Relative Values of the Rate Constants of Elementary Reactions, Defined by Solving the Inverse Kinetic Problem for Isoprene Polymerization on the $NdCl_3 \cdot nROH$-$Al(i\text{-}C_4H_9)_3$-Piperylene Catalyst System at 25 °C (Isopentane is a Solvent).

n	1.45	1.63
$k_{c12rel.}^{\;*}$	1.75	5.00
$k_{c21rel.}$	0.71	1.21
$k_{trel.}$	0.58	0.48
$k_{p1rel.}$	2.09	2.89
$k_{p2rel.}$	1.61	2.02
$k_{tm1rel.}$	1.63	2.09
$k_{tm2rel.}$	1.60	1.97
$k_{ta1rel.}$	1.23	1.35
$k_{ta2rel.}$	0.72	0.39
$k_{tp2rel.}$	0.20	0.19

* **Here** and below, the reaction rate constants are given relative to values of the constants in case of similar isoprene polymerization on $NdCl_3 \cdot 0.75ROH$-$Al(i\text{-}C_4H_9)_3$-piperylene catalyst system at 25 °C.

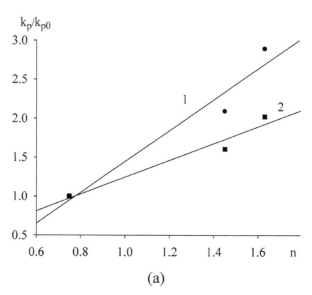

(a)

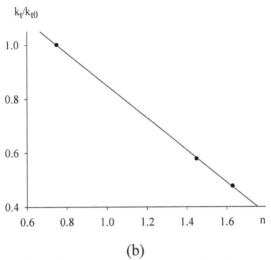

(b)

FIGURE 10.3   Relative chain propagation rate constant (a) on 1 type active site (curve 1) and 2 (curve 2) and relative rate constant of active site destruction, (b) as a functions of the n number, where n is a molecule number of ROH in the composition of $NdCl_3 \cdot nROH$ $Al(i\text{-}C_4H_9)_3$-piperylene catalyst system (isopentane is a solvent, temperature is 25 °C).

These effects are caused by correlation of ROH amount (included into catalyst complex) to particle size of microheterogenous catalyst. Average particle diameter of the formed catalyst is decreased during increase in ROH content, according to Ref. 4. As shown in Figure 10.3, it increases the speed of the process and reduces the probability of active site destruction.

Step 3: Numerical experiment

The developed model allows to investigate the influence of two independent control parameters of the process on the molecular weight of the polymer, that is, catalyst concentration and monomer concentration. Intervals of initial catalyst concentration values (determined by neodymium) and initial monomer concentration values for ion-coordination polymerization of isoprene on $NdCl_3 \cdot nROH\text{-}Al(i\text{-}C_4H_9)_3$-piperylene catalytic system, were as follows (we selected the widest range of concentrations):

- the initial concentration of catalyst $[Nd]_0$ is 10^{-5}–10^{-2} mol/L;
- the initial concentration of monomer $[M]_0$ is 0.5–10 mol/L.

10.2.1   INITIAL CONCENTRATION OF CATALYST

It is found that, with other equal conditions, by increasing the initial concentration of catalyst, $NdCl_3 \cdot nROH\text{-}Al(i\text{-}C_4H_9)_3$-piperylene, polyisoprene number average molecular weight decreases (Fig. 10.4).

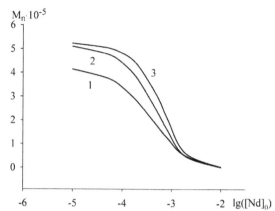

FIGURE 10.4   Calculated M_n of polyisoprene as a function of the $NdCl_3 \cdot nROH$-$Al(i$-$C_4H_9)_3$-piperylene catalyst system initial concentration ($[Nd]_0$) for isoprene polymerization ($[M]_0 = 1.5$ mol/L) at 25 °C, isopentane is a solvent, 120 min is duration time of polymerization process, n is 0.75 (1), 1.45 (2), and 1.63 (3).

10.2.2   INITIAL CONCENTRATION OF MONOMER

It has been established that increasing monomer concentration is followed by an increase in polyisoprene number average molecular weight (Fig. 10.5).

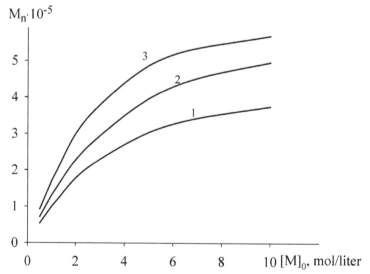

FIGURE 10.5   Calculated M_n of polyisoprene as a function of monomer initial concentration ($[M]_0$) for isoprene polymerization ($[Nd]_0 = 0.0007$ mol/L) at 25 °C, isopentane is a solvent, 120 min is duration time of polymerization process, n is 0.75 (1), 1.45 (2), and 1.63 (3).

The found regularities of influence of the initial concentrations of the catalyst system and those of monomer on number average molecular weight of polyisoprene are fully consistent with the known theoretical concepts,[1] which is a good proof of predictive capability of the developed mathematical formalism.

10.3   CONCLUSION

Thus, this chapter presents the results of mathematical formalism, which describes kinetics of ion-coordination isoprene polymerization on $NdCl_3 \cdot nROH$-$Al(i\text{-}C_4H_9)_3$-piperylene catalytic system. The results of this work demonstrate that the reaction rate constants of chain propagation and active site destruction are almost linearly dependents on the n, where n is ROH molecule number in composition of $NdCl_3 \cdot nROH$-$Al(i\text{-}C_4H_9)_3$-piperylene catalyst system. Macrokinetics of this process, including determination of macrokinetic parameter effects (the local nonuniformity of concentration, degree of turbulence in the reaction mix) on the polymerization kinetics and molecular weight characteristics of polyisoprene remained uninvestigated in this study. Conducting this type of investigation in the future makes it possible to consider the influence of the above parameters, arising from the nonideal conditions of the actual polymerization processes on the molecular characteristics of diene polymers. The influence pattern of parameters, controlling the process, on molecular mass characteristics of the obtainable polyisoprene (characteristics which are obtained using the model developed in the study) can provide the basis for high-performance technology of polyisoprene synthesis with controlled properties.

This research was supported by grant of the President of the Russian Federation MD-4973.2014.8, grants RFBR (projects 14-03-97027 and 14-33-50025).

KEYWORDS

- mathematical modeling, molecular weight characteristics of the inverse kinetic problem, polyisoprene, Ziegler–Natta catalyst.
- mathematical modeling, molecular weight characteristics of the inverse kinetic problem, polyisoprene, Ziegler–Natta catalyst. radical polymerization

REFERENCES

1. Friebe, L.; Nuyken, O.; Obrecht, W. Neodymium based Ziegler Catalysts—Fundamental Chemistry; Springer: Munich, 2006; p 287.

2. Jende, L. N.; Maichle-Mössmer, C.; Anwander, R. *Chem. Eur. J.*, **2013**, *19 (48)*, 16–32.
3. Morozov, Y. V.; Mingaleev, V. Z.; Nasyrov, I. S.; Zakharov, V. P.; Monakov, Yu. B. *Doklady Chem.*, **2011**, *440 (2)*, 286.
4. Zakharov, V. P.; Mingaleev, V. Z.; Zakharova, E. M.; Nasyrov, I. Sh.; Zhavoronkov, D. A. *Russ. J. Appl. Chem.*, **2013**, *86 (6)*, 911.
5. Kumar, A.; Gupta, R. K. *Fundamentals of Polymer Engineering, Revised and Expanded*; Marcel Dekker: New York, 2003; p 712.
6. Quarteroni, A.; Sacco, R.; Saleri, F. *Numerical Mathematics*; Springer-Verlag: Berlin, 2006; p 655..
7. Yeow, Y. L.; Guan, B.; Wu, L.; Yap, T. M.; Liow, J. L.; Leong, Y. K. *AICHE J.*, **2013**, *59 (3)*, 912.
8. De Leon, A. R.; Chough, K. C. Analysis of Mixed Data: Methods and Applications, CRC Press: USA, 2013; p 262.

CHAPTER 11

FILMS OF POLY(3-HYDROXYBUTYRATE)—POLYVINYL ALCOHOL BLENDS

A. A. OL`KHOV[1,2,] A. L. IORDANSKII[2], and G. E. ZAIKOV[3]

[1]Plekhanov Russian University of Economics, Stremyanny per. 36, Moscow117997, Russia, E-mail: aolkhov72@yandex.ru
[2]Semenov Institute of Chemical Physics, Russian Academy of Sciences, Kosygin Str. 4, Moscow 119991, Russia
[3]Emanuel Institute of Biochemical Physics, Russian Academy of Sciences, Kosygin Str. 4, Moscow 119991, Russia

CONTENTS

ABSTRACT

The morphology of extruded films based on blends of polyvinyl alcohol and poly(3-hydroxybutyrate) (PHB) was studied for various compositions. The methods of DSC and X-ray analysis were used. The phase-sensitive characteristics of the composite films, diffusion, and water vapor permeability were also investigated. Process linkages of water and swelling cause the first areas, the second—by processes of a relaxation and transition of structure of composites to an equilibrium condition. In addition, the tensile modulus and relative elongation-at-break were measured. Changes in the glass transition temperature of the blends and constant melting points of the components show their partial compatibility in intercrystallite regions. At a content of PHB in the composite films equal to 20–30 wt.%, their mechanical characteristics and water diffusion coefficients are dramatically changed. This fact, along with the analysis of the X-ray diffractograms, indicates a phase inversion in the above narrow concentration interval. The complex pattern of the kinetic curves of water vapor permeability is likely to be related to additional crystallization, which is induced in the composite films in the presence of water.

11.1   INTRODUCTION

To vary the physicochemical, mechanical, and diffusion characteristics of the polyvinyl alcohol (PVA) and to widen the area of practical application,[1-3] mixed compositions with a moderately hydrophilic polymer were proposed.[4]

To improve the mechanical behavior of poly(3-hydroxybutyrate) (PHB) and simultaneously to depress an expenditure in its production, the modification can be made through the PHB blending with other relevant polymers. Resulting polymer blends are potentially able to gain properties to be different from properties of parent blend-forming polymers.

PHB was used as a modifying polymeric component. The selection of this polymer is due to its biocompatibility with animal tissues and blood. Taking into account the similar properties of PVA, one may expect that a new class of polymer materials for medical purposes will be created.[5, 6]

A widespread procedure for regulating the drug release rate involves controlled changes in the balance of hydrophilic interactions in the polymer matrix at the molecular level. Therefore, regulation of structural organization at the molecular and supramolecular levels makes it possible to control the rate of drug delivery and hence to improve the therapeutic efficacy of new medicines.

11.2   EXPERIMENTAL

We studied the PVA (trade mark 8/27, Russia) containing 27% VAc; $M = 3.8 \times 10^4$ g/mol. The corresponding differential scanning calorimetry (DSC) curve shows two melting peaks at 130 and 170 °C. As the other component, we used powdered PHB (Lot M-0997, Biomer, Germany) with $M = 3.4 \times 10^5$ g/mol; its melting point is 176 °C, and the degree of crystallinity is equal to 69% (X-ray) or 78% (DSC).

The mechanical mixtures were prepared in the following proportions PVA: PHB = 100:0, 90:10, 80:20, 70:30, 50:50, and 0:100 wt.%. The as-prepared mixtures were processed into films with a thickness of 60 ± 5 pm using an ARP-20 single-screw extruder (Russia) with a screw diameter of 20 mm and a diameter-to-length ratio of 25. In the extruder zones, the temperature varied from 150 to 190 °C.

The structure of the as-prepared films was characterized by DSC methods on a Mettler TA-4000 indium-calibrated thermal analyzer; the scanning rate was 20 K/min. The structure of the test samples was also characterized by X-ray analysis on an automated X-ray diffractometer (CuK$_\alpha$-radiation, $\lambda = 0.154$ nm) with a linear coordinate detector (Joint Institute for Nuclear Research, Chernogolovka).[7]

The morphology of the films was studied by scanning electron microscope Hitachi S-570 (Japan) with an accelerating voltage of 20 kV. The film sample was cooled in liquid nitrogen for 30 min. Chips on films were made perpendicular and parallel to the direction of extrusion. The sample is then glued on dissecting table with a special adhesive—carbon or silver. Next, the sample was deposited on thin gold film with a thickness of 200–300 Å.

The tensile modulus and relative elongation-at-break of the composite films were estimated using a ZE-40 tensile machine (Germany) at a cross-head speed of 100 mm/min; the gauge width of the test samples was 10 mm. The water vapor permeability of the composite films was measured according to the standard beaker method[8] at 23 ± 1 °C and a saturated water vapor pressure of 21 mmHg; the results were taken from five parallel experiments. The accuracy of weighing was equal to ±0.001 g.

11.3   RESULTS AND DISCUSSION

The results of DSC studies of the composite films based on the PVA copolymer and PHB are summarized for various film compositions in Figure 11.1 and Table 11.1. As is seen, the average glass transition temperature T_g of the blends lies between 24 and 54 °C, the values corresponding to the glass transition temperatures of the starting PHB and PVA. This fact indicates a mixing of segments of these macromolecules in amorphous regions. The changes in T_g of the composites with respect to T_g of the starting polymers clearly indicate the compatibility of the components. However, the presence of a crystalline phase in both components at various content ratios suggests a limited interaction between the polymers.

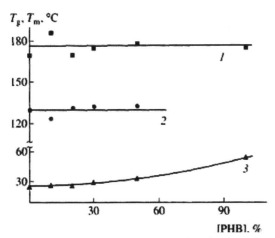

FIGURE 11.1   Dependence of (1) high-temperature and (2) low-temperature melting points and (3) glass transition temperature of the composite films from concentration of PHB.

Analysis of the nature of the double peak in the DSC melting curve of the PVA samples is beyond the scope of this work. However, one may attempt to explain this fact on the basis of published data.[9–11]

This behavior may be related to a wide size distribution of crystallites, as well as to their metastable character. As was asserted in Ref. 10, the typical DSC curves of crystalline polymers show several melting peaks; their height and position on the temperature axis are controlled by the temperature–time conditions of the sample processing in the extruder.

The presence of the double peak in the DSC curves of the PVA films may also be explained by the effect of extrusion-induced orientation, which may lead to the development of various morphological structures with different temperatures of phase transitions. As was stated in Ref. 11, polymers may have various crystalline modifications with identical crystallographic parameters of the unit cell but different energy levels.

TABLE 11.1   Characteristics of the Composite Films Based on PVA and PHB.

PVA:PHB (%)	T_m (°C)	$P_w{}^* \times 10^8$, [g·cm/ cm²·h·mmHg]	$C_w{}^{**} \times 10^3$, [g/ cm³·mmHg]
100:0	129/170	280.0	37.0
90: 10	123/186	75.0	9.0
80:20	130/170	105.0	3.0
70:30	132/175	115.0	1.0
50:50	132/178	125.0	1.4
0:100	176	0.33	0.1

Note: Two melting temperatures (T_m) of the composite films correspond to the two peaks in the DSC healing scans.
*Water vapor permeability coefficient.
**Water solubility coefficient.

As was found, the positions of the high-temperature and low-temperature peaks in the DSC curves, which correspond to the melting points of the starting components, are almost independent of the blend composition and remain invariable in the whole concentration range under study (Fig. 11.1). However, the transient region characterizing the glass transition temperature of the PVA–PHB system assumes different positions on the temperature axis depending on the concentration of PHB. This situation is vividly illustrated in Figure 11.1, where T_g of the blend is seen to increase with the content of PHB.

To reveal the morphology of crystalline regions in the composite films, the DSC measurements were combined with an X-ray study. As was shown for the entire composition range under study, the quasi-crystalline phase of the PVA copolymer is characterized by a well-pronounced orientation of the macromolecular axes along the texture axis, which coincides with the extrusion direction: the (110) X-ray reflection is located on the equator (Fig. 11.2a). In the crystal lattice of the PVA copolymer, the normal to the (110) plane is perpendicular to the *c*-axis, along which the axes of macromolecules are oriented. In samples containing 10 and 20% PHB, the crystalline phase of PHB is almost fully oriented. In this case, the (020) reflection is located on the equator. The normal to the (020) plane is the *b*-axis, and the axis coincides with the direction of the extrusion of PVA–PHB blends. The *c*-axis of the unit cell of PHB, which is parallel to the directions of macromolecules, is perpendicular to the extrusion direction.

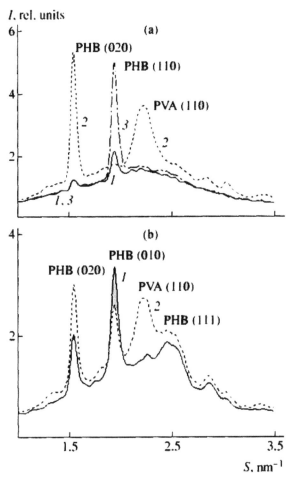

FIGURE 11.2   X-ray diffractograms of the films with a composition of (a) 80:20 and (b) 70:30 wt.% recorded (1) alone the orientation axis and (2, 3) at an angle of (2) 90° or (3) 20° to the orientation axis. $S = 2 \sin\theta/\lambda$, where θ is the X-ray scattering angle and λ is the wavelength.

X-ray diffractograms of films containing 30% and more PHB show the presence of the isotropic phase of PHB in appreciable amounts (Fig. 11.2b). The parameters of the unit crystal cells of the polymers in the blends were estimated. The crystalline phase of PHB is shown to be characterized by an orthorhombic unit cell: $a = 0.576$, $b = 1.32$, and $c = 0.596$ nm.[10] The crystalline phase of the PVA in the quasi-crystalline modification includes regions with a close packing of parallel chains (the γ modification); the parameters of the unit cell are $a =$

0.78, $b = 0.253$, and $c = 0.549$ nm.[12] In the corresponding X-ray diffractograms, these regions are associated with an X-ray reflection at $S = 2.22$ nm^{-1}.

The data of X-ray analysis allow one to conclude that the crystal lattice parameters of the components of the composite films are invariable; only the content of the isotropic component of the crystalline PHB phase is changed in the region of the phase transition.

Figure 11.3 shows the chip surface PVA film (a) and PHB (b). Polymer films are homogeneous materials. In Figure 11.4, PHB presented diameters phase depending upon the concentration of PVA in the mixed films and photo-PHB film by SEM. Mixed films are heterogeneous polymeric materials with a distinct phase boundary.

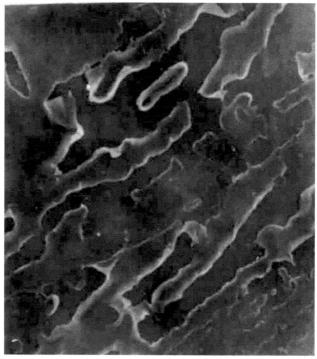

FIGURE 11.3   Morphology of the films PVA (a) and PHB (b).

PHB average phase diameter is 0.2–0.7 microns at a concentration of 10–20% as shown in Figure 11.4a,c. With increasing concentration of PHB up to 30%, PHB average phase diameter is increased by 0.5–1.2 micron as shown in Figure 11.4e. Increasing the concentration of PHB to 50% (Fig. 11.4g) has a bimodal average diameter distribution phase PHB: 0.2–0.8 microns and 1.0–1.5 micron.

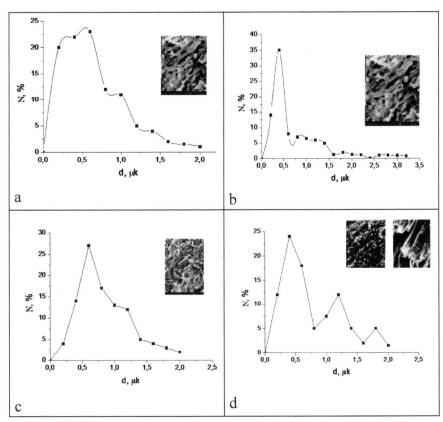

FIGURE 11.4 Distribution of the particle diameters in blended PVA–PHB films according to PHB content: 10% (a), 20% (b), 30% (c), and 50% (d).

Increasing the average diameter of PHB phase at a concentration of more than 20% due to the formation of two continuous phases: PHB polymer and PVA in blends. When there is a change of mechanical (Figs. 11.5 and 11.6) and diffusion (Fig. 11.7), the properties of films are based on blends of PVA–PHB.

Phase PHB films in a mixed fiber. The fibers are oriented along the direction of extrusion. This can be seen in Fig. 11.4c,d. Fiber formation in the mixtures of polymers in the melt flow is common.[11,13,14]

The results of mechanical tests of the PVA–PHB composites (uniaxial tension) fully support the above conclusion that phase inversion takes place in the films containing 20–30% PHB. This reasoning is proved by an abrupt inflection in the dependences of the tensile strength σ_t (Fig. 11.5, curve 1) and elongation-at-break ε_t (Fig. 11.5, curve 2) on the composition of the blends, as well as by the presence of a minimum in the dependence of the tensile modulus E_t (Fig. 11.6) on the composition of the composite films.

According to the data presented in Figs. 11.5 and 11.6, the properties of the composite films are controlled by the PVA matrix when the concentration of PHB is below 20% but determined by the PHB matrix at a concentration of 30% PHB and higher.

Almost invariable σ_t and ε_t values of the composite films before and after the phase inversion region may be explained as follows: particles of the dispersed phase do not serve as stress-concentrating sites in the matrix. This situation is possible when the dispersed phase is uniformly distributed within the matrix and the dimensions of its particles are rather small. This is usually observed in the case of a complete or partial miscibility of the polymer components.[15,16]

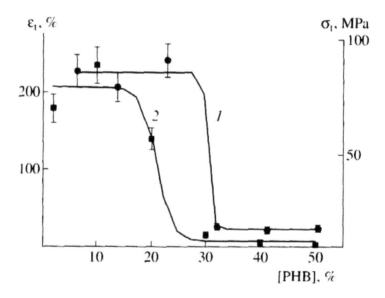

FIGURE 11.5   Dependence of (1) tensile strength σ_t and (2) relative elongation-at-break ε_t of the composite films from content of PHB.

Let us consider the process of water vapor transfer through the PVA–PHB composite films.

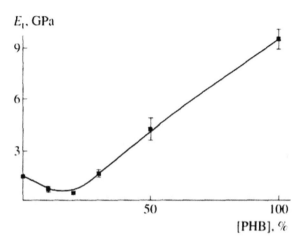

FIGURE 11.6   Dependence of tensile modulus E_t of the composite films from content of PHB.

Figure 11.8 presents the kinetic curves of vapor permeability for various PHB contents. One may distinguish three characteristic portions in these curves. The initial portion refers to a non-steady-state transport mechanism (Fig. 11.8b). In this region, the diffusion flow depends on time because the diffusion is related to the physical–chemical process of the binding of water molecules on functional groups of PHB, which show a marked affinity to water (carbonyl groups,[16] acetate and hydroxyl groups of the PVA[17,18]).

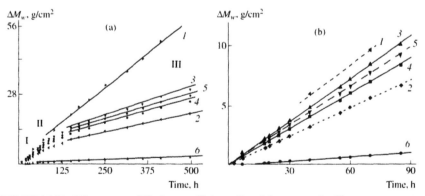

FIGURE 11.7   Water vapor diffusion coefficients D_w of the composite films versus content of PHB.

The next (middle) portion in the kinetic permeability curves corresponds to quasi-steady-state transport, where the segmental mobility of PVA increases under the action of the diffusing water and an additional crystallization of the hydrophilic component is likely to occur.[19] Such phenomena were described in detail and analyzed in Ref. 20. As the degree of crystallinity in the composite films increases, the rate of the effective flow of the diffusing component must decrease, and this trend is well seen in the last portion of the kinetic curves (Fig. 11.8a). Since newly formed crystallites are impermeable to water and therefore create an additional diffusion resistance, the slope of the last portion of the kinetic curves markedly decreases compared with the slope of the curves in the preceding portion.

The dependence of diffusion coefficients on the composition of the polymer blend (Fig. 11.7) shows a characteristic inflection at a content of 30% PHB, which corresponds to the phase inversion of the polymer matrix. In this region, one may observe an abrupt decrease in the tensile strength and relative elongation-at-break (Fig. 11.5). Strictly speaking, the traditional interpretation of diffusion equations in this region seems to be poorly justified and requires a detailed analysis, including consideration of the convective transfer mechanism.[21] The correlation between the transport and mechanical characteristics of the films indicates the important role of structural and morphological elements of the PVA–PHB mixed compositions.

The table lists the vapor permeability and solubility coefficients for the initial polymers (PVA and PHB) and composite films of various compositions estimated by the Daynes–Barrer method.[22] As seen from the table, the solubility coefficient C_W monotonically decreases with the increasing concentration of PHB. This result seems to be quite evident, since the amount of sorbed water depends on the nature and concentration of functional groups in the polymer.

The coefficient of water vapor permeability slightly depends on the composition of the blends. Since this parameter is controlled by diffusion, equilibrium water sorption, and structure of the films at the crystal level, the interpretation of the results is ambiguous. For example, the above decrease in the solubility may be compensated by a structural amorphization of the components of the composite films near the phase inversion point.

As the content of the PVA copolymer in the polymer blend is increased, the concentration of hydroxyl groups evidently rises. The hydrophilization of the PHB matrix (with increasing content of the PVA component) provides a monotonic increase in the water solubility coefficient without any visible inflection points and extrema (Table 11.1). Even though the films may experience various structural rearrangements at the crystal level, as is evidenced by the DSC data, X-ray analysis and mechanical tests, the water solubility in the PVA–PHB composite films is still sensitive only to changes in the ratio between the contents of the hydrophilic and moderately hydrophilic components. Therefore, the content of dissolved water is controlled by the nature and concentration of functional groups in the polymer blend.

As the content of PHB in the blends is increased, the concentration of hydroxyl groups decreases; that is, at the molecular level, hydroxyl groups with their high group contribution (according to van Krevelen[23]) are substituted by ester groups of PHB with a much lower group contribution.[24]

11.4   CONCLUSION

Blends based on polymers with different hydrophilic groups are of scientific and practical interest from the point of view of wide variation coefficients of water sorption, diffusion and permeability.

In spite of limited concentration interval of partly miscible blending (no more than 30 wt.% of PHB), these PHB–PVA blends are of great interest as novel biodegradable films and coatings. Additionally, the blends based on water-soluble and biocompatible PVA and friendly environmental and biocompatible PHB could be treated as new generation of environmentally protected materials in packaging industry, agricultural application as well as biomedicine areas.[25–27]

KEYWORDS

- films
- morphology
- poly(3-hydroxybutyrate)
- blends, polyvinyl alcohol
- diffusion, mechanical parameters
- SEM, DSC, X-ray analysis

REFERENCES

1. Sudesh, K.; Abe, H.; Doi, Y. *Prog. Polym. Sci.* **2000**, *25*, 1503–1555.
2. Holmes, P. A. In *Developments in Crystalline Polymers*; Bassett, D. C., Ed.; Elsevier Applied Science: London, 1988; Vol. 2, p 1.
3. Barak, P.; Coqnet, Y.; Halbach, T. R.; Molina, J. A. E. *J. Environ. Qual.* **1991**, *20*, 173–179.

4. Mergaert, J.; Webb, A.; Anderson, C.; Wouters, A.; Swings, J. *Appl. Environ. Microbiol.* **1993**, *93*, 3233–3238.
5. Luo, S.; Netravali, A. N. *Polym. Composites* **1999**, *20*, 367–378.
6. Orlando, G. *Regenerative Medicine Applications in Organ Transplantation;* Elsevier: London, 2014.
7. Iordanskii, A. L.; Razumovskii, L. P.; Krivandin, A. V.; Lebedeva, T. L. *Desalination* **1996**, *104*, 27–35.
8. Iordanskii, A. L.; Ol`khov, A. A.; Pankova, Yu. N.; Kosenko, R. Yu.; Zaikov, G. E. In *Chemical Reaction in Condensed Phase: The Quantitative Level;* Zaikov, G. E. Nova Science Publication: New York. 2006; p 139.
9. Godovskii, Yu. K. *Teplofizicheskie metody issledovaniya polimerov (Thermophysical Methods for Polymer Investigation);* Khimiya: Moscow, 1976.
10. Paul, D.; Newman, S. M. *Polymer Blends;* Academic: New York, 1979.
11. Ol`khov, A. A.; Iordanskii, A. L.; Zaikov, G. E. *J. Balk. Tribol. Assoc.* **2014**, *20*, 101–110.
11. Brule B., Flat, J. J. In *Macromolecular Symposia. Special Issue: Fillers, Filled Polymers and Polymer Blends;* Wiley-VCH: Weinheim, 2006; Vol. 233, p 210.
12. Zoldan, J.; Siegmann, A.; Narkis, M. In *Macromolecular Symposia. Special Issue: Fillers, Filled Polymers and Polymer Blends;* Wiley-VCH: Weinheim, 2006; Vol. 233, p 123.
13. Iordanskii, A. L.; Rudakova, T. E.; Zaikov, G. E. Interaction of Polymers with Bioactive and Corrosive Media. Ser. In *New Concepts in Polymer Science*; VSP Science Press: Utrecht, Tokyo, 1994.
14. Iordanskii, A. L.; Olkhov, A. A.; Zaikov, G. E.; Shibryaeva, L. S.; Litvinov, L. A.; Vlasov, S. V. *J. Polym. Plastics Technol. Eng.* **2000**, *39*, 783–792.
15. Iordanskii, A. L.; Olkhov, A. A.; Kamaev, P. P.; Wasserman, A. M. *Desalination* **1999**, *126*, 139–145.
16. Hassan, C. M., Peppas, N. A. Biopolymers. PVA Hydrogels. In *Advances in Polymer Science*; Berlin: Springer-Verlag, 2000.
17. Rozenberg, M. E. *Polimery na osnove vinilatsetata (Vinyl Acetate-Based Polymers)*; Khimiya: Leningrad, 1983.
18. Rowland, S. P. *Water in Polymers*; American Chemical Society: Washington, 1980.
19. Pankova, Yu. N.; Shchegolikhin, A. N.; Iordanskii, A. L.; Zhulkina, A. L.; Ol`khov, A. A.; Zaikov, G. E. *J. Mol. Liquids* **2010**, *156*, 65–69.
20. Iordanskii, A. L.; Bonartseva, G. A.; Pankova, Yu. N.; Rogovina, S. Z.; Gumargalieva, K. Z.; Zaikov, G. E.; Berlin, A. A. In *Current State-of-the-Art on Novel Materials*; Balköse, D., Horak, D., Šoltés, L., Ed.; Apple Academic Press: New York, 2014; Vol. 1. p 450.
21. Iordanskii, A. L.; Ol`khov, A. A.; Pankova, Yu. N.; Kosenko, R. Yu.; Zaikov, G. E. In *Polymer and Biopolymer Analysis and Characterization;* Zaikov, G. E., Jimenez, A., Ed.; Nova Science Publication: New York, 2007; p 103.
22. Van Krevelen, D. W.; Nijenhuis, K. Te. *Properties of Polymer. Their Correlation with Chemical Structure, Their Numerical Estimation and Prediction from Additive Group Contributions;* 4th ed.; Elsevier: Amsterdam, 2009.
23. 2009Vogler, E. A. *Water in Biomaterials. Surface Sciences;* Morra, M., Ed.; Wiley: Chichester, 2001.
24. Panov, A. A.; Beloborodova, T. G.; Anasova, T. A.; Zaikov, G. E. *J. Balk. Tribol. Assoc.* **2008**, *2*, 243–247.
25. Bonartsev, A. P.; Iordanskii, A. L.; Bonartseva, G. A.; Zaikov, G. E. Biodegradation and Medical Application of Microbial Poly(3-hydroxybutyrate). *J. Balk. Tribol. Assoc.* **2008**, *15*, 359–395.
26. Iordanskii, A. L.; Chvalun, S. N.; Shcherbina, M. A.; Karpova, S. G.; Lomakin, S. M.; Shilkina, N. G.; Rogovina, S. Z.; Zaikov, G. E.; Chen, X. J.; Berlin, A. A. *J. Balk. Tribol. Assoc.* **2013**, *19*, 144–150.

CHAPTER 12

DESORPTION OF THE PLASTICIZER (DIALKYLPHTHALATE) IN PVC

I. G. KALININA, K. Z. GUMARGALIEVA, and S. A. SEMENOV

N. N. Semenov Institute of Chemical Physics, Russian Academy of Sciences, Moscow, Russia

CONTENTS

ABSTRACT

It was shown that desorption of the plasticizer (dialkylphthalate) in PVC is accelerated by bio-overgrowth of plasticized PVC samples. Thus, the transfer of the plasticizer is controlled by its diffusion.

Biodegradation is governed by different factors that include polymer characteristics, type of organism, and nature of pretreatment. The polymer characteristics such as its mobility, tacticity, crystallinity, molecular weight, the type of functional groups and substituents present in its structure, and plasticizers or additives added to the polymer, all play an important role in its degradation.

PVC is a strong plastic that resist abrasion of chemicals and has low moisture absorption. There are many studies about thermal and photodegradation of PVC but there only few reports available on biodegradation of PVC.[1]

In this study, the long-term effects of free and encapsulated 2-n octyl-4-iso-thiazolin-3-one (OIT) incorporation into PVC was determined on properties of the plastic following burial for 20 months in microbially active soil. In general, fewer bacteria and fungi were isolated from samples containing OIT when compared with biocide-free controls. No patterns in fungal colony predominant or range was observed between the PVC formulations, OIT containing samples retained more plasticizer than the controls and hence underwent reduced biodegradation by microorganisms.

According to Ref. 2, PVC having low molecular weight can be exposed to biodegradation by the use of white-rot fungi.

Problems of biodegradation were not discussed or considered in the works devoted to the study of plasticized PVC plasticates aged under model or climate conditions or during long-term exploitation of 15–30 years.[3,4] This is partly caused by the fact that desorption of corrosion inhibitors, as well as increased temperature regimes are not suitable conditions for grouping of microorganisms. But aging of PVC plasticate-made articles made of plasticized PVC under real conditions of their exploitation is often bound to the processes of diffusional desorption of plasticizing additives from the material.[5,6]

Because biodegradability claims often appear in poorly documented promotional literature only, a comparative biodegradability survey of some novel highly plasticized PVC, traditional PE, and PP wraps, as well as coated or impregnated paper products was made.[7] All materials were exposed in soil under conditions generally favorable for biodegradation. The measurement of CO_2 evolution in Biometer flask, decline in tensile strengths, in combination with newly developed residual weight determinations, and other residue analysis techniques were utilized in assessing biodegradation. During 3 months of exposure, traditional PE and PP films did not undergo measurable deterioration, but a group of newly formulated and heavily plasticized PVC films underwent extrusive biodeterioration and up to 27.3% of their carbon was converted to CO_2. However, gas chromatography, residual weight determination, chloride release, and viscosity measurements indicated that only the plasticizer but not the PVC resin was mineralized.

The desorption of phthalate plasticizer accumulated by microorganisms, widely used in PVC may lead to biodegradation of the plasticized PVC material.

In this connection, we have studied the features of microscopic fungal effects on diffusion desorption of the plasticizer (dialkyl phthalate) from plasticized PVC of I-40-13 type (GOST 5960-72).

Microscopic fungus *Aspergillus niger*, obtained from PVC insulation of wires exploited in the regions of tropical climate, was used as a bioagent.

Cultivation of microscopic fungus on plasticized PVC was performed under optimal conditions for growth of the microorganism ($T = 29\,°C$, $\varphi = 100\%$) during 12 months.

Quantitative estimation of dialkyl phthalate (DAP) concentration in the wire isolation was conducted with the help of UV-spectrophotometric analysis. DAP concentration in the samples was calculated by intensity of the absorption bend $\lambda = 230$ nm, characteristic for that substance.

It was shown[8-9] that regularities of the change of plasticizing additive concentration in plasticate is determined by the ratio of diffusion and desorption rates. For most of system containing a migrating low molecular component one of the following two variants of the process take place. If the rate of desorption from the surface is lower than that of low molecular component transfer in the material volume, the total rate of the process is defined by the intensity of additive desorption into the environment (so-called desorptional sphere of proceeding process).

Thus, if desorption of the component proceeds much faster than its diffusion in the material, the total rate of the process and consequently the amount of migrated low molecular additive will be limited by diffusion (diffusional sphere).

Diffusional desorption of low molecular components from polymer materials at the process proceeding in various spheres were calculated from the expressions.[8-9]

In diffusional domain

$$m_t/m_o = (8/\pi)\exp(\pi^2 \_ x \underline{D} \_ xt)\, 4\, I^2 \tag{12.1}$$

In the desorption domain

$$m_t/m_o = \exp(\underline{W} \_ xt)\, S \cdot I \tag{12.2}$$

where m_0 and m_t are initial and current concentrations of low molecular component in the polymer material at the moment of time t, respectively, W is the desorption rate of components form the material surface, D is the diffusion coefficient in the material, S is the solubility of the component in the material, I is the half thickness of the material film.

Equations 12.1 and 12.2 may be generalized as follows:

$$m_t/m_0 = \exp(-k\ t) \tag{12.3}$$

where k is the effective rate constant of low molecular component transfer from the material.

Based on eq 12.3 and the values of plasticizer concentration in the plasticized item, obtained by UV-spectrometric analysis, one can calculate the effective rate constants of DAP transfer in the samples studied, k. Obtaining these constants, diffusion coefficients (D) and desorption rates (W) for DAP from the test samples of PVC plasticate and those treated by bio-overgrowing are easily calculated. The parameters k, D, and W, obtained with the help of our experimental data, are shown in the Table 12.1.

Comparison of these values with corresponding sorption–diffusional parameters, determined by other (independent) techniques, allows us to make a valid conclusion about the type of the plasticizer transfer.

D_p and W_p were calculated using the literature values of physicochemical and sorption–diffusional characteristics of DAP and PVC-plasticate.[8,10]

TABLE 12.1 Sorption–Diffusional Parameters for the PVC Plasticate Dialkyl Phthalate System.

Parameter	Experimental k, D, W	Experimental k, D, W	Calculated D_p, W_p
	Micromycete influence	Influence of $T = 30$ °C, $\varphi = 100\%$	Influence of $T = 30$ °C
k, s^{-1}	1.00×10^{-8}	1.60×10^{-9}	
D, cm^2/s	0.64×10^{-10}	0.82×10^{-11}	1.10×10^{-10}
W, g/cm^2•s	1.50×10^{-10}	2.40×10^{-11}	4.30×10^{-11}

There are k—effective rate constant (s^{-1}), D—diffusion coefficient (cm^2/s), W—desorption rate (g/cm^2•s).

Calculation of the diffusion coefficient by the eq 12.1 accepts DAP solubility in PVC plasticate as 0.3 g/cm^2.

The rate of dialkyl phthalate desorption from the insulation surface was calculated by the Herz equation:

$$W_p = P_T/(2\pi MkT)^{1/2},\qquad(12.4)$$

where P_T is the pressure of plasticizer vapors at temperature T, M is the molecular mass of DAP (390 g/mol).

The pressure of DAP vapors at 30 °C (P_{30}) was calculated by interpolation dependences obtained from the Clausius–Clapeyron equation:

$$\lg P_T = (-A/T) + B,\qquad(12.5)$$

where A and B are constants.

To determine constants A, B, and P_{30}, the system of two equations of the (12.5) type for 293 and 333 K was solved.

Plasticizers' vapor pressures at current temperatures are shown in special literature and are 1.3×10^{-4} and 4.0×10^{-3} dyn/cm^2, respectively. Calculations have shown that the pressure of DAP vapors at 30 °C is 3.4×10^{-4} dyn/cm^2. The desorption rate (W_p), obtained from eq 12.4, is shown in Table 12.1.

The coefficient of dialkyl phthalate diffusion from plasticized PVC (D_p) at 30 °C were calculated with the help of data from the work[10], which show the coefficient D values for DAP in plasticized insulation at various temperatures. Applying an equation of the Arrhenius type, the effective activation energy was calculated and found to be 58.4 kJ/mol. Then, based on the expression mentioned, D_p at 30 °C as shown in the table was obtained.

The analysis of the data from this table shows close values of W and W_p for test samples. The agreement between diffusion coefficients D and D_p was less than satisfactory. Consequently, the plasticizer loss at complex temperature—humidity influence on PVC—insulation ($T = 30$ °C and $\varphi = 100\%$) is mainly determined by its desorption from the material surface. This conclusion correlates with the data reported by Bezveliev.[4]

Satisfactory coincidence of D and values are observed on microscopic fungus influence on the plasticized PVC. In this case, the values of W and W_p are different at least by an order of magnitude.

Thus, the analysis performed shows that transfer of the plasticizer under influence of a microscopic factor on the PVC plasticate is limited by diffusion of plasticizing additive in the material volume.

KEYWORDS

- **diffusion**
- **biodegradation**
- **plasticized PVC**

REFERENCES

1. Coulthwaite, L.; Bayley, K.; Liauw, C.; Craig, G.; Verran, *J. Int. Biodeter. Biodegrad.* **2005,** *56* (2), 86–93.
2. Shah, A.; Hasan, F.; Hameed, A.; Ahmed, S. *Biotechnol. Adv.* **2008,** *26* (N3), 246–265.
3. Moiseev, Yu. V.; Malunov, V. V.; Gumargalieva, K. Z.; Soloviev, A. I.; *Plastmassy* **1985,**9, 60.
4. Gumargalieva, K. Z.; Ivanov, V. B.; Zaikov, G. E.; Moiseev, Yu. V.; Pokholok, T. V. *Khimicheskaya Fizika 14,* 753–763.
5. Tinius, K. M. *Plasticizers*; Khimiya: Moscow-Leningrad, 1964.
6. Bezveliev, A. G. *Plastmassy,* **1983,** *2*, 8–9.
7. Yabannavar, A.; Bartha, R. *Soil Biol. Biochem.,* **1993,** *25* (*11*), 1469–1475.
8. Borisov, B. I.; Colloid, Zh. **1973,** *35* (*11*), 140.
9. Borisov, B. I. and Gromov, N. I. *Corrosion and Protection in Oil and Gas Industry*; Vol. 3, pp19–22.
10. Borisov, B. I.; Colloid, Zh. Vol. 40, *3*, 533–535.

CHAPTER 13

THERMOPHYSICAL CHARACTERYSTICS OF CHLORINATED ETHYLENE–PROPYLENE–DIENE

I. A. MIKHAYLOV[1], YU. O. ANDRIASYAN[2], G. E. ZAIKOV[2], and A. A. POPOV[1,2]

[1]Plekhanov Russian University of Economics (PRUE), 117997 Moscow, Stremyanny per. 36, Russia
[2]Emanuel Institute of Biochemical Physics of Russian Academy of Sciences (IBCP RAS), 119991 Moscow, Kosygina Str. 4, Russia

CONTENTS

ABSTRACT

Macromolecular structure of chlorine-containing ethylene–propylene–diene caoutchoucs (EPDM) and their probable changes, which happen in the process of caoutchoucs vulcanization, were studied by method of infrared spectroscopy. Caoutchoucs CEPDM-2, CEPDM-4, and CEPDM-16, consequently containing 2, 4, and 16% (mole) chlorine, were taken for the investigation. As a sample for comparison, initial ethylene–propylene–diene caoutchouc EPDM without halogen was used. To determine thermostability of macromolecular structures of investigated polymers under vulcanization temperature, the specimens of caoutchoucs were subjected to heating vulcanizing press under temperature 151 °C during 5, 10, 20, 30, 40, 50, and 60 min.

To study thermophysical characteristics of chlorine-containing EPDM, methods of deferential thermal analysis (DTA) and thermogravimetric analysis (TGA) were used. As a result, spectral investigation of specimens were established, that in the process of obtaining CEPDM-2, chlorine combine with fragment of ethylidene norbornene (ENB), but the structure of the main chain was not touched, as a result of which, haloid modified double linkage and halogen in α-position to this linkage, that has to result in increase of rubber vulcanization rate based on CEPDM-2.

When obtaining CEPDM-4 and CEPDM-16 chlorine associated as to fragment ENB as to the main chain of polymer, where in consequence of dehydrochlorination reaction, another unsaturation can be afforded. This process will permanently result in decrease of ozone resistance of rubbers from these caoutchoucs.

Data of DTA and TGA confirm structural transformation of caoutchoucs CEPDM, established by infrared spectroscopy method. So data of DTA show that oxidizing of caoutchoucs EPDM and CEPDM in terms of degree of their chlorination begins in temperature interval 210–240 °C, which shows their high stability. Effective destruction of polymeric chains was observed in field of temperatures 410–420 °C, which has all types of investigated rubbers. Pyrolysis of caoutchoucs EPDM and CEPDM are observed under 480 °C, but under higher temperatures, polymers are carbonified.

Data of TGA show that investigated caoutchoucs in case of stability to weight loss on 10% under thermal influence is possible to range: CEPDM-2 (350 °C) > EPDM (315 °C) > CEPDM-4 (277 °C) > CEPDM-16 (250 °C). In buckles show, the temperature under which polymer weight loss proceeds on 10%.

13.1 INTRODUCTION

It is known[1-7] ethylene–propylene–diene caoutchoucs (EPDM) and rubbers, which based on them possess very valuable properties, such as thermostability, weather stability, ozone stability, chemical stability to range of aggressive environments, high dielectric indexes, high tensile strength and elasticity; caoutchoucs have high filler loading capacity and are mixed well with the ingredients.

Combination of EPDM with all-purpose caoutchoucs permits to obtain composition with wider complex valuable properties.[1-5] But widespread use of these compositions are limited by degraded rate of vulcanization of EPDM and, as a rule, its bad co-vulcanization with high unsaturated all-purpose caoutchoucs.[8]

It is known that all above-mentioned complex of EPDM properties are explained by special tips of structural framework of its macromolecules. The haloid modification of EPDM was carried out with the aim of saving these properties and trying to remove defects, appropriate of EPDM. Among other factors, chlorination of ethylene–propylene–diene caoutchouc with diene comonomer of ethylidene norbornene (ENB) was carried out:

$$CH{=}CH$$
$$CH{-}CH_2{-}CH \qquad \text{(SKEPTE).}$$
$$CH_2{-}C{=}CH{-}CH_3$$

It was interesting to carry out haloid modification of this caoutchouc to study structure of modified caoutchouc and also probable structural changes occurring in the process of its heating under vulcanization temperature. As specimens of investigation, EPDM with ethylidene norbornene (SKEPTE-40) and its chlorinated derivatives (CEPDM) with content of chlorine 2, 4, and 16% mole were taken, marked in this work as CEPDM-2, CEPDM-4, and CEPDM-16, respectively. Also, it is well known from literature that the addition of big amount of halogens in macromolecule of polyolefin caoutchouc may result in destruction of macromolecular chain and decrease of polymer thermostability. So, for example, destruction of polymer under chlorination, more than 2%,[2] was observed. Therefore, it was interesting to study thermostability of chlorinated ethylene–propylene–diene caoutchoucs, which have different content of halogen.

13.2 EXPERIMENTAL

In this work, the forthcoming time intervals of heating 5, 10, 20, 30, 40, 50, and 60 min were accepted, considering that optimum of vulcanization of chlorinated

EPDM (under 151 °C) in the different systems of accelerators can rage in wide time diapason.

Investigated EPDM and CEPDM were preliminary extracted by hot acetone during 24 h, then exposed to drying before permanent weight with the purpose of removing volatile impurity. Polymer chain was obtained from purifying the caoutchoucs by hot-pressing method.

Structural investigations were carried out by infrared of spectroscopy method. The temperature of specimens, 151 °C, that correspond to the temperature of rubber vulcanization based on EPDM with different duration of heating time under 151 °C: EPDM (heat 5–60 min); CEPDM-2 (heat 5–60 min); CEPDM-4 (heat 5 min); CEPDM-4 (heat 10–40 min); and CEPDM-4 (heat 50–60 min).

Differential thermal (DTA) and thermogravimetric analysis (TGA) of investigated polymers were carried out for studying the influence of chlorination extent on thermostability of macromolecules of CEPDM.[3] Tenderness of DTA and TGA was changed depending on size of heat effect and amount of emitting of volatile products.

13.3   RESULTS AND DISCUSSION

Data obtained by infrared spectroscopy show that structure of macromolecules of EPDM, depending on heating time, doesn't change. In all the interval of heating time, the spectrum of EPDM has distinctive bands of absorption in the field 720, 800, 1150, and 1680 cm^{-1}. The absorption band in the field 720 cm^{-1} characterizes pendulum oscillation sequentially combined with three or five methylene groups. The absorption band in the field 800 cm^{-1} characterizes no plane deformation oscillation of CH group in the presence of double linkage in the compound CRR' = CHR''. The absorption band in the field of 1150 cm^{-1} is determined by the oscillation of propylene unit $-CH_3\!\!-\!\!\overset{\displaystyle CH_3}{\underset{\displaystyle |}{CH}}\!\!-\!\!CH_2\!\!-$ in case of nonconjugated double linkage.

The bands of absorption characterize EPDM, which has the structure as follows:

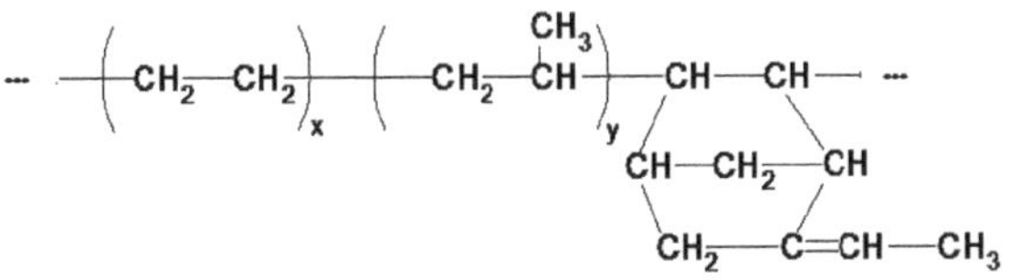

When considering spectrum CEPDM-2, we can see that absorption band disappears in the fields 800 and 1680 cm^{-1} and two new absorption bands appear in the fields 1510 and 1570 cm^{-1}, characterizing oscillation of cyclical compound. Then, under heating in all time interval in infrared-spectrum of CEPDM-2, no changes were observed, which shows us the stability of its structure.

Based on obtained data, we can suggest the next scheme of changes of macromolecular structure of EPDM in the process of obtaining CEPDM-2.

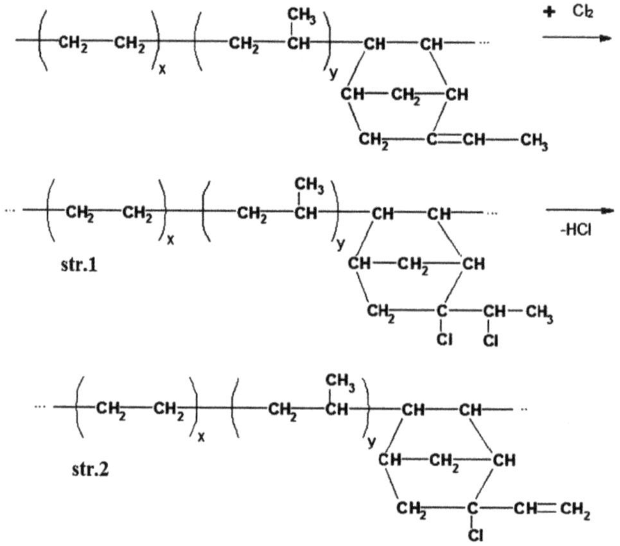

In the process of obtaining CEPDM-2, the attaching of chlorine proceeds after double linkage of ethylidene norbornene fragment, which explains the disappearance of absorption band at fields 800 and 1650 cm^{-1}.

The next dehydrochlorination of formed structures result in formation of double linkage. The chlorine atom in this structure is situated in α-condition to double linkage.

When considering spectrum CEPDM-4, we can observe new absorption band, such as band in the field 687 cm^{-1}, characterizing of oscillation of CCl group; 820 cm^{-1}, characterizing of oscillation of CH group in aromatic cycle; and 1265 cm^{-1}, characterizing of oscillation of double linkage. Under heating polymer during 10 min, the absorption band in field 1150 cm^{-1} disappears, all others absorption bands remain without changes.

While heating at 20, 30, and 40 min in spectrum CEPDM-4, we do not observe changes, but when heated for 50 min, wide absorption band in the field 1040 cm^{-1} is observed, characterizing the presence of ester group. When heating for 60 min, the spectrum of CEPDM-4 does not change.

When considering spectrum CEPDM-16 and its changes depending on heating time, we found similarity in spectrums of CEPDM-4 and CEPDM-16, with difference in absorption band 1040 cm^{-1} observed after 20th min of polymer heating.

Based on obtained experimental data, we can assume the next structural changes in macromolecule of EPDM in the process of obtaining CEPDM-4 and CEPDM-16:

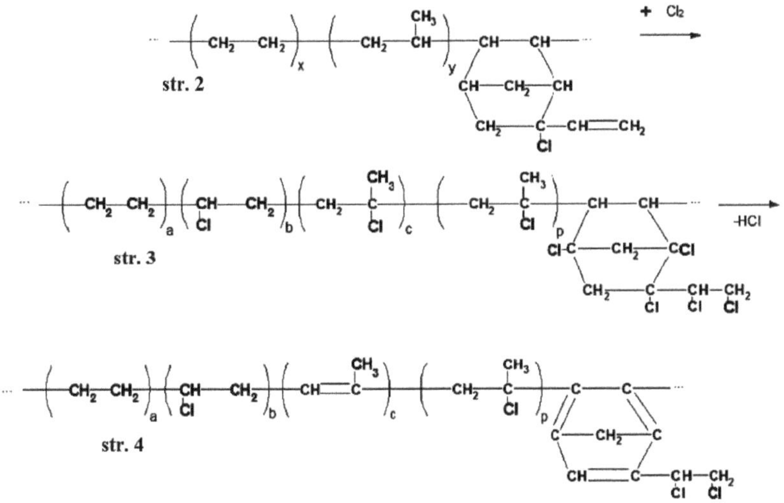

where $x = a + b$, $y = c + p$; $a, c = 8/12$; $b, p = 2/3$

In the first stage, similar structural changes were observed when obtaining CEPDM-2 (structure 2). In the second stage, we can see chlorine adjoining to cyclical fragment CEPDM-2 and to the main chain of polymer. In the process of chlorination of the structure 2, the intermediate structure 3 is formed, which in consequence of dehydrochlorination reaction modifies to structure 4. The structure 3 characterizes building of macromolecule of CEPDM-4 and CEPDM-16 with difference that in macromolecule of CEPDM-16, more the content of units –(–CHCl–CH$_2$–)$_b$– and units –CH=CH–, obtained in the process of dehydrochlorination of part of units –CHCl=CH$_2$–.

Content of chlorine in the final product, which determines by its amount, adjoined to ethylene macromolecule fragments, and also partially its content of fragments

From shown structural formulae, we can see that fragment $\left(-CH_2-\underset{Cl}{\overset{CH_3}{C}}-\right)_p$

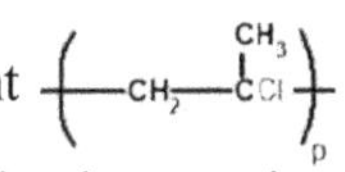

in structure 3 has more stability in the process of dehydrochlorination reaction

$$\left(-CH_2-\overset{\overset{\displaystyle CH_3}{|}}{\underset{|}{C}}Cl-\right)_c,$$ which we can explain, so that macromolecular chains of EPDM,

except alternating short block (8–12) monomer units of ethylene and propylene, also contain some percent of blocks with 2–3 monomer units in each one and also alternating singular units of ethylene and propylene. The reaction of dehydrochlorination in big propylene blocks proceeds easier because of elimination of molecule of HCl in one of propylene units, which results in development of "neighbors effect," because elimination of next molecule HCl in adjacent units is attended by formation of conjugated double linkage. In the little blocks and isolated units of propylene, the proceeding of dehydrochlorination reaction is very difficult.

In the process of heating of CEPDM-4 and CEPDM-16, which have structural formula 4, and on the 10th min, the dehydrochlorination of fragment

$$\left(-CH_2-\overset{\overset{\displaystyle CH_3}{|}}{\underset{|}{C}}Cl-\right)_p$$ begins, it explains why the absorption band in field 1150 cm^{-1}

disappears. Thereof, that percent of unsaturation in CEPDM-4 is less than that for CEPDM-16. Oxidation of the CEPDM-4 begins on 50th min of heating and CEPDM-16 at 20th, as is evidenced by appearance of band 1040 cm^{-1} in spectrums of these caoutchoucs.

The investigation associated with influence of extent of chlorination on thermostability of macromolecule of CEPDM were then carried out. Data of investigation of EPDM and CEPDM are given in Figure 13.1.

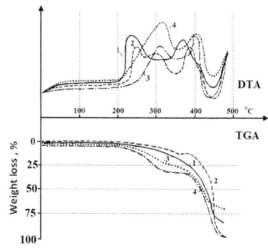

FIGURE 13.1 Curve of differential thermal and thermo-gravimetric analysis of investigated caoutchoucs: (1) EPDM, (2) CEPDM-2, (3) CEPDM-4, and (4) CEPDM-16.

We can see from represented data of DTA, oxidation of investigated polymers depending on extent of their chlorination begins in interval of temperatures from 210 to 240 °C, that is, polymers possess high thermostability. So, for example, the beginning of oxidation is observed in EPDM and CEPDM-19, (curves 1 and 4) under 210 °C; in CEPDM-2, under 230 °C (curve 2); and in CEPDM-4, under 240 °C (curve 3).

Effective destruction of polymer chains of all types of investigated polymers begins in the field of temperatures 410–420 °C.

Pyrolysis of polymers is observed under 480 °C, when the further increase in temperature takes place, polymers are carbonized.

Curves of DTA for EPDM and CEPDM have the same character. Each curve has two maximums, which, depending on type of polymers, are stated in direct temperature fields. So, for EPDM (curve 1), the first maximum is observed under temperature 230 °C, the second—under 380 °C. It shows that macromolecule of EPDM has two distinguished by its capability to oxidation of structure; apparently, they are structural blocks of polyethylene and propylene. Gradual removal of maximum in the field of high temperatures occurs with increase of content of chlorine in polymer. So, little removal in CEPDM-2 (curve 2) is observed, that is, the first maximum is under 250 °C, the second—under 390 °C. The maximums were observed under similar temperatures in CEPDM-4 and CEPDM-16 (curves 3 and 4): the first under 315 °C and second under 405 °C. It is necessary to note that curve of DTA, depending on type of polymers, are different angles of inclination, characterizing pyrolysis rate. So, curves 1 and 2, characterizing the behavior of EPDM and CEPDM-2, which have similar inclination angle, which can consider approximate to 90 °C, and CEPDM-4 and CEPDM-16 have inclination angle, which are similar and approximate to 45 °C. The proximity of temperatures under which we can observe maximum on curve of DTA, and also the proximity of inclination angles of these curves show us the proximity of structural parameters of polymers macromolecules. Here, it is necessary to pay attention to proximity of maximums and inclination angles of EPDM and CEPDM-2 (curves 1 and 2), on the curve 2, characterizing thermal changes CEPDM-2, third maximum in the temperature interval 300–310 °C appears, which shows us formation of new structure in consequence of chlorine modification. We can suggest that this changed fragment of ethylidene norbornene:

The above-mentioned assumption confirms data about investigation of structures of these polymers by infrared spectroscopy method. When analyzing the data obtained in the process of conducting of thermogravimetric analysis (TGA), we can see that the method is less sensitive to structural changes of investigated polymers, however, on thermostability of polymer. So, weight loss of investigated polymers on 10% occurs in EPDM under 315 °C; in CEPDM-2, under 350 °C; in CEPDM-4, under 277 °C; and in CEPDM-16, under 250 °C. This shows us that thermostability of CEPDM-2 is more than that of all other types of investigated ethylene–propylene caoutchoucs. Lower stability of CEPDM-4 and CEPDM-16 is explained by processing them under heating of dehydrochlorination and oxidation reaction.

13.4 CONCLUSION

So, carried-out investigation of structure of chlorinated ethylene–propylene–diene caoutchoucs (CEPDM) showed that in the process of obtaining chlorinated EPDM with the content of 2% (mole) chlorine, adjoining of halogen occurs on fragment of ENB, in consequence of which, new double linkage is formed, in relation to which, chlorine is in α-position. The main chain of macromolecule does not change in the process of chlorination.

When obtaining CEPDM-4 and CEPDM-16, chlorine adjoins to fragment of ENB and to the main chain of polymer, where, in consequence of dehydrochlorination reaction, additional unsaturation is formed as double linkage.

Data obtained when investigating polymers by DTA and TGA method confirm hypothesis that when we added in ethylene–propylene–diene caoutchouc of 2% (mole) chlorine, the halogen adjoins to fragment of ENB, did not touch the main chain when adding chlorine in amount of 4 and 16 mole, the last one adjoins to fragment of ENB and to the main chain of polymer, where, in consequence of dehydrochlorination reaction, additional unsaturation is formed.

Destructions of macromolecular chains of EPDM in consequence of chlorination and dehydrochlorination reaction, which lay in base of haloid modification, apparently, are not observed, because chlorine containing ethylene–propylene–diene caoutchoucs have high thermostability.

Thus, carried out structural and thermophysical investigation of chlorinated ethylene–propylene–diene caoutchoucs (CEPDM) allows to estimate macromolecular structures of ethylene–propylene–diene caoutchoucs, which is formed in the process of haloid modification and possible changes of these structures under thermal condition.

KEYWORDS

- **halide modification**
- **caoutchouc**
- **mechanical chemistry**
- **chlorine-containing ethylene–propylene–diene caoutchouc (CEP-DM)**
- **rubber, elastomer.**

REFERENCES

1. *Sadhan, K. De.; White, J. R. Rubber Technologist's Handbook*; Rapra Technology Limited: UK, 2001; p 61–64.
2. *Brennan, P. Y.* First Look at Terpolymer Rubber Syntesis. *Chem. Eng.* **1965**, *72* (*14*), 94–96. *Sutton, M. S.* Blends of Royane with Other Polymers. *Rubber World* **1964**, *149* (*5*) 62–68.
3. Livanova, N. M.; Popov, A. A.; Karpova, S. G.; Shershnev, V. A.; Ivashkin, V. B.; *Vysokomolek. Soed* **2002**, *44* (*1*), 71–77
4. Karpeles, R.; Grossi, A. V. EPDM Rubber Technology. In *Handbook of Elastomers*, 2nd ed.; Bhowmick, A. K., Stephens, H. L., Eds.; Marcel Decker, Inc.: New York, 2001; pp 845–876.
5. John, A. Vander Laan R. R. Ethylene Propylene Rubbers. In *The Vanderbilt Rubber Handbook*; 13th ed.; R. T. Vanderbilt Co., Inc.: Norwalk, 1990; pp 123–148.
6. Ver Strate, G. Ethylene Propylene Elastomers. In *Encyclopedia of Polymer Science and Engineering*; Springer, 1986; Vol. 6, pp 522–564.
7. Ronkin, G. M., Andriasyan, Yu. O. Investigation of Ethylene-Propylene-Diene Co-Polymers Chlorination Process and Properties of Obtained Materials. *Promyshlennost SK* **1981**, *6*, 8–11.

CHAPTER 14

SIMULATION OF CO-POLYMERIZATION PROCESSES

T. A. MIKHAILOVA[1], E. N. MIFTAKHOV[2], and S. A. MUSTAFINA[1]

[1]Sterlitamak Branch of Bashkir State University, Deparment of Mathematical Modelling, Republic of Bashkortostan, Sterlitamak City, Lenin Avenue 37, 453103 Russia, E-mail: Mustafina_SA@mail.ru, Tel.: +7 917 404 45 58 (Mustafina S.A.), T.A.Mihailova@yandex.ru, Tel.: +7 961 047 97 99 (Mikhailova T.A.)
[2]Ufa State Aviation Technical University, Ishimbay Branch, Deparment of Physics and Mathematics, Republic of Bashkortostan, Ishimbay City, Gubkina Str. 15, 453200 Russia, Tel.:+7 917 451 88 10

CONTENTS

ABSTRACT

Statistical approach to the description of mathematical model of butadiene–styrene copolymerization process has been considered in the paper. The algorithm, which is based on the Monte Carlo method, has been described. The mathematical model allows predicting molecular characteristics depending on conversion and carrying out the calculation of copolymer's molecular weight distribution.

14.1 INTRODUCTION

At the present time, the range of synthetic rubbers, which are made at the domestic industrial enterprises, are rather wide. Rubbers that represent a product of radical copolymerization of butadiene with styrene in an emulsion are the most widespread. The combination of monomers in a chain of successive reactions of connection with each other influences qualitatively important properties of rubbers. But their production represents difficult technological process, which studying becomes simpler at building of mathematical model. Simulation allows one not only to predict the properties of a product but also optimize the production process.[1]

14.2 EXPERIMENTAL PART

In describing the mathematical model of the copolymerization processes, there are two approaches: kinetic and statistical. Kinetic approach is classical in solving problems of chemical kinetics and has successfully established itself not only in the study of physical and chemical phenomena but also in the optimization of technological processes in the chemical industry. This method for the simulation of polymerization processes involves composing and numerical solution of kinetic equations for the concentrations of all types of particles involved in the process. These equations are derived from the conditions of the material balance for each component reactions involving the law of mass action, which determines the rate of formation and disappearance of this component.[2]

As the number of monomer molecules can reach tens of thousands, the kinetic scheme, which includes all the main reaction occurring in the system, is reduced to an almost infinite system of differential equations. Directly to solve such a system is not possible, so at writing the model equations, it is converted to moments of the molecular weight distribution.[3] Using such a simplification of the system allows us to calculate the average molecular characteristics of the resulting product,[4] as well as under conditions of uncertainty of the rate constants of some elementary reactions successfully formulate and solve the inverse problem.[3] However, to get the picture changes, the molecular weight distribution and studying the composition of the product obtained in this approach is no longer possible.

When using statistical approach, the polymer chain's simulation is carried out by means of random variables. The basis of this approach is the simulation of each polymer molecule's propagation. Each link of the growing polymeric chain is considered as concrete random process of the conditional motion along the polymer molecule, and the probability of random processes realization is considered to an equal portion of the molecules corresponding to it among all others in a reaction system. As the method works at the level of particles, it gives the chance to accumulate information on structure and length of the formed polymer chains and to receive molecular characteristics of polymerization's product at any moment.

For realization a statistical approach to the simulation of copolymerization process we apply the method proposed in 1977 by the American physicist Gillespie. Describe the algorithm of the following sequence of steps.[5]

1. For building a model of butadiene–styrene copolymerization, let us assume that the reactivity of the active center at the end of a growing chain is determined by the nature of the terminal unit. Then the kinetic scheme of butadiene–styrene copolymerization can be described by the following steps:

    initiator decomposition

    $$I \xrightarrow{k_d} 2R,$$

    initiation of active centers

    $$R + M^\beta \xrightarrow{k_{i\beta}} P^\beta_{A(\beta),B(\beta)},$$

    chain propagation

    $$P^\alpha_{n,m} + M^\beta \xrightarrow{k_{p\alpha\beta}} P^\beta_{n+A(\beta),m+B(\beta)},$$

    chain transfer

    $$P^\alpha_{n,m} + S \xrightarrow{k_{re\mu a}} Q_{n,m} + S_0,$$

    chain termination by disproportionation

    $$P^\alpha_{n,m} + P^\beta_{r,q} \xrightarrow{k_{da\beta}} Q_{n,m} + Q_{r,q},$$

chain termination by recombination

$$P^{\alpha}_{n,m} + P^{\beta}_{r,q} \xrightarrow{k_{r\alpha\beta}} Q_{n+r,m+q},$$

where α, $\beta = \overline{1,2}$; M^1, M^2 are the monomers of the first and second type; $P_{n,m}$ and $Q_{n,m}$ are the active and inactive polymer chains with length $m+n$, comprising m units of the M^1 monomer and n units of the M^2 monomer, respectively; k_i, k_p, k_{reg}, k_d, k_r are the reaction rate constants of initiation, growth, chain propagation, disproportionation, and recombination elementary stages, respectively; $A(\alpha) = \{1$, if $\beta = 1$; else $0\}$; $A(\beta) = \{1$, if $\beta = 2$; else $0\}$.[2]

2. Transform the experimental rate constant of elementary reactions to stochastic rate constants according to the following equations:
    $\tilde{k} = k$ for first order reactions;
    $\tilde{k} = \dfrac{k}{V \cdot N_A}$ for bimolecular reactions between different species (V is the reaction volume, N_A is the Avogadro's number).

3. Then calculate the reaction rate for every reaction according to the equation:

$$R_i = \tilde{k}_i \cdot X_A \cdot X_B, \tag{14.1}$$

where $\tilde{k}_i$ is the rate constant of the i th reaction in which reagents A and B participate; X_A, X_B are the numbers of reagent's molecules.
The total reaction rate is then calculated as the summation of the individual reaction rates:

$$R_{sum} = R_1 + R_2 + \dots + R_n, \tag{14.2}$$

where n is the number of elementary reactions forming kinetic scheme of the copolymerization process.

4. Then the probability of any reaction taking place at a given time is calculated by the following equation:

$$p_i = \frac{R_i}{R_{sum}}, \quad i = 1..n \tag{14.3}$$

It is apparent that $p_1 + p_2 + \dots + p_n = 1$.

5. Generate a random number r, uniformly distributed between 0 and 1, and pick up value k such that the inequality takes place:

$$\sum_{i=1}^{k-1} p_i < r < \sum_{i=1}^{k} p_i. \tag{14.4}$$

Consequently, reaction under an index k has to result from an imitating choice.

6. Continuing reasoning similarly, we will build all scheme of carrying out reaction.

14.3   RESULTS AND DISCUSSION

On the basis of the developed algorithm, the program complex was created in the integrated development programming environment Visual Studio in the C# language. It allows to carry out calculation of process of emulsion-type butadiene–styrene copolymerization in the reactor.

To illustrate the operation of the program, complex computational experiment has been performed to study the process brought to the 70% monomer conversion. The following compounding of process was used: polymerizer volume—10.8 m³, the total weight of the monomer mixture—3 t., dosage of butadiene—70 weight parts, styrene—30 weight parts, initiator (pinane hydroperoxide)—0.048 weight parts, chain–transfer agent (tertiary dodecyl mercaptan)—0.28 weight parts. The obtained number average, weight average molecular masses, and intrinsic viscosity showed a satisfactory agreement with the results from the kinetic model of the process (Figs. 14.1–14.3).

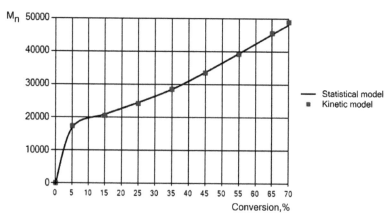

FIGURE 14.1 Dependence of number average molecular mass values of copolymer on conversion.

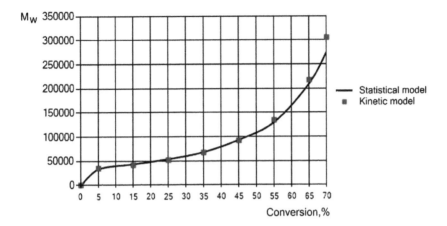

FIGURE 14.2 Dependence of weight average molecular mass values of copolymer on conversion.

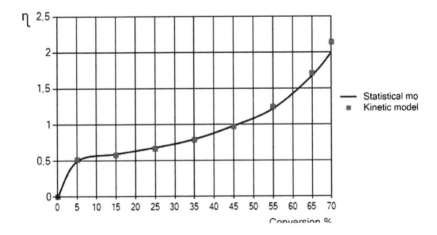

FIGURE 14.3 Dependence of intrinsic viscosity values of copolymer on conversion.

For an assessment of quality of the received product, the absolute impor-tance is made by molecular weight distribution as at any sample of polymer, there are macromolecules of the different sizes. It shows a ratio of quantities of macromolecules of different molecular masses. Figure 14.4 shows the plot of copolymer's molecular weight distribution at 70% conversion, which represents the dependence between molecular weight and polymer mass fractions. Molecu-lar weight distribution is narrow, that is, a copolymer prevails the fraction with a certain molecular weight and the part of fractions with smaller or larger values of molecular weights is much lower. In this example, there is predominance of fractions having a molecular weight about $5 \times 10^4 - 6 \times 10^4$. The peak of a curve corresponds to value of number average molecular mass of copolymer.

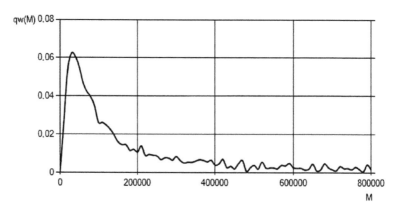

FIGURE 14.4   Molecular weight distribution of copolymer at 70% conversion.

14.4   CONCLUSION

The statistical approach considered in this paper allows studying properties of the product of emulsion-type butadiene–styrene copolymerization. As simulation of each macromolecule's growth and tracing of the processes happening to it is the basis of this approach, it allows accumulating information on composition and length of the formed copolymer chains. On the basis of this information, carrying out the calculation of molecular weight distribution and prediction dependences of change of the main copolymer's molecular characteristics on conversion is possible at any moment.

KEYWORDS

- copolymerization
- butadiene
- styrene
- statistical approach
- Monte Carlo method

REFERENCES

Podval'nyi, S. L. *Simulation of Industrial Polymerization Processes*; Khimia (Chemistry, in Rus.) Publishing House: Moscow, p 1979–256.

Mustafina, S. A.; Miftakhov, E. N.; Mikhailova, T. A. Solving the direct problem of butadiene-styrene copolymerization. *Int. J. Chem. Sci.* **2014**, *12* (2), 564–572.

Miftakhov, E. N.; Nasyrov, I. Sh.; Mustafina, S. A. Simulation of Butadiene–Styrene Emulsion Copolymerization Process *Bashkir Chem. J. Ufa* **2011**, *18* (*1*) 21–24.

Miftakhov, E. N.; Mustafina, S. A.; Nasyrov, I. Sh. *Simulation of Butadiene–Styrene Free Radical Copolymerization Process in the Production of Emulsion-Type Rubbers*; Herald KTU: Kazan, 2012; Vol. 24, p 78–82.

Gillespie, D. Exact Stochastic Simulation of Coupled Chemical Reactions. *J. Phys. Chem.–* **1977**, *81* (*25*), 2340–2361.

CHAPTER 15

EFFECTS OF LOWMOLECULAR ELECTROLYTES ON PARTICLE SIZE OF POLYELECTROLYTE–SURFACTANT COMPLEXES

M. S. BABAEV and S. V. KOLESOV

Institute of Organic Chemistry, Ufa Scientific Centre of RAS, pr. Oktyabrya 71, Ufa, Russia, E-mail: b.marat.c@mail.ru

CONTENTS

ABSTRACT

By turbidity measurement and laser light scattering method the effects of the inclusion of lowmolecular electrolytes (NaCl, HClandNaOH) on the process of formation and particle size of polyelectrolyte–surfactant complexes based on poly-N, diallyl-N, N-dimethylammonium chloride, and ionogenic surfactant—sodium dodecyl sulfate have been studied.It has been shown that the presence of lowmolecular electrolytes exercises a significant influence upon the limiting values ofthe phase of polyelectrolyte–surfactant complexes' emergence and their particle sizes.

15.1   INTRODUCTION

At this point, a great number of unique researches and surveys devoted to amphiphilicpolyelectrolyte–surfactant complexes(PSC), which are the products of the interreaction of oppositely charged polyelectrolytes and surfactants(SF), have been published.[1,2] These complexes are generatedas a result of condensation of amphiphilicions of SFon the oppositely charged macroionsunder the SF concentrations, which are significantly lower than critical micelle concentration (CMC).[3,4,5]An intense interest in these systems is due to their unusual properties and practical applications opportunities in the fields of ecology, medicine, and pharmaceutics. A feature of these complexes is the presence of intramolecular micellar phase with significant solubilizing capacity in relation to various organic compounds. Among the ionic polymers,which appear as a basis of PSCthere are such extremely promising polymers as water-soluble polymers based on N,N-diallyl-N,N-dimethylammonium chloride, most of them are nontoxic and have antibacterial properties.[6,7]

    The significant contribution to the process of binding of polyelectrolyte–SF has been made by electrostatic interaction of SFions and ionized polyelectrolyte groups.[8]The introduction of lowmolecular electrolytes reduces the binding constants ofpolyelectrolyte–surfactants (SF) and increasesthe cooperativeof the processes. Currently, the data concerning the effect of the inclusion of lowmolecular electrolytes on PSC particle sizeis not enough. The study of these questions is topicaland significant for the development of methods for the preparation and management of characteristics and properties of PSC, both scientifically andfor all practical purposes.The objective of this research is to determine the effects of the presence of lowmolecular electrolytes' additives on the formation and particle size of PSCbased on poly-N,N-diallyl-N,N-dimethylammonium chloride (PDADMAC) with a micelle-forming ionogenicSF–sodiumdodecyl sulfate (SDS).

15.2 EXPERIMENTAL PART

As a cationic polyelectrolyte (PE) ofPDADMAC,the sample of molecular mas-sof 131,600 has been used. The molecular mass () has been determined on the basis of viscosimetric measurements according to the equation of Mark–Kuhn–Houwink: $[h] = KM_\eta^a$, where $[h]$ stands for the characteristic viscosity (cm^3/gm), K and a are constants (according to Ref. 9,$K = 4.83 \times 10^{-3}$ and $a = 0.88$, the solvent−1 mol/L of NaCl,$T = 30 \pm 0.01°C$). Sodium dodecyl sulfate (SDS) of the company "Aldrich" has been further rectifiedthree times by recrystalliza-tion from ethanol.

The conditions of phase separation of mixtures of PE and SFin solution have been studied by turbidity titration. A total of 0.003 mol/L of SDS's aqueous solution have been addedby portionswhile stirring to the prefabricated in water solution of 0.02 mol/L of PDADMAC. After the addition of each new portion of the SFsolution for5 min (the time, exceeding the limit of which does not lead to a noticeable change in the optical density of the solution of PSC), the optical density of D dispersionhas been measured on the UV-spectrometer Shimadzu UV–VIS–NIR 3100 at the wavelength of 500 nm and atthe absorbing layer thickness of 1 cm. PDADMACaqueous solutions have beenprepared with con-stant stirringthe day before turbidity titration. The titration's result is presented in terms of the dependenceof the optical density of the solution Don the molar ratio of components $z = [SF]/[PE]$. The composition of the reaction mixture has been taken for $z_1$and beyond the limit ofthis composition in the systems along with the aggregative resistant colloidal particles, the insoluble precipitate of PSCappears by storing them at room temperature within 7 days.On the titration curve, the rangeof the reaction mixture's composition of$z > z_1$complies with the sharp rates of increased optical density.

To determine the particles' size of PSC, the particlesize analyzer SALD-7101 Shimadzu has been applied;the operating principle of which is based on a static laser light scattering at the wavelength of$\lambda = 375$ nm. Calculations of the particles' size were carried out automatically with the help ofsoftware program complex,that is,WingSald II. The result of calculations is represented in terms of the dependence of the particles' diameter$d$ (nm) on their numeric quotient$q$ (%). Thecalibration of instruments was conducted using a special calibration sample MBP 1 in such a way that the determined values of particles' sizeof MBP 1–10 differentiate by not more than 3% from passport data.

15.3   RESULTS AND DISCUSSION

Bindingof oppositely charged SF ions by means of polyelectrolytes occurs as a result of electrostatic interactions of charged groups and due to the hydrophobic interactions ofSF alkyl radicals with hydrophobic units of the polymer chain.[10,11] The electrostatic binding of SDS bypolyelectrolyte lies in the formation of salt compounds between positively charged quaternary nitrogenatom, that each componentof the polycation consists of, andsulfonate group of SDS. Ion exchange reaction can be represented by the scheme:

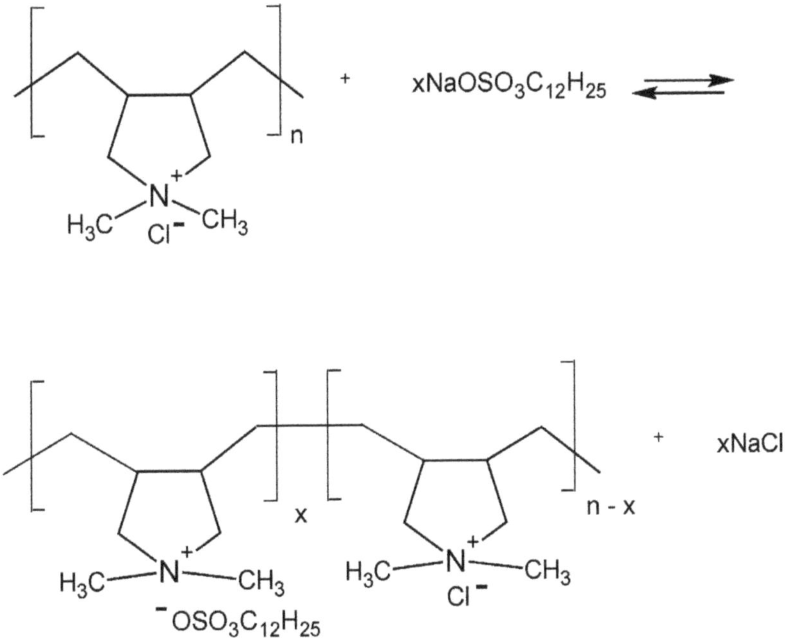

The introduction of lowmolecular electrolyte in the chemistry of polyelectrolytes is a well-known method of the regulation of conformational conditions of the polyelectrolytes.The presence of lowmolecular electrolyte ions leads to the charge compensation ofpolyion, to the reduction of the electrostatic repulsion between them, and as the consequence, to the extent of polyion'sunfolding, depending on theconcentration of lowmolecular electrolyte. Probably polyelectrolyte–surfactant bindingwill occur more proportionallyin a salt-free solution than in the case of suppression of the polyelectrolyte's effect by means of a lowmolecular salt that eventually will affect the particle size ofPSC and these particles' persistent quality.

The presence of chloride of sodium in the solution has a considerable impact on thelyophilizing ability of polyelectrolyte. The results of turbidity titration of PDADMACsolutionsby SDS solution under the different amounts of NaClare shown in Figure 15.1. The figure demonstrates thatthe increase in concentration of sodium chloride in a solution leads to theregular titration curves shifting toward the lower values of z, that is,to thedecline in the lyophilizing ability of polyelectrolyte. The values of z_1 for the systems of PSC, obtainedunder the lack of salt and at 0.01, 0.05, 0.1, and 0.25 mol/L ofNaCl solution, are 0.78, 0.7, 0.6, 0.55, and 0.53,respectively. The decrease in thelyophilizing ability of polyelectrolyte along with the increase inlowmolecular electrolyte contenthas probably been initiated due to the strengthening of cooperative interactionsof polyelectrolyte–SF, as well as thesuppression of electrostatic interactions between the polyelectrolyte chains at high salt concentrations.

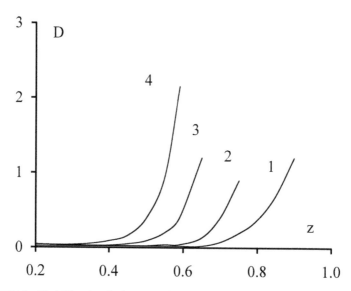

FIGURE 15.1 Turbidity titration'scurve of 0.02 mol/L of the PDADMACsolution by0.003 mol/L of the solution of SDS at different concentrations ofNaCl(mol/L): (1)0,(2) 0.01,(3) 0.05,and (4) 0.25.

The presence of salt at the moment of formation of PSCsubstantiallyaffects the particle size of the complexes based on PDADMAC (Fig. 15.2). Under the lack of salt additives, the average particle size of PSC is 153 nm. During the formationof PSCin 0.01, 0.05, and 0.25 mol/L ofNaCl solution, the size of the formed particles in these complexes comprise on the average 193, 814, and 1484 nm.

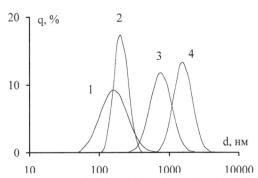

FIGURE 15.2   The distribution of PSC particles due to the size of these particles and on the basis of PDADMACsolutions with different concentrations of NaCl.The concentration of NaCl (mol/L): (1) 0,(2) 0.01,(3) 0.05,and (4) 0.25. The particle sizes are determined at z_1. For illustration purposes, the horizontal axis is represented in a logarithmic scale.

The additives of hydrogen chloride and sodium hydroxide also decreasethe lyophilizing ability of PE. The values of $z_1$for the systems of PSC,formed in 0.1 mol/L ofNaOH and 0.1 mol/L of HCl solutions, equal to 0.70 and 0.62, respectively. NaCl has the effect on lyophilizing ability of PEout of the three tested electrolytes. As we can see (Fig. 15.3), turbidity titration curve of PE bySF solution is in the range of smaller values ofzin the presence ofNaCl additives. This is probably due to the fact that NaCl contains counterions of both SF andpolyelectrolyte, and this increases their cooperative bindingmore than other polyelectrolytes. HClhas somewhat lesser effect, the anions of which are counterionsonly ofpolyelectrolyte. NaOH has the least effect on the lyophilizing ability of polyelectrolyte,cations of which arecounterions only of SF.

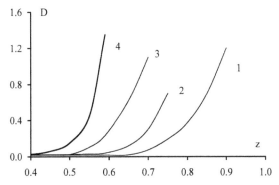

FIGURE 15.3   Turbidity titration's curve of 0.02 mol/L of PDADMACsolutionby the0.003 mol/L of SDS solution: (1) aqueous solutions of the components,(2) in 0.1 mol/LofNaOH solution,(3) in 0.1 mol/L of HCl solution,and (4) in 0.1 mol/L of NaCl solution.

Additives of hydrogen chloride and sodium hydroxidealso lead to anincreasein the generatedPSC particles' sizes. In these conditions and in the presence of NaOH, the particles' sizes have the largestvalues (Fig. 15.4, curve 1). High particle sizes of PSCin the presence ofNaOH in comparison with other lowmolecular electrolytes can be explained by following way. Sinceout of the three electrolytes, NaOHhas the least effect on the lyophilizing ability of polyelectrolyte, formation of the dispersionof PSC in the presence of NaOH occurs at thehigh content of SFin the system. The larger amounts of SF and the larger amounts of additives of lowmolecular electrolyte provided, then, probably, the aggregation processes proceedmore intensely, and thisleads to the generation of larger particles.

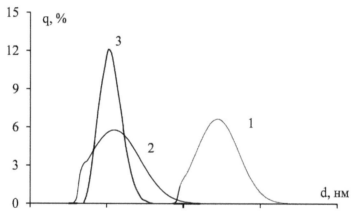

FIGURE 15.4 The distribution of PSC particlesdue to the size on the basis of PDADMAC,generated in 0.1 mol/L of lowmolecular electrolytes' solution: (1) NaOH,(2) HCl,and (3) NaCl. For illustration purposes, the horizontal axis is represented in a logarithmic scale. The average particle size is equal to 31.2 (1), 1.4 (2),and 1.3 (3) microns. z_1 values are equal to: 0.70 (1), 0.62 (2),and 0.55 (3).

15.4 CONCLUSION

Thus, the presence of lowmolecular electrolyte exercises a significant effect on the lyophilizing ability of polyelectrolyte and particles' sizes of polyelectrolyte–surfactant complexes. Those ions of lowmolecular electrolytes, which are counterions of the both components of the reaction mixture surfactant–polyelectrolyte.In this respect,counterions that are components of the polyelectrolyte have the greatest effectson the lyophilizing ability of polyelectrolyte, than those counterions, which form the parts of the surfactant.

KEYWORDS

- polyelectrolytes
- self-assemble
- polyelectrolyte–surfactant complexes
- poly-N,N-diallyl-N,N-dimethylammonium chloride, ionic force
- microparticles

REFERENCES

1. Kogej, K. Association and Structure Formation in Oppositely Charged PolyelectrolyteSurfactant Mixtures. *Adv. Colloid Interface Sci.* **2010,** *158(1–2),*68–83.
2. Bain, C. D.; Claesson, P. M.; Langevin, D.; Meszaros, R.; Nylander, T.; Stubenrauch, C.; Titmuss, S.; von Klitzing, R. Complexes of Surfactants with Oppositely Charged Polymers at Surfaces and in Bulk. *Adv. Colloid Interface Sci.* **2010,***155(1–2),*32–49.
3. Ibragimova, Z. K.; Kasaikin, V. A.; Zezin, A. B.; Kabanov, V. A. Non-Stoichiometric Polyelectrolyte Complexes of PolyacrylicAcid and Cationic Surfactants. *Polym. Sci.* **1986,** *28(8),*1826.
4. Kasaikin, V. A.; Efremov, V. A.; Zakharova, J. A.; Zezin, A. B.; Kabanov, V. A. Formation of Intramolecular MicellarPhase as a Necessary Condition for Binding of AmphiphilicIons with Oppositely Charged Polyelectrolytes. *Rep. Russ. Acad. Sci.* **1997,***354,*126.
5. Kasaikin, V. A.; Litmanovich, E. A.; Zezin, A. B.; Kabanov, V. A. *Rep. Russ. Acad. Sci.* **1999,** *367(3),*359.
6. Vorob'eva, A. I.; Sultanova, G. R.; Bulgakov, A. K.; Zaynchkovsky, V. I.; Kolesov, S. V. Synthesis and Biological Properties of the Copolymer Based on N,N-diallyl-N,N-DimethylammoniumChloride. *Chem. Pharmaceutic. J.* **2012,***46(11),*21–23.
7. Vlasov, P. S.; Chernyi, S. N.; Domnina, N. S. FunctionalyzedPolyampholytes on the Basis of Copolymers of N,N-dialkyl-N,N-dimethylammoniumChloride and Maleic Acid. *Russ. J. Gen. Chem.* **2010,***80(7),*1314–1319.
8. Ruso, J. M.; Sarmiento, F. The Interaction between TrimethylAmmonium Bromides with Poly(L-aspartate): AThermodynamics Study. *Colloid Polym. Sci.* **2000,***278,*800–804.
9. Wandrey, C.; Hernández-Barajas, J.; Hunkeler, D. Radical Polymerisation Polyelectrolytes Advances in Polymer Science. **1999,**145,*123 (10)*Kabanov V. A. *Pure Appl. Chem.* **2004,***76(9),* 1659.
10. Kabanov, V. A. From Synthetic Polyelectrolytes to Polymer-Subunit Vaccines. *Pure Appl. Chem.* **2004,***76 (9),* 1659.
11. Shilova, S. V.; Tret'yakova, A. Ya.; Barabanov, V. P. Association of QuaternizedPoly(4-Vinylpyridine) and Sodium Dodecyl Sulfate in Water-Alcohol Solvents. *Polym. Sci. Series A.* **2010,***52 (12),* 1283.

CHAPTER 16

CHAIN PROPAGATION IN METHYL METHACRYLATE AND STYRENE POLYMERIZATION

S. V. KOLESOV, N. N. SIGAEVA, I. I. NASIBULLIN, and A. K. FRIESEN

Institute of Organic Chemistry, Ufa Scientific Center, Russian Academy of Sciences, pr. Oktyabrya 71, Ufa 450054, Russia, E-mail: gip@anrb.ru

CONTENTS

ABSTRACT

Within radically initiated MMA and styrene polymerization systems, metallocenes are shown to be actively participating in formation of two kinds of chain propagation active centers—free radicals and living coordination active centers, and the polymerization takes place as radical coordination one. Relative contribution of each kind of centers into the observed kinetic peculiarities of the polymerization depends much on metallocene kind and the process conditions.

16.1 INTRODUCTION

Radical polymerization of vinyl monomers—methyl methacrylate and styrene—in the presence of some metallocenes together with increasing the process rate is characterized by some kinetic peculiarities. Basic of them are already mentioned in scientific literature, that is, conversion change of average molecular weight and MMD of polymers[1–5]; ability of polymers derived in the presence of metallocenes to initiate living polymerization of new monomer portions[2]; prolonged until complete consumption of monomer, and post-effect of polymerization after short-term UV irradiation of the polymerization system.[3] Furthermore, advanced yield of stereoregular polymethyl methacrylate (PMMA) is observed as compared with free radical polymerization of monomer under the same temperature.[6] Consideration of these peculiarities suggested that in the course of polymerization, there appeared coexistence of several, namely, two kinds of active centers for macromolecule growth. Availability of two kinds of active centers under styrene polymerization in the presence of ferrocenes, titanocene, and zirconocene dichloride has been proved by solution of the inverse MMD problem.[5]

16.2 EXPERIMENTAL PART

Methyl methacrylate (MMA) and styrene (St) were washed with 10% aqueous NaOH solution, dried over $CaCl_2$, and distilled twice under reduced pressure. For polymerization, used fraction of MMA with $T = 39\ °C$ ($p = 100$ mmHg.), and $n_D^{20} = 1.4130$, $d_4^{20} = 0.936$ g/mL; for St, fraction with $T = 66\ °C$ ($p = 40$ mmHg.), $n_D^{20} = 1.5449$, and $d_4^{20} = 0.903$ g/mL. The purity of the monomer was controlled by ^{1}H NMR and ^{13}C-spectroscopy.

Benzoyl peroxide (BP) and azobisisobutyronitrile (AIBN), recrystallized twice from methanol and dried at room temperature in vacuum to constant weight.

Ferrocene, acetyl ferrocene, titanocene dichloride, and zirconocene dichloride (99.9%, Aldrich, USA) $(C_5Me_5)_2ZrCl_2$ и $(C_5Me_5)_2Fe$) were synthesized in the Institute of Applied Physics of RAS (Nizhny Novgorod).

The kinetics of polymerization is studied by the dilatometric method.[7] For bulk polymerization, the reaction mixture (monomer, initiating system) was poured into the vial and degassed triplicate cycles of freezing–thawing to a residual pressure of 1.3 Pa. The vial was then sealed, placed in a thermostat to maintain the temperature with precision ±0.1 °C, and maintained at the selected temperature until the desired conversion.

Computational Details. The quantum chemical calculations were performed with the program Priroda-06.[8–10] The generalized gradient approximation for the exchange-correlation functional by Perdew, Burke, and Ernzerhof was employed.[11] The electronic configurations of the molecular systems were described by the orbital basis sets of contracted Gaussian-type functions of size (5s1p)/[3s1p] for H, (11s6p2d)/[6s3p2d] for C and O, and (17s13p8d)/[12s9p4d] for Fe, which were used in combination with the density-fitting basis sets of uncontracted Gaussian-type functions of size (5s2p) for H, (10s3p3d1f) for C and O, and (18s6p6d5f5g) for Fe. All structures were optimized without symmetry restriction. Analytical second derivatives were used to determine the type of a stationary point. The enthalpies of reactions ($\Delta_r H°$) and the activation enthalpies ($H°_{act}$) were determined at 298 K (their values are expressed in kJ/mol). The particles $CH_3–CH_2–C^•(CH_3)(COOCH_3)$ and $CH_3–CH_2–CH^•Ph$ were adopted as models of poly(methyl methacrylate) and polystyrene growing radicals (further $R^•$).

16.3   RESULTS AND DISCUSSION

Analysis of quantum chemical simulation results of possible interactions within monomer – metallocene – radical initiator – free radical systems suggested that under usual fashion of radical polymerization of vinyl monomers together with free radicals, the active centers of coordination polymerization could be formed[12,13] different by their structure for polymerization of both nonpolar (styrene) and polar (methyl methacrylate) monomers.

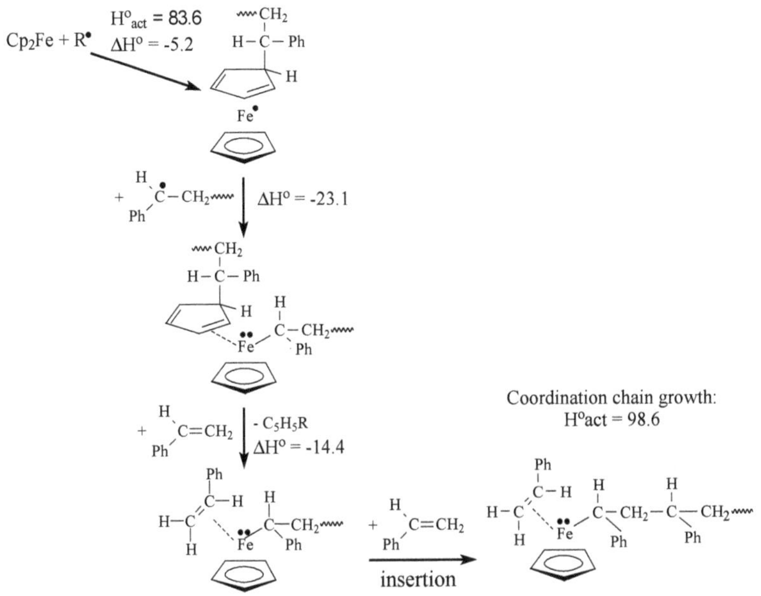

The scheme of active center formation and functioning in the coordination polymerization of styrene in the presence of ferrocene.

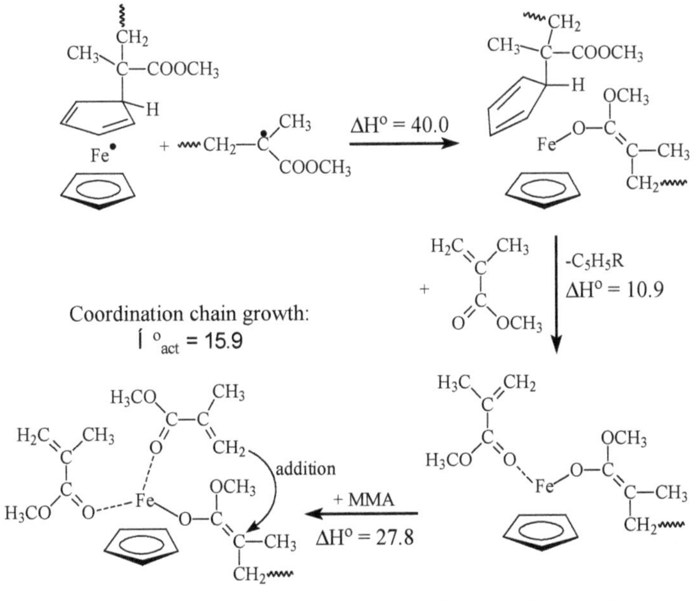

The scheme of active center formation and functioning in the coordination polymerization of MMA in the presence of ferrocene.

It is the availability of the coordination active center and chain propagation on the coordination mechanism that causes kinetic characteristics of living polymerization—prolonged post-effect of polymerization, conversion change of average molecular weights and MMD of polymers as well as effect of stereoregularity during polymerization.

The existence of two kinds of active center is also observed in kinetic peculiarities of monomer consumption in the presence of different metallocenes (ferrocene—Cp_2Fe, decamethylferrocene—$(C_5Me_5)_2Fe$, acetyl ferrocene—AcCp-FeCp, titanocenedichloride—Cp_2TiCl_2, zirconocene dichloride—Cp_2ZrCl_2, and decamethyl zirconocene dichloride—$(C_5Me_5)_2ZrCl_2$) depending on the ratio of metallocene/initiator, the process temperature, and kind of the radical initiator. Kinetic curve character revealing gel effect due to diffusion limitation of quadratic macroradicals growth termination, qualitatively testifies the availability of free radical component in the general polymerization process. The contribution of free radical channel to the polymerization process can be evaluated by the severity of gel effect on the polymerization kinetic curves in the systems with metallocene and radical initiator depending on the experiment conditions. Kinetic data analysis has revealed that over differential curves of MMA polymerization in the presence of metallocenes, if initiated by BP within temperature range 40–80 °C, there is some gel-effect severity decrease and the maximum of its manifestation is shifted toward the range of lower polymerization times, irrespective of the temperature. Gel effect for Cp_2Fe–BP system at 40–50 °C is flattening, and at 55–60 °C, is absolutely disappearing (Fig. 16.1).

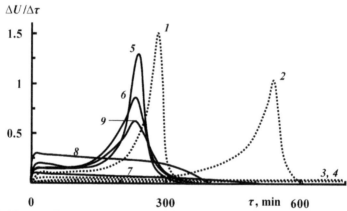

FIGURE 16.1 Differential curves of MMA polymerization in the presence of BP (1–4) and BP–Cp_2Fe (5–9). [BP] = [metallocene] = 1.0 (mmol/L); polymerization temperature, °C. 40 (1, 5), 50 (2, 6), 55 (7), 60 (3, 8), and 70 (4, 9). Dashed curves—BP only initiated polymerization.

Gel-effect degeneration occurs at polymerization temperature of 40 °C in case of using $(C_5Me_5)_2Fe$ and Cp_2TiCl_2 (Figs. 16.2 and 16.3). Gel-effect degeneration in systems with Cp_2TiCl_2 was also mentioned in paper.[4] Under further temperature increase, the gel effect is again evident, and the more the temperature, the more the manifestation thereof. The share of macroradicals participating in the process is determined by rate of their formation in the initiation reactions and rates of consumption for the account of quadratic termination and jumping into coordination propagation centers. Balance of these rates under certain conditions of the experiments provides for stationary character of the process. Absolute gel-effect degeneration is likely to demonstrate such a significant decrease of macroradicals share participating in the process that their concentration over the general kinetic curves for the account of the quadratic termination resistance provides no visible self-acceleration of the process.

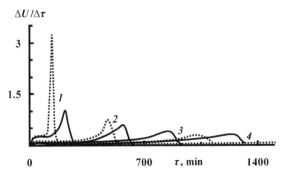

FIGURE 16.2   Differential curves of MMA polymerization in the presence of BP and BP–$(C_5Me_5)_2Fe$. [BP] = [metallocene] = 1.0 (mmol/L); polymerization temperature, °C: 80 (1), 60 (2), 50 (3), and 40 C (4). Dashed curves—as in Figure 16.1.

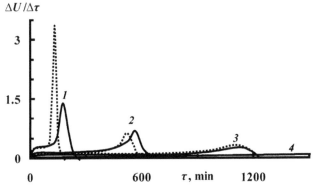

FIGURE 16.3   Differential curves of MMA polymerization in the presence of BP and BP–Cp_2TiCl_2. [BP] = [metallocene] = 1.0 (mmol/L); polymerization temperature, °C: 80 (1), 60 (2), 50 (3), and 40 C (4). Dashed curves—as in Figure 16.1.

Change of relative input of free radical component of polymerization depends on metallocene/BP mole ratio. In the case of Cp_2Fe, the absolute gel-effect degeneration occurs when the equimole composition of the initiator system. Increase of Cp_2Fe/BP mole ratio is enhancing manifestation thereof, which is likely due to metallocene affecting initiator decomposition. In the case of $(C_5Me_5)_2Fe$, the gel-effect manifestation is moving toward longest reaction time, with the metallocene/BP mole ratio increasing to 2. The presence of even small amounts of acetyl ferrocene within the initiator system composition results in significant increase of MMA polymerization process, but when its concentration is reaching 10–20 mmol/L, gel-effect manifestation decreases.

Polymers obtained in the presence of metallocenes and carefully refined from radical initiator residue are able to initiate polymerization of new monomer portion, that is, to act as "polymer metallocene catalysts." Comparison of conversion dependencies of MMA polymerization over such PMMA catalysts obtained at 50–70 °C (Fig. 16.4) shows that "polymer metallocene catalysts" activity obtained through polymerization at 60 °C is higher than that of the polymers synthesized at either 50 or 70 °C. These results are well correlated with Figure 16.1 data showing that it is at 60 °C and equimole Cp_2Fe/BP ratio over polymerization kinetic curve that gel effect is fully degenerated, that is, coordination active centers are formed at most.

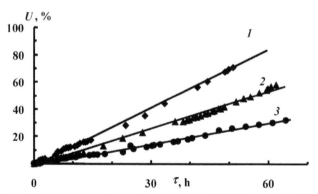

FIGURE 16.4 MMA polymerization in the presence of PMMA-catalyst (3% of monomer weight), obtained using BP–Cp_2Fe initiating system, at, °C: 60 (1), 70 (2), and 50 (3). T_p = 60 °C.

Formation of active centers of polymerization is also affected by some kinds of metallocenes and the process conditions for MMA–AIBN–metallocene polymerization systems. When polymerization took place at 40 °C, there was actually no gel effect when using $(C_5Me_5)_2Fe$, Cp_2Fe, and Cp_2ZrCl_2 (Fig. 16.5).

Small manifestation of gel effect occurred at polymerization temperature of 50 °C for all metallocenes analyzed herein. Kinetic polymerization curves obtained at 60–70 °C testify primary input of free radical active centers to the process. Gel-effect manifestation starts earlier at 60 °C for metallocene–AIBN systems, than at 50 °C, and its manifestation degree increases greatly. However, coordination active centers are formed under these conditions as evidenced by ability of the obtained polymers to act as polymer catalysts (Fig. 16.6).

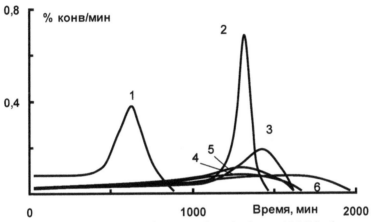

FIGURE 16.5 Differential curves of MMA polymerization at 40 °C in the presence of [AIBN] = [metallocene] = 1.0 (mmol/L). Metallocene: Cp_2TiCl_2 (1), AcCpFeCp (2), without metallocene (3), $(C_5Me_5)_2Fe$ (4), Cp_2Fe (5), and Cp_2ZrCl_2 (6).

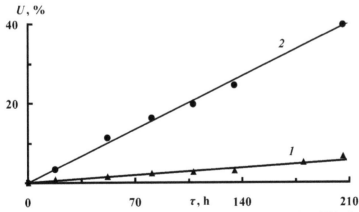

FIGURE 16.6 MMA polymerization in the presence of PMMA-catalyst (3% of monomer weight), obtained at 60 °C in metallocene—AINB initiating system. Metallocene: Cp_2Fe (1), AcCpFeCp (2). T_p = 20 °C.

Summarizing the above results, one can say that under certain conditions of MMA and styrene radical polymerization in the presence of metallocenes, the channel of coordination chain propagation is advancing in such a way that it fully masks manifestation of free radical chain propagation.

16.4 CONCLUSION

Therefore, the obtained experimental data bring us to the conclusion that within radically initiated MMA and styrene polymerization systems, metallocenes are actively participating in formation of two kinds of chain propagation active centers—free radicals and living coordination active centers, and the polymerization takes place as radical coordination one, accordingly. Relative contribution of each kind of centers into the observed kinetic peculiarities of the polymerization much depends on metallocene kind and the process conditions. Under certain conditions, the channel for living (coordination) chain propagation is advancing so much that fully masks manifestation of free radical chain propagation flags.

KEYWORDS

- methyl methacrylate and styrene polymerization
- metallocenes, post-effect of polymerization

REFERENCES

1. Grognes, E. L.; Claverie, J.; Poli, R. *J. Am. Chem. Soc.* **2001**, *123*, 9513.
2. Kotlova, E. S.; Pavlovskaya, M. V.; Grishin, I. D.; Semeikin, O. N.; Ustinyuk, N. A.; Grishin, D. F. *Polym. Sci. Ser. B.* **2011**, *53* (*3*), 456.
3. Sigaeva, N. N.; Zakharova, E. M.; Garifullin, R. N.; Utyasheva, G. V.; Kolesov, S. V. *Polym. Sci. Ser. B.* **2010**, *52* (*3–4*), 214.
4. Grishin, D. F.; Semenycheva, L. L.; Telegina, E. V.; Smirnov, A. V.; Nevodchikov, V. I. Herald of Russian Academy of Sciences. *Seriya Chem.* **2003**, *2*, 482 (in Russian).
5. Sigaeva, N. N.; Kolesov, S. V.; Abdulgalimova, A. U.; Garifullina, R. N.; Prokudina, E. M.; Spivak, S. I.; Budtov, V. P.; Monakov, Yu. B. *Polym. Sci. Ser. A.* **2004**, *40* (*8*), 1305.
6. Puzin, Yu. I.; Prokudina, E. M.; Yumagulova, R. Kh.; Masluhov, R. R.; Kolesov, S. V. The Reports of Russian Academy of Sciences. *Russ. Acad. Sci.* **2002**, *386* (*1*), 69 (in Russian).
7. Gladyshev, G. P.; Popov, V. A. *Radical Polymerization at High Degrees of Conversion*; Nauka: Moscow, 1974; p 244 (in Russian).
8. Laikov, D. N., PRIRODA, Electronic Structure Code, Version 6, 2006.

9. Laikov, D. N. The Development of the Economical Approach to the Calculations of Molecules by Density Functional Theory, its Application for Solving the Complex Chemical Tasks. Ph. D. (Phys–Math) Thesis, M.V. Lomonosov Moscow State University, 2000 (in Russian).

10. Laikov, D. N.; Ustynyuk, Yu. A. PRIRODA-04: A Quantum Chemical Program Suite. New Possibilities in the Study of Molecular Systems with the Application Parallel Computing. *Rus. Chem. Bull., Intern. Ed* **2005**, *54* (*3*), 820.

11. Perdew, J. P.; Burke, K.; Ernzerhof, M. Generalized Gradient Approximation Made Simple. *Phys. Rev. Lett* **1996**, *77* (*18*), 3865–3868.

12. Friesen, A. K.; Khursan, S. L.; Kolesov, S. V.; Monakov, Yu. B. *Chem. Fizika* **2009**, *28* (*8*), 87 (in Russian).

13. Monakov, Yu. B.; Friesen, F. K.; Kolesov, S. V. *J. Characterisation Develop. Novel Mater.* **2011**, *2* (*3–4*), 375.

CHAPTER 17

POLYMERIC COMPOSITIONS BASED ON STYRENE–BUTADIENE–STYRENE

M. I. ABDULLIN, A. A. BASYROV, S. N. NIKOLAEV, A. S. GADEEV, N. V. KOLTAEV, and YU. A. KOKSHAROVA

Bashkir State University, 100 Mingazhev Str., Ufa 450074, Russia, E-mail: koksharova.yulya@yandex.ru

CONTENTS

ABSTRACT

Dependence of MFI of plasticizer on polymeric composition was studied. Polymeric compositions with specific for using in 3D-printing were received.

17.1 INTRODUCTION

There are lots of problems in modern chemistry, which can be decided only by using polymeric composite materials. Their using increases thanks to broad field of possibility of changing and improvement properties. Conductive polymeric composites are usually consisting of a polymer or so-polymer, as a matrix, and as a conductive powder.[1–7]

Perspective filler for conductive polymeric composites is a technical nanocarbon Printex XE-2B. Chain structure of such technical carbon makes broad field of it when used in polymeric composites. However, rheological properties of polymeric compositions filled with Printex XE-2B technical carbon and their recycling are not studied.

The object of the chapter is receipt of polymeric materials based on SBS and filled with Printex XE-2B technical carbon (filling 15 mass% and more) which could be recycled using three prototyping technologies.

17.2 EXPERIMENTAL PART

Basic materials:
1. Styrene–Butadiene–Styrene block so-polymer LG-501 granules;
2. Technical carbon (TC) Printex XE-2B;
3. Oil plasticizer OP-6 (*PN-6 oil*)

Analyzed compositions (mass%):
- SBS LG-501: Oil plasticizer OP-6*(PN-6 oil)*—55:45
- SBS LG-501: Oil plasticizer OP-6*(PN-6 oil)*: TC Printex XE-2B—50:45:5
- SBS LG-501: Oil plasticizer OP-6*(PN-6 oil)*: TC Printex XE-2B—45:45:10
- SBS LG-501: Oil plasticizer OP-6*(PN-6 oil)*: TC Printex XE-2B—40:45:15
- SBS LG-501: Oil plasticizer OP-6*(PN-6 oil)*: TC Printex XE-2B—35:45:20
- SBS LG-501: Oil plasticizer OP-6*(PN-6 oil)*: TC Printex XE-2B—30:55:15
- SBS LG-501: Oil plasticizer OP-6*(PN-6 oil)*: TC Printex XE-2B—20:65:15
- SBS LG-501: Oil plasticizer OP-6*(PN-6 oil)*: TC Printex XE-2B—15:70:15

The specified compositions were compounded with filaments receipt (diameter 3 mm).

Rheological properties of polymers were studied by capillary viscometer in the temperature interval 130–170 °C, pressure 98.9 N. The method's principle is to determine in grams the mass of material that was extruded in 10 min. Melt flow index (g/10 min) was calculated according to the formula:

$$MFI = \frac{600.m}{t},$$

where m—mass of extruded polymer's part (grams), t—time of polymer's flow.

17.3   RESULTS AND DISCUSSION

It was established earlier that the optimum ratio of electric conductance technological effectiveness is observed at extent of filling with polymeric composition TC Printex XE-2B from 15 to 20%.[8] Nevertheless, recycling of polymeric composites with the specified TC content is the lowest. The flow of polymer melts was estimated in the melt flow index parameter (MFI). Processing of polymeric compositions on the 3D printer requires value of MFI more than 15 g/10 min.

Two series of polymeric compositions were prepared:
1. With TC content from 0 to 20 mass% and constant content of plasticizer PN-6 oil 45 mass%.
2. With the content of PN-6 oil from 45 to 70 mass% and constant TC content of 15%.

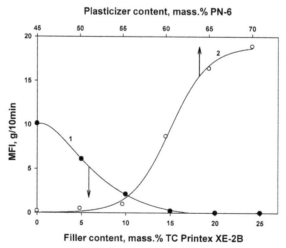

FIGURE 17.1   Dependence of MFI of the modified SBS-compositions on the contents: (1) TC Printex XE-2B (150 °C, 98.9 N, PN-6 45 mass%) and (2) PN-6 oils (150 °C, 98.9 N, TC Printex XE-2B 15 mass%).

It was established that processing of the SBS compositions filled with technical carbon of the Printex XE-2B brand is impossible to carry out in the lack of plasticizer at TC content of more than 5 mass%. At introduction of PN-6 to structure of SBS-compositions in quantity less than 45 mass%, melt does not show flow. A melt flow index increases from 0.2 to 18.9 g/10 min as PN-6 oil into SBS-composition increases from 45 up to 70 mass% (Fig. 17.1). This change is connected with the fact that PN-6 oil works as SBS plasticizer, which increases mobility of macromolecules of polymer and, therefore, SBS-composition melt flow.

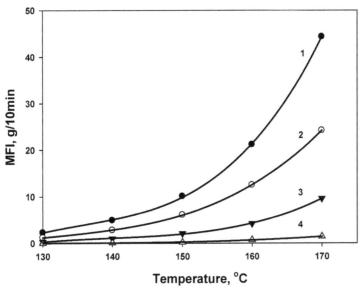

FIGURE 17.2   Dependence of the MFI of compositions on the basis of SBS and PN-6 (45 mass%), filled with TC PrintexXE-2B, on temperature 98.9 N. Content of TC Printex XE-2B, mass%: (1) 0, (2) 5, (3) 10, and (4) 15.

The comparative analysis of the received MFI dependent on temperature, which showed that MFI value decreases with increase in TC content. For composition 4, with the highest TC content (15%), MFI changes slightly with increase in temperature from 130 to 170 °C (Fig. 17.2).

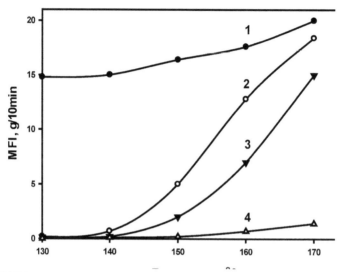

FIGURE 17.3 Dependence of the MFI of compositions on the basis of SBS and TC Printex XE-2B (15 mass%), filled with PN-6 oil, on temperature 98.9 N. Content of PN-6 oil, mass%: (1) 70, (2) 65, (3) 55, and (4) 45.

It was established that MFI value increases with increase in the content of oil at composition (TC-15%). Composition 1 containing PN-6 oil in the amount of 70 mass% is characterized by the highest MFI value (20 g/10 min); however, the composition melt index changes to a lower extent with increase in temperature range from 130 to 170 °C, in comparison with compositions 2 and 3. For compositions 2 and 3 in the same temperature range, there is a spasmodic increase in melt flow. The sample 4 shows the lowest MFI values—less than 5 g/10 min (Fig. 17.3).

Thus, polymeric compositions with optimum electric conductance and with the corresponding technological conditions of processing on 3D—the printer was received ($T \geq 170$ °C; $P = 98.9$ N; MFI ≥ 15 g/10 min), mass%:

1. SBSLG-501: PN-6 oil: TC Printex XE-2B-30:55:15;
2. SBSLG-501: PN-6 oil: TC Printex XE-2B-20:65:15;
3. SBSLG-501: PN-6 oil: TC Printex XE-2B-15:70:15.

17.4 CONCLUSION

The electroconductive polymeric compositions on the basis of butadiene of styrene block so-polymer of the LG-501 brand possessing necessary technological properties for using in the 3D printing are developed. It is established that MFI value increases with increase in the content of PN-6 oil in composition. The increase in the TC content reduces a tendency to grow with MFI compositions.

KEYWORDS

- **technical carbon**
- **Printex XE-2B**
- **SBS**
- **melt flow index**

REFERENCES

1. Encyclopedia of Polymers. Soviet Encyclopedia: Moscow, 1974; Vol. 2, pp 269–275.
2. Agayev, U. H.; Shikhaliyev, K. S.; Zeynalov, E. B. et al. Stabilization of Polypropylene`s Structure and Chemical Properties. *Plastics* **1988**, *12*, 51–52.
3. Zaikin, A. E.; Galikhanov, M. F. *Bases of Creation of Polymeric Composite Materials*; KGTU: Kazan, 2001; p 140.
4. Karpova, S. G.; Lednev, O. A.; Nikolaeva, N. Yu.; Lebedeva, E. D.; Popov, A. A. Physical and Chemical Properties of the Modified Polyethylene. *High Mol. Mater.* **1994**, *36* (*5*), 788–793.
5. Marikhin, V. A.; Myasnikova, L. P. Supramolecular Structure of Polymers. Khimia: Moscow, **1977**.
6. Bagirov, M. A.; Abbasov, T. F.; Kerimov, F. Sh. Dependence of Modification Fillers of Structure and Electro-physical Properties of High-Pressure Polyethylene. *Plastics* **1989**, *3*, 71–73.
7. Marino, K; Kuleznev, V. N. *Functional Fillers for Plastics*; Publishing House/Scientific Bases and Technologies: Moscow, 2010; p 462.
8. Abdullin, M. I.; Basyrov, A. A.; Koltaev N. V. et al. Comparison of Conductive Polymeric Composites Filled with Technical Carbon and Carbon Fibers Electrical Conductivity *J. Postgrad. Doc. Students Sci. Public.* **2014**, *8*, 95–99.

CHAPTER 18

POLYVINYLCHLORIDE COMPOUNDS; RESISTANT TO AGGRESSIVE ORGANIC MEDIA

G. F. AMINOVA[1], A. R. MASKOVA[1], R. F. NAFIKOVA[2],
L. B. STEPANOVA[3], A. K. MAZITOVA[1], and R. R. DAMINEV[2]

[1]Federal State Educational Institution of Higher Professional Education Ufa State Petroleum Technological University, Kosmonavtov Str. 1, Ufa 450062, Republic of Bashkortostan, Russian Federation, E-mail: asunasf@mail.ru
[2]Federal State Educational Institution of Higher Professional Education Ufa State Petroleum Technological University, Prospect Str. 2, Sterlitamak 453118, Sterlitamak, Republic of Bashkortostan, Russian Federation, Email: Taiffa27@mail.ru
[3]Federal State Educational Institution of Higher Professional Education Kazan State Technological University, Karl Marx Str. 68, Kazan 420015, Republic of Tatarstan, Russian Federation, Email: Lenadez@mail.ru

CONTENTS

ABSTRACT

This paper presents the test results of synthesized butyl alkyl phenoxy alkyl phthalates as plasticizers in PVC formulations of the top and intermediate linoleum layers. Laboratory tests of butyl alkyl phenoxy alkyl phthalates showed that PVC formulations of the top and intermediate linoleum layers based on our developed plasticizers on all parameters meet the requirements of existing standards, and in indicators of benzene resistance and oil resistance, even surpass standard samples.

18.1 INTRODUCTION

Development of the production of film materials and polymeric coverings are inseparably linked to the development of production of plasticizers. The use of plasticizers provides purposeful change of the properties of polymeric materials.[1] Simplification of processing, increase of elasticity and plasticity, improvement of the resistance to frost and fire—is not the full list of properties of polymeric materials regulated by plasticizers. In most cases, only the selection of plasticizers manages to give the polymer good formability, high elasticity, and resistance to frost.[1] Successful selection of plasticizer expands the polymers scope of use and extends their lifespan. Most plasticizers produced in industry are used for plasticizing of polyvinyl chloride (PVC).[2,3]

Among all polymers, polyvinylchloride (PVC) has the broadest application in construction. The main advantages of PVC are the ability of production of various kinds of products with different properties. Areas of its application vary from building products such as rigid pipe, siding, and profiles, to semi-rigid floor and wall coverings, and then to elastic wires and cables, and single-ply roofing materials.

The main qualities of PVC in construction: durability, mechanical strength, rigidity, light weight, resistance to corrosion, chemical, weather, and temperature influence. PVC is excellent fire-resisting material. It hardly gives in to ignition. PVC does not conduct electricity, and thus, is ideal as an insulating material. Also, the main characteristic of construction materials from PVC is their durability. Generalizing all aforesaid, we can say that the unique properties of PVC and its low price make it a material capable to compete with any of the polymers in many fields.[3] It is caused by development of manufacturing processes, improving the development of equipment for the manufacture of PVC products, the quality of raw materials, and application of new compositions to achieve the necessary consumer properties of products.[1,4]

18.2   THE EXPERIMENTAL PART

The plasticizers are investigated and applied various classes of chemical compounds: alcohols, simple and compound ethers, acetals, organic compounds of nitrogen, sulfur, chlorine, and others. There are more than 300 marks of plasticizers, most of which are phthalic acid ethers, share of which is 95% of total consumption of ether plasticizers.

Despite the fact that a large amount of various plasticizers for PVC is available currently, mainly dioctyl phthalate (DOP) is used, which allows to obtain a material with set physical, mechanical, and performance characteristics. DOP is considered as an international standard PVC plasticizer, meeting the requirements of processing. Requirements for other plasticizers are generally consistent with the requirements of the DOP, but it is relatively expensive and scarce. Therefore, dibutyl phthalate (DBP) is used widely due to its low cost. Plasticizers DOP and DBP are used to plasticize the cable plastic; linoleum; plastic construction profiles; decorative materials; artificial leather; technical films; products for food, medical, and coatings industry. However, DBP does not provide long-term operation of PVC compounds. DBP has a high volatility, which leads to intensive losses of plasticizer from plastic, the deterioration of physical and mechanical properties and reduce the term of operation of the products received on its basis.[2,4–9]

Butyl benzyl phthalate is also used for plasticizing polyvinyl chloride and its copolymers. It possesses high plasticizing properties and is known as a plasticizer of suspension and emulsion polyvinyl chloride, polyvinyl butyral, and chlorinated rubber. Butyl benzyl phthalate is used in the manufacturing of polymeric construction materials, artificial leather, and other benzene-resistant and oil-resistant PVC materials. However, due to the scarcity of benzyl alcohol, its release is limited. Also, it should be noted that it's possible to use phenols as certain polymeric materials plasticizers, such as polyamides and cellulose acetate. However, they are not suitable for plasticizing PVC.[5,10,11] Therefore, expanding of the range of plasticizers improving the performance properties of composite PVC materials for construction application is relevant and practically significant task.

Previously,[12–14] we describe methods of synthesis and the physicochemical and physicomechanical properties of butyl alkyl phenoxy alkyl phthalates.

This article presents the results of tests of the developed compounds as plasticizers in PVC formulations of the top and intermediate layers of linoleum.

The choice of the plasticized composition for a particular product is determined by the operating conditions. The selection criteria are varied and depend on purpose of the product.

Durability is one of the main characteristics of polymer material, which characterizes its resistance to degradation by mechanical effects. The low strength causes a rapid destruction of the material during operation.

Another important characteristic of the PVC plastic is the elongation at break, characterizing the resizing polymeric body under the effect of the applied load.

Melt flow index (MFI) is a parameter that largely determines the choice of processing conditions of the polymer composition. Furthermore, the magnitude of the MFI may be used for quality control of raw materials (components) of the polymeric composition, that is, MFI is an important parameter that determines the processability of PVC compounds.

Quite often, choosing a polymer material account of its thermal behavior is more important than the assessment of the strength characteristics. Possibility of operating the plasticized composition in the low-temperature region depends upon the temperature brittleness.

Plasticizers have a significant impact on the structure and properties of polymers. With their help, it is possible to directly change the physical, mechanical, and rheological properties of the polymer material: durability, hardness, frost resistance and brittleness, toughness and processability, melt flow index as well as the thermal, electrical, and other properties of the polymers.[1–3,15–17]

18.3   RESULTS AND DISCUSSION

Considering results of determination of physicochemical and physicomechanical properties, 11 prototypes of butyl alkyl phenoxy alkyl phthalates were tested as PVC plasticizers. The received compounds were tested as plasticizers in PVC formulations of top and intermediate linoleum layers. Prototypes of plasticizers (Table 18.1) were entered into the PVC formulation of top (Table 18.2) and intermediate (Table 18.3) linoleum layers at ratios with DOP 2:1 or instead of it. In appearance, prototypes plasticizers are oily yellowish transparent liquid.

TABLE 18.1   Prototypes of Plasticizers.

Title	Degree of oxyalkylation of butanol (prototype number)
Butoxyethylphenoxyethylphthalates	1.5 (I), 2.0 (II), 2.2 (III), 2.4 (IV), 3.0 (V)
Butoxypropylphenoxypropylphthalates	1.5 (VI), 2.0 (VII), 2.2 (VIII), 2.4 (IX), 3.2 (X)
Butoxyethylphenoxypropylphthalates	1.5 (XI), 2.0 (XII), 2.2 (XIII), 2.4 (XIV), 3.0 (XV)
Butoxypropylphenoxyethylphthalates	1.5 (XVI), 2.0 (XVII), 2.2 (XVIII), 2.4 (XIX), 3.2 (XX)

The degree of oxyethylation of phenol in the prototypes is 1.0; degree of oxypropylation is 2.1.

The prepared composition was rolled in laboratory mill at a temperature of 160–162 °C within 10 min, while rolling compositions no difficulties arise, films did not adhere to the rolls, the obtained plastic samples do not have holes, chips, and cracks.

PVC films of the top and intermediate linoleum layers were analyzed according to the norms of standard 00203312-100-2006.

Technological parameters: melt flow index was determined according to All Union State standard 11645-73; strength and elongation at break All Union State standard11262-80; thermal stability according to All Union State standard 14041-91; brittleness temperature according to All Union State standard16782-92.

TABLE 18.2   Results of Tests of Experimental Films in Industrial Compounding of the Top Linoleum Layer.

Indicator		Prototypes													Butyl benzyl phth-alate	Norms of standard 00203312-100-2006.
		PVC plastics with proposed plasticizers						PVC plastics with proposed plasticizers + DOP in ratio 2:1								
		I	II	III	X	XVII	XVIII	DOP + I	DOP + II	DOP + III	DOP + X	DOP + XVII	DOP + XVIII		
Tensile strength, kgf/cm²	Along	285	273	269	272	271	270	288	273	270	282	286	284	299	Not less than 175
	Across	249	225	237	240	239	238	256	251	259	251	257	255	273	Not less than 175
Elongation at break (%)	Along	260	283	267	254	264	260	257	264	247	251	259	257	216	Not less than 100
	Across	241	261	254	225	249	240	232	261	254	215	238	231	201	Not less than 100
Linear change, (%)		2.4	2.2	1.9	2.3	2.1	2.0	2.3	2.1	2.0	2.4	2.3	2.2	2.6	Not more than 3.0
Performance parameter															
Thermal stability at 180 °C		1 h 48 min	1 h 33 min	1 h 54 min	1 h 43 min	1 h 47 min	1 h 46 min	1 h 42 min	1 h 43 min	1 h 44 min	1 h 41 min	1 h 43 min	1 h 42 min	1 h 05 min	With DOP 1 h 45 min
MFI, g/10 min T = 170 °C, P = 16.6 kgf		8.0	7.5	8.2	7.1	7.9	7.7	8.9	9.5	9.4	8.1	8.6	8.4	7.4	7.1
Brittleness temperature, (°C)		Maintain													−25
Water absorption ability, (%)		0.437	0.474	0.485	0.482	0.484	0.483	0.473	0.499	0.502	0.498	0.499	0.501	0.204	0.195
Extractibility by petrol, (%)		1.26	1.44	1.48	1.53	1.40	1.49	1.51	1.76	1.53	1.86	1.83	1.85	11.72	18.00
Extractibility by oil, (%)		10.5	10.7	10.8	10.2	10.5	10.3	10.8	10.8	10.9	11.1	10.9	10.9	4.35	11.0

TABLE 18.3 Results of Tests of Experimental Films in Industrial Compounding of the Intermediate Linoleum Layer.

Indicator		PVC plastics with proposed plasticizers					PVC plastics with proposed plasticizers + DOP in ratio 1:2					Butyl-benzyl-phthalate	Norms of standard 00203312-100-2006.
Prototypes		VI	VIII	IX	XIX	XX	DOP +VI	DOP+ VIII	DOP+ IX	DOP +XIX	DOP +XX		
Tensile strength, kgf/cm²	Along	135	143	169	174	171	168	183	174	176	173	179	Not less than 100
	Across	108	112	128	135	132	134	142	147	156	151	159	Not less than 100
Elongation at break, (%)	Along	192	221	237	246	244	256	246	234	249	247	251	Not less than 100
	Across	173	216	229	234	231	240	226	234	239	237	249	Not less than 100
Linear change, (%)		2.4	2.2	1.9	2.5	2.4	2.3	2.1	2.0	2.6	2.5	2.6	Not more than3.0
Performance parameter													
Thermal stability at 180 °C, min		1h 48 min	1 h 33 min	1 h 54 min	1 h 57 min	1 h 55 min	1 h 42 min	1 h 43 min	1 h 44 min	1 h 59 min	1 h 57 min	1 h 05 min	With DOP 37 min
MFI, g/10min T = 170 °C, P = 16.6 kgf		9.0	9.5	10.2	9.8	9.7	9.9	9.5	9.4	9.8	9.8	9.4	9.9

TABLE 18.3 (*Continued*)

	Maintain											-25
Brittleness temperature, (°C)												
Water absorption ability, (%)	0.437	0.474	0.485	0.457	0.449	0.473	0.499	0.502	0.462	0.453	0.204	0.195
Extractibility by petrol, (%)	1.56	1.54	1.58	1.53	1.51	1.59	1.80	1.85	1.71	1.69	11.72	18.00
Extractibility by oil, (%)	10.8	10.9	10.9	10.8	10.8	11.0	10.9	11.1	11.0	11.0	4.35	11.0

Tables 18.2 and 18.3 show that the PVC compounding of the upper and intermediate linoleumlayers on the basis of our developed plasticizers by all parameters meet the requirements of existing standards, and in terms of petrol and oil resistance, even surpass standard samples. It should also be noted that the use of butyl alkyl phenoxy alkyl phthalates in relation to DOP 1:2 and without it in the formulation of PVC films promotes improvement of operational characteristics of finished product—unbonded multilayered PVC linoleum, namely reduce change in linear dimensions, electrical surface resistivity, the absolute residual strain, and abrasion.

18.4 CONCLUSION

Thus, by results of tests the synthesized butyl alkyl phenoxy alkyl phthalates have sufficiently high efficiency as PVC plasticizers and are recommended for wide tests.

KEYWORDS

- **benzene resistance**
- **butyl alkyl phenoxy alkyl phthalates**
- **linoleum, oil resistance**
- **PVCformulations, PVC plasticizers**
- **melt flow index**
- **thermal stability**
- **brittleness temperature**

REFERENCES

1. Wilkie, Ch.; Summers, J.; Daniels, Ch. Polyvinyl Chloride. SPb.: Profession, 2007; 728.
2. Barshteyn, R. S.; Kirilovich, V. I.; Nosovskiy, Y. E. Plasticizers for Polymers; Khimia: Moscow, 1982; p 196.
3. Gutkovich, S. A. Features of Reception and Application of PVC with Different Physicochemical Characteristics. PhD Dissertation, Moscow, 2011, 314.
4. Aminova, G. F.; Gabitov, A. I.; Maskova, A. R.; Khusnutdinov, B. R.; Abdrakhmanova, L. K.; Nafikova, R. F. Obtaining of Linoleum with Improved Physical and Mechanical Properties. *Electron. Sci. J. Oil Gas Bus.* **2013**, *6*, 508–537. URL: http://ogbus.ru/eng/authors/AminovaGF/AminovaGF_2.pdf
5. Hamaev, V. H. Synthesis and Investigation of Properties of Ether Compounds and Development on its Basis Plasticizers and Components for Synthetic Oils. PhD Dissertation, Ufa, 1982, 486.
6. Maskova, A. R.; Stepanova, L. B.; Aminova, G. K. Testing of Formulations of PVC Materials for Construction Application using Symmetric and Asymmetric Alkoxylated Alcohols Phthalates. *Izv. KSUAE Kazan* **2012**, *2* (*20*), 177–182.
7. Mazitova, A. K.; Nafikova, R. F.; Aminova, G. K. Polyvinyl Chloride Plasticizers. In *Science and Epoch: Monograph;* Ageeva, G. M., Aminova, G .K., Balyaeva, S. A., Kirikova, O. I., Ed.; VSPU: Voronezh, 2011; Vol. 7, Chapter XVII, T. 500, p 276–296.
8. Aminova, G. K.; Maskova, A. R.; Buylova, E. A.; Stepanova, L. B.; Mazitova, A. K. Phthalates of Alkoxylated Alcohols—Plasticizers of PVC Compositions Construction Application. *Bashkir Chem. J.* **2012**, *19* (*3*), 118–121.
9. Tinius, K. *Plasticizers*; Khimia: Moscow, 1964; p 915.

10. 938 533 (USSR). Butylphenoxyethylphthalate as Plasticizer of Polyvinyl Chloride. Hamaev, V. H.; Bikkulov, A. Z.; Mazitova, A. K.; Khannanov, R. N.; Svinuhov, A. G.; Smorodin, A. A.; Pavlychev, V. N.; Teplov, B. F. Publication Prohibited.

11. Mamedov, R. I.; Ismailov, R. A.; Sadik-zade, O. D.; Guseinov, F. I.; Sadyhov, Sh. F. Plastification of Polyvinyl Chloride by Esters of Alkosinaphthenic acids. *Plastics* **1972**, *10*, 39–40.

12. Aminova, G. F.; Gabitov, A. I; Maskova, A. R.; Rysaev, D. U; Goryushinsky, I. V. New Types of Composite PVC Materials of Finishing Application. *News KSUAE* **2013**, *3* (*25*), 80–85.

13. Aminova, G. K.; Maskova, A. R.; Sayapova, R. G.; Yagafarova, G. G.; Gorelov, V. S. PVC Linoleum with Improved Performance Characteristic. In *Science, Technology and Higher Education [Text]: Materials of the II International Research and Practice Conference*, Accent Graphics communications: Westwood, Canada, 2013; Vol. II, p 16–19.

14. Aminova, G. F.; Gabitov, A. I.; Maskova, A. R.; Abdrakhmanova, L. K.; Nafikova, R. F. Composite PVC Materials of Finishing Application Based Butoxypropylphenoxypropylphthalates. *Sci. Peace* **2013**, *2* (*2*), 40–42.

15. Aminova, G. K.; Maskova, A. R.; Buylova, E. A.; Gorelov, V. S.; Mazitova, A. K. Plasticizers for Polyvinyl Chloride Compositions of Construction Application. *Ind. Prod. Use Elastom.* **2012**, *4*, 29–32.

16. Kryzhanovsky, V. K.; Kerber, L. M.; Burlov, V. V.; Panimatchenko, A. D. Manufacture of Polymeric Materials: Textbook. SPb.: Profession, 2004. 464.

17. Aminova, G. K.; Maskova, A. R.; Nafikova, R. F.; Mazitova, A. K. PVC Linoleum. In *Science and Epoch: Monograph;* Aminova, G. K., Bagnetova, E. A., Barsukova, S. A., Kirikova, O. I., Ed.; VSPU: Voronezh, Science Inform: Moscow, 2013; Vol. 11, Chap. IX, T. 500, p 187–212.

CHAPTER 19

CALCIUM–ZINC STABILIZING SYSTEM INTEGRATED ACTION FOR PVC COMPOUNDS

L. B. STEPANOVA[1], R. F. NAFIKOVA[2], T. R. DEBERDEEV[1],
R. J. DEBERDEEV[1], and R. R. DAMINEV[2]

[1]Federal State Educational Institution of Higher Professional Education Kazan State Technological University,Karl Marx Str. 68, Kazan 420015,Republic of Tatarstan,Russian Federation,Email: Lenadez@mail.ru
[2]Federal State Educational Institution of Higher Professional Education Ufa State Petroleum Technological University,Prospect Str.,Sterlitamak453118, Republic of Bashkortostan, ,Russian Federation, Email: Taiffa27@mail.ru

CONTENTS

ABSTRACT

New liquid complex stabilizers for PVC compounds based on calcium and zinc salts of organic acids are created. It is shown that the secondary stabilizers epoxidized soybean oil (ESO), phosphite NF, and dipenterythritol (DPET) significantly increase their effectiveness in providing static and dynamic thermal stability, melt PVC color stability of compositions and articles.

19.1   INTRODUCTION

Polyvinyl chloride (PVC) is the basis of many composite materials and occupies a leading position in terms of production of thermoplastic polymers. Currently, 69% of PVC provides market of plastic construction materials, prevails in manufacturing pipes and fittings, exterior cladding of walls, windows, and materials for decoration.

In the development of the production of PVC and the continuous expansion of the areas of application leading place success in creating stabilizing additives, used in the processing and exploitation of polymer processing as PVC because of its anomalously low thermal stability is impossible without effective stabilization.

Modern range of PVC stabilizers is very wide, the main share of which are the metal-containing additives—divalent metal carboxylates, preferably calcium stearate, barium, zinc, and lead. Their main functionis binding depleting the decay of hydrogen chloride PVC and weakening the destructive action of mechanical forces, especially intense during thermomechanical processing of PVC.[1–8]Currently, competing with traditional stabilizer is nontoxic stabilizing system, derived from calcium and zinc salts of the higher organic acids. This is primarily due to the growth requirements of ecological safety of polymer products.[9–11]We obtained the calcium–zinc stabilizers (CSR) containing calcium oleate, zinc oleate,and monooleate glycerol in a molar ratio of 1.5:0.5:0.5, and stabilizer (CSE) with a molar ratio of 2-ethylhexoate calcium:2-ethylhexoate zinc:monooleate glycerol to 1 5:0.5:1, respectively.

Increased requirements for stabilizers to provide high color stability, static and dynamic thermal stability, low initial yellowness in products, and good processability of materials with maximum performance processing equipment requires improvedefficiency of calcium–zinc stabilizers. It is known that the efficacy of calcium–zinc stabilizers largely achieved by combined use of various secondary stabilizers.[12–14]

In this context, the aim of this work was to study the effect of different secondary stabilizer efficiency of CSR, CSE, and the creation of multifunctional action of stabilizing systems.

19.2   THE EXPERIMENTAL PART

Preparation of PVC composition and PVC films for analysis. Ingredients PVC composition, polymer stabilizers, plasticizers, lubricants, fillers, - a two-stage stirred laboratory mixer TGHK 5 for 60 m and chickpeas for uniform distribution of the components in the mixture.

The performance characteristics of PVC materials were evaluated by standard methods.

Thermal stability of time PVC was determined by the induction period of the time change, indicator color "congo red," when allocating HCl PVC during aging (175 °C) in accordance with GOST 14041-91 (determination of tendency to release hydrogen chloride and other acidic products at high temperature in the compositions and product-based polymers and copolymers of vinyl chloride). The method of the Congo-red on the thermostat «LAUDA» PROLINE P 5.

To assess the impact of complex stabilizers on the technological properties of PVC compounds,plastograph«Brabender» was used. Tests were conducted at a temperature of 180°Cinmixing chamber and stirrer speed of 35 rev/min. In chamber,plastographwas charged with 60 g precooked PVC composition.

To determine the dynamic thermal stability,the resultant mixture was charged into a mixing chamber plastograph «Brabender» at a melt temperature of 190 °C and speed of rotation of the cams was 35 rev/min. From plastography,thermal stability and dynamic torque value in steady statewas determined.

19.3   RESULTS AND DISCUSSION

We used stabilizers, acting by different mechanisms x, namely ESO, DPET—functioning as acceptors HCl; phosphite NF—inhibiting thermal degradation of PVC interaction with unstable carbonyl groups.[15]

Influence of secondary stabilizers and congestion thermostabilizing effectiveness of CSR, CSE studied in powdered PVC compositions with 1 part by weight of the stabilizer content 100 parts PVC 7059M. Figures 19.1 and 19.2 show the dependence of the thermal stability of PVC powder compositions of the content as a part of CSR, CSE secondary stabilizers.

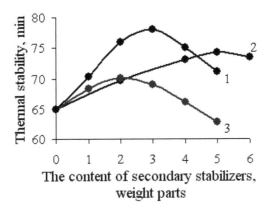

FIGURE 19.1 Dependence of thermostable STI PVC composition of the content of the secondary stabilizers on CSR (160 °C). (1)DPET,(2)ESO,and (3)phosphite NF.

With the introduction of the CSR, CSE secondary stabilizers termostabil s of PVC composition increases, the greatest effect is achieved when dosed in Re: 5 parts by weight—ESO, twoparts— NF phosphite, threeparts by weight—DPET on 100 parts by weight complex calcium–zinc stabilizer.

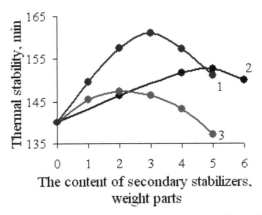

FIGURE 19.2 Dependence of thermostable STI PVC composition of the content of the secondary stabilizers on CSE (160 °C). (1) DPET,(2) SEM,and (3)phosphite NF.

The effectiveness of secondary stabilizers also evaluated in terms of color stability and the rate of release of NA l PVC, which indirectly can judge the time of saving material original color with accelerated aging, as well as maintaining the physical, mechanical, and other parameters (Figs. 19.3and 19.4).

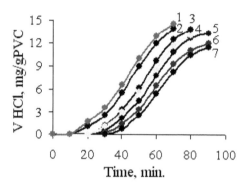

FIGURE 19.3 Effect of different secondary stabilizers HCl evolution rate of the PVC at a temperature chamber 165 °C.(1) CSR,(2) CSR + DPET,(3) CSE,(4) CSR M,(5) CSE + ESO,(6) CSE + phosphite NF,and (7) CSE M.

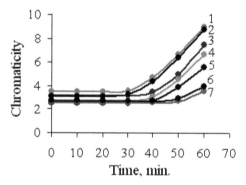

FIGURE 19.4 Effect of different secondary stabilizers to change the color stability of the films during heat treatment at 180 °C. (1) CSR,(2) CSR + DPET,(3) CSE,(4) CSE + ESO,(5) CSE + phosphite NF,(6) CSR M,and (7) CSE M.

Research results (Figs. 19.3 and19.4) show a positive effect of using the secondary stabilizers in improving the color stability of films and reduce the rate of dehydrochlorination of the polymer. More pronounced effect on reducing the number of released HCl occur while using phosphite NF and ESO, they also have a more stabilizingaction.

In view of the data modernized versions of stabilizers CSR, CSE were created, while p-containing soda as part of secondary stabilizers: 5 parts by weight— ESO, twoparts—NF phosphite, threeparts by weight—DPET on 100 parts by weight of the stabilizer complex, and labeled them as complex stabilizers CSR-M, M-CSE. The main characteristics of the developed complex stabilizers are given in Table 19.1.

CSR (CSE), in conjunction with SEM, phosphite NF, DPET enhances the action of liquid complex stabilizers. Thermal and color stability of CSR-M (CSE-M) above 1.5 times the rate of release of hydrogen chloride substantially below that determines the appropriateness of their use in complex stabilizers.

TABLE 19.1   Characteristics of the Calcium–Zinc Complex Art and Stabilizer.

Indicator	Integrated stabilizer			
	CSR	CSE	CSR-M	CSE-M
Appearance	Movable homogeneous oily liquid of light brown color			
Mass fraction of calcium, (%)	3.03	4.22	2.55	3.74
Mass fraction of zinc,(%)	1.64	2.3	1.31	1.97
Density at 20°C, g/cm³	0.992	1.007	0.990	0.982
Flash Point, °C	194	197	197	199
Acid number, mg KOH/g	6.2	6.0	5.8	5.6
Mass fraction of volatiles,(%)	1.6	1.4	1.3	1.2
The thermal stability of PVC at 160 °C, min	65	140	94	183
Color stability of the film at 180 °C, min	30	35	40	50

To study the influence of complex stabilizers on the technological properties of the PVC composition in conditions close to the processing, plastograph«Brabender»was used. Action liquid complex stabilizers, compared to foreign analogs, were studied in a model uPVC window profile composition (Table19.2).

TABLE19.2   **Test** Results Unplasticized PVC Compositions Plastograph «Brabender» (Temperature 180 °C, Rotational Speed 35 rev/min).

Quality indicators	CSR	CSE	CSR-M	CSE-M	Stabiol CZ 2818	ArStab KC 317
Time of onset of melting, s	30	28	26	24	32	25
Melting start temperature, °C	165	158	160	154	162	154
Maximum torque, Nm	35.2	42.1	35.5	39.1	48.1	46.8
Melting time, s	58	39	35	22	86	26
The equilibrium torque, Nm	27.2	29.3	26.9	28.9	38.8	27.3

TABLE19.2   (*Continued*)

The energy spent on melt-ing, kNm	7.1	4.3	6.4	4.1	12.1	4.5
Dynamic stability and thermostats, min	17.13	17.23	19.42	19.55	7.67	14.8

As seen from the data, using complex stabilizers, there is a significant torque reduction, energy expended on melting, and increasing the dynamic thermal stability of PVC composition. The effectiveness of the new complex stabilizers are at a level similar to imported and some indicators surpass them.

19.4   CONCLUSION

In general, conducted complex research shows versatility of action of new liquid calcium–zinc stabilizers complex. Secondary stabilizers: ESO, phosphite NF, DPET significantly increases their effectiveness in providing static and dynamic thermal stability, color stability, and melt PVC compositions.

KEYWORDS

- PVC
- stabilizers.

REFERENCES

1. Minsker, K. S.;Fedoseyeva, G. T. Degradation and Stabilization of PVC. *Chemistry*1979, 272.
2. Wilkie, C.;Summers,J.;Daniels,C. SPb: Polyvinyl Chloride Ave about Occupations, 2007;p 728.
3. Gorbunov, B. N.*Chemistry and Technology of Stabilizing Materials Polymerase p/BN Gorbunov*;Gurvich,Y. A., Maslov,I. P.Khimia:Moscow, 1981; p 283.
4. Liquid Stabilizer Comprising Metal Soar and Solubilized Metal Perchlorate/DF Anderson; Applicant and Patentee Akcros Chemical America. Pat number 55575951 US MPK6 C 09 K 15/32, C 08 K 5/09. № 418057; appl. 4/6/1995; publ. 11.19.1996.
5. Jia-you, X. Gaofenzicailiaokexieyugongeheng/X. Jia-you, G. Shao-yun, W. Wei-lai*Polym. Mater. Sci. Technol.*2005,*21*(*2*),241–244.
6. Brinkmann, S. *Stabilisatoren und Füllstoffe*; Kunststoffe,Brinkmann,S.,Regel,K.,Ed.;RWTH Publication: Moscow, 1995;12,pp 2186–2189.
7. Fisch, M.;Loeffler,O.Liquid Polyvinyl Chloride Resin Stabilizer Systems Based On Complexes of Organic Triphosphites with Zinc Chloride/KJ Bae.Applicant and Patentee Argus Chemical Corp. Pat 4,782,170 US MKI C 07 F 3/06,№ 862882; appl. 13/05/86; publ. 11.01.88.

8. Hiroyuki, G. Synergistic Effects Esters Pentaerythritol and Hydro and Rowan Fatty Acids Fish Fat with Metal Soap on Stabilization PVC/G. Hiroyuki, S. Akinori, N. Hirosi, H. Yuzou, I. Takeo. *Jap. J. Polym. Sci. and Technol.*1994,*51*(*8*),511–517.

9. Available at http://www.polymery.ru/ etter.php? n_id = 2613.

10. Available at http://www.stroybox.ru/index .php? id = 882511004.

11. Available at http://veretelnikov.prom.ua/a70795-stabilizatory-pvh-profilej.html

12. Homogenous Stabilizer Compositions for Vinyl Halide Polymers/ER Quinn; Applicant and Patentooblad and Tel The Lubrizol Corp. Pat 4,743,397 US MKI4 C 09 K 15/32,№ 931770; appl. 17/11/86; publ. 05.10.88.

13. Application number 3708711 FRG MKI4 C 08 L 27/6. Stabilisierungsmittelfür Vinylchloridpo-lymerisate.G. Marx; and declared Tel and patentee BASF LACKE & FARBEN.№ 3708711.8; appl. 03/18/87; publ. 06.10.88.

14. Choi, J. H.;Fortner,L.E.;Mottine,J.J. Stabilization of PVC BodiesApplicant and Patentee AT & Technologies Inc.Pat 4584241 USA. In MKI 32 15/00, H 01 B 7/00,№ 597 130; appl. 04/06/84; publ. 04.22.86.

15. *Chemical Additives to Polymers*;Maslova, I. P.,Ed.;Khimia:Moscow, pp 1981–264.

CHAPTER 20

DEVELOPMENT OF LIQUID STABILIZING SYSTEMS PVC PROCESSING

L. B. STEPANOVA[1], R. F. NAFIKOVA[2], T. R. DEBERDEEV[1], R. J. DEBERDEEV[1], and R. N. FATKULLIN[2]

[1]Federal State Educational Institution of Higher Professional Education, Kazan State Technological University, Republic of Tatarstan, 420015 Kazan, Karl Marx Str. 68, Russian Federation, E-mail: Lenadez@mail.ru
[2]Federal State Educational Institution of Higher Professional Education, Ufa State Petroleum Technological University, Republic of Bashkortostan, 453118 Sterlitamak, Prospect Str. 2, Russian Federation, E-mail: Taiffa27@mail.ru

CONTENTS

ABSTRACT

In this chapter, a process for preparing liquid calcium–zinc stabilizers complex PVC stepwise reaction of 2-ethylhexanoic (oleic acid) with mixtures of oxides of calcium, zinc, oleic acid, glycerin,and ester in the presence of plasticizers (DINP, DOTF) and phenolic antioxidant (ionol, Irganox 1010) is investigated. Studies on technological and operational properties of PVC compounds, the optimal ratio of components and integrated systems producing liquid stabilizers with at least 98% yield, presented.

20.1 INTRODUCTION

Polyvinyl chloride (PVC) is one of the most tonnage polymers, produced both in Russia and abroad, world production of this polymer has reached 50 million tons/year. In Russia, the production volume for 2012 amounted 642 tons PVC.[1] Depending on the method of preparation, formulation, and processing technology, this polymer gives a large range of materials and products, which are characterized by different properties.

Meanwhile, the distinctive feature of PVC is its anomalously low thermal stability. Under the action of many chemical, physical, biological, and mechanical factors,PVC easily decomposes with the release of HCl.[2] From the outset, the use of PVC has arisen a need to increase its stability for processing and use.

For the stabilization of PVC compositions during processing and improving the performance and processing characteristics of PVC materials,4–5 kinds of specific additives were administered: stabilizers, lubricants, antioxidants, photostabilizers, etc.[3]

Modern range of PVC stabilizers is very wide, the main share occupied by metal-containing stabilizers. Their main function-binding emitted during the decay of hydrogen chloride PVC. When processing the most commonly used PVC barium, cadmium, lead-containing compounds, they are effective stabilizers, but in view of the toxicity, data are metal-containing stabilizers and cannot be used in the manufacturing of materials and products in contact with medical and food.[4]

In recent years, due to environmental considerations,nontoxic calcium–zinc (Ca/Zn) complex stabilizeris preferable to use, despite the fact that these systems are expensive. In Russia, the complex range of stabilizers is very limited, they are mostly purchased abroad. In this regard, a very urgent work aimed at creating a nontoxic stabilizer systems for PVC compounds.

Previously, to develop a simple method for producing environmentally safe nontoxic complex PVC stabilizers, some features of the process of interaction of fatty monocarboxylic acids with glycerol in the presence of divalent metal oxides (MgO, CaO, ZnO) were studied and investigated the effect of the resulting glycerol monooleate on the technological properties of PVC compounds.[5-6]

20.2   EXPERIMENTAL PART

For preparation of PVC, it's composition as well as PVC films for analysis should be considered. The main ingredients of PVC composition are: polymer stabilizers, plasticizers, lubricants, fillers. Meanwhile a two-stage stirred laboratory mixer TGHK 5 for 60 min for uniform distribution of the components in the mixture was applied.

The performance and characteristics of PVCmaterials were evaluated by standard methods.

The thermal stability time of PVC was determined by the induction period of the time change indicator color "congo red" when allocating HCl PVC during aging (175 °C) in accordance with GOST 14041-91 (Determination of tendency to release hydrogen chloride and other acidic products at high temperature in the compositions and product-based polymers and copolymers of vinyl chloride. The method of the congored) on the thermostat «LAUDA» PROLINE P 5.

Plastograph «Brabender» was used to assess the impact of complex stabilizers on the technological properties of PVC compounds.Tests were conducted at a temperature of the mixing chamber and 180 °S stirrer speed of 35 rev/min. In chamber, plastograph charged with 60 g precooked PVC composition.

To determine the dynamic thermal stability, the resultant mixture was charged into a mixing chamber plastograph «Brabender» at a melt temperature of 190 °C and speed of rotation of the cams 35 rev/min. From plastogram,thermal stability and dynamic torque value are determined in steady state.

20.3   RESULTS AND DISCUSSION

In practice, increasing use is being made of liquid metal-containing stabilizers, which are typically formed by reacting carboxylic acids with the oxides and hydroxides of metals in a solvent or plasticizer at elevated temperatures.[7]

However, this method of synthesis of stabilizers has a significant drawback. At synthesis temperatures hydrolysis of ester plasticizers, the accumulation of acid products, accelerating, subsequently, the processing of PVC compositions thermal decomposition of the plasticizer and polymer.

Meanwhile, the efficacy of calcium–zinc stabilizers largely determines the possibility of using them in combination with known synergists. Among them, the lubricantsare of particular interest as lubricating stabilizers can improve manufacturability of processing PVC compounds, increase productivity, and reduce production lines, the need for separate lubricants, and reduce the total cost of the stabilizer system. It may be important quantitative ratio of the components in the mixture as unplasticized PVC compositions have strict limitations on the content of primary plasticizers.

The foregoing process has identified the need to study the stepwise interaction of liquid organic monocarboxylic acid mixtures of oxides of calcium, zinc, in the first step to give coprecipitated salts Me^{2+}—stabilizers, the second—oleic acid with glycerol, to produce mechanochemical stabilizer.

To create an integrated liquid calcium–zinc stabilizers of PVC, commercially available nontoxic raw components were used, such as oleic acid, 2-ethylhexanoic acid, glycerol,trinonilfenilfosfit (phosphite NF),epoxidized soybean oil (ESO),dipenterythritol (DPET),ionol,andIrganox 1010.

Coprecipitated calcium and zinc salts of oleic (2-ethylhexanoic acid) was obtained at an equimolar ratio of organic acid, calcium oxide, and zinc. In the initial period of the synthesis, in both cases, for a certain time, a sharp decrease in the acid number due to the formation of salts of organic acids is observed, simultaneously increasing the viscosity of the reaction mass. To reduce the viscosity of the reaction mass, plasticizer (DINP, DOTF) and antioxidant (Ionol, Irganox 1010) were added. Antioxidants inhibit the thermal decomposition of the plasticizer in the course of synthesis, then the PVC compositions enhances the thermal stability of the polymer.

Delayed addition of a plasticizer and an antioxidant for the introduction stage of the synthesis of calcium–zinc salts based on oleic (2-ethylhexanoic acid) allows to intensify the process of salt formation and improve the quality of the stabilizer by reducing the degree of hydrolysis and thermal degradation of the ester plasticizer. Based on the identified characteristics of the synthesis, calcium–zinc salts of oleic (2-ethylhexanoic acid) at different molar ratio of the oxides of calcium and zinc were obtained. The results are shown in Tables 20.1 and 20.2.

TABLE 20.1 Preparation of Calcium–Zinc Salts with Oleic Acid in the Presence of a Plasticizer DOTF at 130 °C.

Reaction time, min	The molar ratio CaO:ZnO					
	1:1	1.25:0.75	1.5:0.5	1:1	1.25:0.75	1.5:0.5
	Acid number, mg KOH/g			Exit salts, (%)		
10	148.6	140.6	144.6	23.4	27.5	25.5
20	108.9	101.6	104.3	43.9	47.6	46.2

TABLE 20.1 (*Continued*)

30*	69.5	64.2	67.6	64.2	66.9	65.2
40	32.9	25.3	27.6	83.0	87.0	85.8
50	8.8	5.6	7.6	95.5	97.1	96.1
60	4.6	4.7	3.6	97.6	97.6	98.1

* **Input** into the reaction mixture DOTF (in an amount of 50 wt.% by weight of oleic acid) Ionol (0.5% by mass DOTF).

TABLE 20.2 Preparation of Calcium–Zinc Salts with 2-Ethylhexanoic Acid in the Presence of a Plasticizer DOTF at 140 °C.

Reaction time, min	The molar ratio CaO:ZnO					
	1:1	1.25:0.75	1.5:0.5	1:1	1.25:0.75	1.5:0.5
	Acid number, mg KOH/g			Exit salts, (%)		
10	268.8	258.3	267.3	27.4	30.2	27.8
20*	184.2	168.1	183.6	50.2	54.6	50.4
30	132	116.2	128.3	64.3	68.6	65.3
40	84	68.9	79.3	77.3	81.4	78.6
50	43.5	28.9	37.4	88.2	92.2	89.9
60	11.2	7.6	8.6	97.0	97.9	97.7
70	4.2	4.2	3.7	98.9	98.9	99.0

* **Input** into the reaction mixture DOTF (in an amount of 70 wt.% by weight of oleic acid) Ionol (0.5% by mass DOTF).

Experimental results show that the highest yield of calcium–zinc salts of organic acids are obtained in all cases. Meanwhile, the known synergistic effect on the thermal stability of mixtures of Ca–Zn, involving calcium salt exchange reaction of labile atoms of CI due to the catalytic action of zinc chloride, quantifies the ratio of calcium and zinc. Evaluation of the thermal stability of PVC compounds (PVC 7059M 100 parts by weight; calcium–zinc stabilizer 1 wt. H.) GOST 14041-91 in isothermal conditions shows that all investigated samples increases the thermal stability of PVC. It was established that the highest possible value of the thermal stability of PVC compositions obtained at a molar ratio of calcium and zinc is 1.5:0.5 (Fig.20.1), therefore, for the further application, this ratio is used.

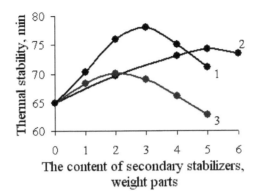

FIGURE 20.1   Effect of the molar ratio of calcium and zinc on the thermal stability of PVC at 160 °C: (1) Ca salt:salt Zn = 1.25:0.75,(2)Ca salt:salt Zn = 1.5:0.5,and (3)Ca salt:salt Zn = 1:1.

To enhance the functionality of the stabilizer in the medium obtained in the first step the calcium and zinc salts of oleic and 2-ethylhexanoic acid glycerin monoester obtained (lubricant). Using a lubricant can achieve maximum efficiency in the processing of PVC compositions, as refining processes associated with mechanical action on the polymer, with a probability of discontinuity chemical bond with the release of HCl and the formation of intermolecular crosslinks.

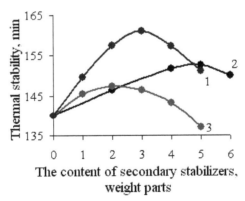

FIGURE 20.2   Effect of calcium–zinc yield glycerol monooleate at a temperature of 165 °C. (1) Salt Me2 + ots.;(2) synthesis of glycerol monooleate in the presence of 2-ethylhexoate, zinc calcium; and (3) synthesis of glycerol monooleate in the presence of calcium–zinc oleates.

It is found that the process of esterification of oleic acid with glycerol in the presence of calcium and zinc salts is more intensive than in the absence, thereof, that gives an indication of their catalytic effect (Fig.20.2). It is shown that in the presence of Me^{2+} oleates, process is effective after 4 h. The acid number is reduced to 5–10 mg KOH/g, the monoester constitutes more than 98% at other equal conditions in the presence of 2-ethylhexoate, calcium, zinc, high output reached after 5 h. The esterification of oleic acid with glycerol at a molar ratio of 1:1 in the presence of calcium and zinc salts of oleic (2-ethylhexanoic acid) showed that the process proceeds effectively at a temperature of 165 °C.

In the esterification of oleic acid with glycerol in the IR spectra, the absorption band gradually disappears with a peak at 1700 cm^{-1}, characteristic of the carboxylic acids. In the products, new characteristic absorption bands in the regions 3200–3600 cm^{-1} and 1420 cm^{-1} are observed, respectively, stretching and deformation vibrations of the OH group, which confirms the formation of the partial esters of glycerol. In the 1400 cm^{-1} absorption bands, the variations for the salts of organic acids are significant.

Kinetics of the reaction of esterification of fatty acids, glycerol organic measured using acid at two different temperatures. The esterification reaction of the acid with glycerol in the presence of calcium and zinc salts include the first order in acid (Fig.20.3, Table 20.3).

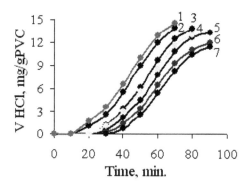

FIGURE 20.3   Dependence of ln K versus the reaction time of esterification of acids with glycerol oleate calcium–zinc: (1) 165 °C,(2) 175 °C; 2 etigeksoat calcium–zinc,(3) 165 °C,and (4) 175 °C.

TABLE 20.3   The Kinetic Parameters of the Reaction of Esterification of Oleic Acid with Glycerol in the Presence of Organic Salts of Calcium, Zinc.

Salts of organic fatty acids	Temperature, K	The reaction rate constant, s^{-1}	The activation energy E, kJ/mol
Oleate Ca–Zn	438	0.000189	21.56
	448	0.000215	

TABLE 20.3   (*Continued*)

2-ethylhexoate Ca–Zn	438	0.000157	25.81
	448	0.000184	

The esterification of oleic acid with glycerol in the presence of milder oleates of calcium–zinc.

Creating complex PVC stabilizers required balance between metal-containing stabilizers and glyceryl monooleate as the lubricant contributes to changing the rheological properties of the polymer melt. When the level exceeds dosage lubrication, certain critical values deteriorate the physicomechanical and performance characteristics of the finished product.

To determine the effect of the content of glycerol monooleate on the technological properties of PVC compounds, a series of laboratory experiments on plastograph «Brabender» were carried out. By plastogram,torques were determined, which characterize conventional melt viscosity and dynamic thermal stability, that is, time from start of the thermomechanical effect to flow in thermal degradation of PVC.

In conducting research, industrial unplasticized composition window profile was selected as a benchmark on the basis of PVC 6669 RV containing hydrophobic chalk, heat stabilizers, titanium dioxide, a mixture of process lubricants, modifiers, processability, and strength in balanced proportions. For the selection of the optimal content of glycerol monooleate in a calcium–zinc stabilizer, a reference sample was replaced with heat stabilizers and technological lubricant on calcium–zinc stabilizer with different content monooleate glycerol.

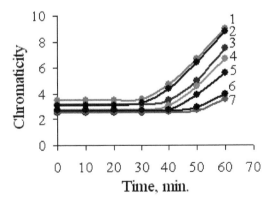

FIGURE 20.4 Dependence of maximum torque and dynamic thermal stability of PVC compositions of the content in the integrated regulator of glycerol monooleate.Maximum torque ($T = 180$ °C; $N = 35$ rev/min).(1) 2-ethylhexoate, calcium–zinc;(2) oleate, calcium–zinc. The dynamic thermal stability ($T = 190$ °C; $N = 35$ rev/min): (3) 2-ethylhexoate, calcium–zinc; and (4) oleate, calcium–zinc

The resulting dependence of the maximum torque and dynamic thermal stability, the content in the composition of complex stabilizer glycerol monooleate show that glycerol monooleate leads to a significant reduction of the maximum torque and increase dynamic thermal stability (Fig.20.4). Among the test samples, a complex stabilizer prepared based on calcium–zincoleate, more conducive to reduction of the maximum torque, is likely to facilitate lubricity oleates of calcium–zinc. When introduced into the PVC formulation complex stabilizer (CSR), prepared at a molar ratio of calcium oleate, zinc oleate, and monooleate glycerol ratio of 1.5:0.5:0.5, respectively, and a complex stabilizer (CSE)—in a molar ratio of 2-ethylhexoate calcium 2-ethylhexoate:zinc monooleate glycerol ratio of 1.5:0.5:1, the maximum torque is close to the level lines of the reference sample containing the optimum amount of external and internal lubricants providing maximum performance of the extruder. Glycerol monooleate in combination with calcium and zinc salts of organic acids in a molar ratio of calcium oleate, zinc oleate, glyceryl monooleate of 1.5:0.5:0.5, respectively, and the complex stabilizer (CSE)—in a molar ratio of calcium 2-ethylhexoate 2-ethylhexanoate, zinc glyceryl monooleate of 1.5:0.5:1 and increase the dynamic heat stability of the PVC composition more than 1.5 times, this is primarily seen as no overheating of the composition and chemical bonding produced during PVC degradation of hydrogen chloride.

Use CSR (CSE) in the recipe profile -pogonajnye not lead to a reduction of its physical and mechanical characteristics, which confirms the balance of complex stabilizer (Table20.4).

TABLE 20.4 Results of the Test Profile—PogonajnyeObtained using Complex Stabilizer (Stabilizer Content of 4 Parts Per 100 Parts by Weight of PVC).

Indicator	TU 5772-215-020-3312-02	Integrated stabilizer	
		CSR	CSE
Deviation from the mass,%	Not morethan +15	13.8	13.5
Hardness, mm	Not morethan0.14	0.12	0.10
Elasticity,%	Not lessthan50	76	74
Longitudinal shrinkage, %	Not more than0.4	0.25	0.23
Water absorption,%, by weight	Not more than1	0.06	0.04
Brittleness temperature, °C	Not higherthan−40	Минус 41	Минус 41

20.4 CONCLUSION

The features of the synthesis of liquid calcium–zinc stabilizers complex:

- Increased formation of coprecipitated calcium–zinc salts of organic acids, adding ester plasticizer (DOTF, DINP) in an amount of 50–70 parts by weight by weight of the loaded organic acid;

- Catalytic activity of the coprecipitated calcium–zinc salts for esterification of oleic acid with glycerol in the second step of the synthesis;

- The reaction of esterification of oleic acid in the presence of oleates (2-ethylhexoate), calcium–zinc is first order in oleic acid.

Thus, studies have shown the possibility of practical implementation of the process to obtain stepwise coprecipitatedoleate and 2-ethylhexoate calcium–zinc and glycerol monooleate to yield complex PVC stabilizers, providing high technological properties of PVC compounds, while maintaining the basic mechanical and operational characteristics of PVC products.

KEYWORDS

- **complex stabilizer**
- **polyvinyl chloride**

REFERENCES

1. Available on http://article.unipack.ru/38511/.
2. Wilkie, C. Polyvinyl ChlorideWilkie, C., Summers, J., Daniels, C, Ed.;SPb.: Profession, 2007, 728 p.
3. Minsker, K. S.; Fedoseyeva, G.T. Degradation and Stabilization of PVC.*Chemistry***1979**, 272.
4. Gorbunov, B.N.*Chemistry and Technology of the Stabilizers of Polymeric Materials*;Gorbunov, B. N., Gurvich, Y. A., Maslov, I. P., Ed.;Khimia: Moscow, 1981;p 283.
5. Stepanova, L. B. Development of Resource-Saving Technology of Integrated Stabilizers for PVC.Aminov, G. K., Nafikov, A. B., Gilmutdinov,A. T., Mazin,L. A., Stepanova,L. B., Ed.;*Industrial Production and Use of Elastomers*; 2010;Vol. 3,pp. 18–21.
6. Stepanova, L. B.Effect of Glycerol Monoesters with Organic Monocarboxylic Fatty Acids on Processing Properties of PVC.Mazin, L. A., Afanasev, F.I., Deberdeev, R. J., Stepanova, L. B., Ed;*Plastics*,**2011**,(*10*),15–16.
7. AS 1325883 USSR. The Stabilizing Composition for Chlorine-Containing Polymers and Method of Producing. K. S.Minsker, M. I. ABDULLIN, Y. D Morozov, Z. G.Rasul, N. V.Davedenko, E. V. Screws, I. N.Mullahmetov, R. Z. Biglova; Applicant and Patentee ACT "Caustic."

CHAPTER 21

MODERNIZATION OF SUSPENSION PVC PRODUCTION

N. A. SHKENEVA, R. N. FATKULLIN, and R. F. NAFIKOVA

Bashkir Soda Company JSC, Tekhnicheskaya Str. 32, The Republic of Bashkortostan, Sterlitamak, The Russian Federation, Email: shkeneva.natalia@gmail.com, fatkullin.rna@kaus.ru, nafikova.rf@kaus.ru

CONTENTS

21.1 INTRODUCTION

Polymeric material's industry at the present time is facing the rapid upsurge due to significant expansion of their application areas. Continuous growth of polymer's consumption volume in Russia brings up a question on increased productiveness of existing companies.

One of the most commonly used plastics nowadays is polyvinylchloride, derived from industrial production as a result of the vinylchloride-monomer polymerization in emulsion, in mass, or in suspension. Suspension method of PVC production is the most widely used method that generally includes stages of vinyl chloride polymerization in autoclave-type reactors equipped with the cooling jacket, polymer suspension deaeration, centrifugal separation, and water removal. Efficiency of the process, as a rule, is determined by polymerization stage. At first glance, the simplest way to increase productiveness is the use of high-performance large-volume polymerization reactors. The reactors' size enlargement results in improved efficiency without increase in the amount of primary and support equipment as well as the workforce, but this, however, has some complexities—it leads to reduction of the jacket cooling surface and the reactor volume proportion that significantly sophisticates elimination of the reaction heat and makes it difficult to keep up required polymerization conditions.[1] Besides, this method is quite expensive, since polymerization reactors are the most expensive components of the device. In this regard, the questions about intensification of working equipment performance becomes very crucial.

Intensification of experimental performance can be achieved in two ways[2]:

1. improvement of reactors design;
2. improvement of technological processes in this type of reactors.

In the present chapter, we will focus on the second way of the problem solution.

21.2 EXPERIMENTAL PART

The stage of suspension polymerization of vinyl chloride can be nominally divided into polymerization process itself and the number of auxiliary procedures such as raw charging, the reactor heating up (commissioning), and discharge of PVC suspension. Specific technology of the reactor charge based on loading of preheated vinyl chloride and warm water allows to speed up the process. Such processing methods enable to exclude heating up stage from the cycle and also have a positive impact on molecular weight uniformity of PVC.[3] However, their realization requires capital investments for production modernization, and therefore, there is much tension around the issue of process safety management.

On condition that other parameters of the process are constant, productivity of the polymerization reactor is determined by kinetics and heat elimination conditions of the reaction. As a result, another widely used method of process optimization is based on activation of vinyl chloride polymerization along with intensification of the reactor cooling. In a general way, this is achieved by selection of more effective initiators, the use of mixed initiating systems, and their loading conditions.

Traditional technology stipulates for one-time batch loading of the whole quantity of initiators or a mixture of initiators with a diverse half-life period. Moreover, specified temperature setting is kept constant by adjusting the cooling of water flow into the jacket of the reactor. This method is defined by a significant difference of heat release rate during polymerization process, as a result of which heat output of the cooling systems is used inefficiently as shown in Figure 21.1.

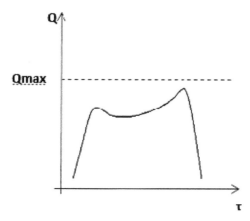

FIGURE 21.1   Quantity of heat relieved by the cooling jacket during vinyl chloride suspension polymerization under traditional technology based on initiator batching.

Alternative to above is a continuous initiator dosage technology according to which a specifically selected initiator with a very short half-life value is loaded into the reactor at the polymerization temperature during the process.[4] Under this method, the temperature inside the reactor is constant due to change of batched initiators amount; as a result, averaging of total heat release of the process with respect to time is observed (Fig. 21.2). This technology enables to significantly increase utilization factor of the cooling systems' thermal output and reduce duration of the polymerization stage by 20–25%.

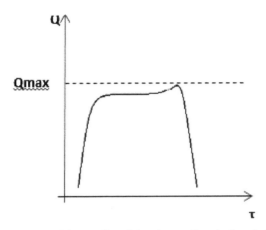

FIGURE 21.2 Quantity of heat relieved by the cooling jacket during vinyl chloride suspension polymerization under continuous initiator dosage technology.

Besides, application of continuous initiator dosage technology has a positive effect on quality attributes of the resulted polymer—contributes to uniformity in particle size distribution, improves homogeneity, and inherent heat stability of PVC due to good distribution of initiator in the system and lack of local hot spots. When using CIB technology, the reactor "hot" loading becomes particularly important—enhanced control over the process and possibility to stop loading of initiator at any moment, guaranteeing proper safety level.

In advanced method of vinyl chloride polymerization reactors, productiveness increase is related to the use of the reflux condensers. Partial cooling process has been derived from the mass process and is known as a very effective method of the reaction heat elimination. In the production of suspension, PVC is used as typical water-cooled tube and shell heat exchangers, whereas a part of the monomer vaporizes from the extent of the reactor, condensates in the reflux condenser tubes, and then the cold monomer comes back to the reaction medium.

21.3   RESULTS AND DISCUSSION

The use of the reflux condensers for suspension polymerization of vinyl chloride predetermines some features of the process conduct connected with foaming of the reaction mass as a result of the monomer boiling that leads to the reflux condenser tubes blockage with the polymer and loss of its heatspreading capability. Yet, gas accumulation and foaming of the reaction mass depend on physical and chemical properties, geometrical parameters of the reactor, and agitation intensity.[3]

Partly, the foaming problem is solved by the use of defoaming agents. In PVC production, as defoaming agents are normally used organosilicon SAS, distinguished by economical efficiency and safety, not influencing polymerization process. High efficiency of their defoaming action is related to low interfacial tension, solvent resistance in foam, and high sputtering ratio.[5]

In the Ref. 6, it is shown that the most dangerous, from the point of view of the reflux condenser blockage, is the starting point of polymerization. For that matter in the beginning of the process base heat quantity of the reaction is removed by means of the cooling jacket, the reflux condenser in this case is not used.

Full use of the reflux condenser potential in vinyl chloride suspension polymerization process is restricted by the fact that rapid increase of heat load onto the reflux condenser leads to enlargement of average particle diameter and deterioration of PVS qualitative properties. This may be caused by desorption of high-molecular stabilizers from the surface of polymerizing suspension drops under intense vinyl chloride vaporization.[1] In the meantime in technical literature, there is no information about negative effect of the reflux condensers working mode on PVC quality; optimum working mode is selected empirically.

Besides, efficiency of partial cooling significantly reduces accumulation of extraneous gases that do not condensate in working mode of the reflux condenser, the major portion of which appears as a result of the initiator decay and is loaded with the feedstock. When using abundant commercial initiators, specific evolution of gas equals to nearly 50 g/kg of the initiator, as a result of which utilization capacity of the reflux condenser heat output during the process decreases by more than 90%. An indicator that allows to evaluate performance of the reflux condenser is the difference between cooling water inlet and outlet temperature. Reduction in the temperature difference tells about low cooling speed. Operational efficiency of the reflux condensers can be restored by means of residual gas removing into recovery system.[7]

21.4 CONCLUSION

Optimization of technological working modes of vinyl chloride reflux condensers in production of suspension PVC by means of minimum technology-based time selection of their plug-in, control during the process, and development of technology of periodic residual gas removal, on condition that qualitative characteristics of the polymer are kept stable, allows to reduce the duration of polymerization stage by more than 30 min; thanks to more efficient use of heat-transfer area. As a result, specific capacity of polymerization reactors increases without extra charges.

Simultaneous implementation of continuous initiator dosage technology, the technology of preheated aqueous phase loading into the reactor, and optimization of heat pick-up in the reflux condensers allows significantly to raise output of the existing industry and increase its efficiency by 40%.

KEYWORDS

- **polyvinyl chloride**
- **suspension polymerization**
- **intensification**
- **initiator dosage**
- **reflux condenser**
- **heat take-off**

REFERENCES

1. *Polyvinylchloride*; Summers, J.; Wilkey, Ch.; Daniels, Ch. St. Professiya ("Profession", in Rus.) Nauka Publishing House: Petersburg, 2007; p 728.
2. Kasatkin, A. G. *Basic Processes and Devices of Chemical Technology*; Khimia ("Chemistry", in Rus.) Publishing House: Moscow, 1971; p 748.
3. Ulyanov, V. M.; Gutkovich, A. D.; Shebyrev, V. V. *Technological Equipment for Suspension Vinyl Chloride Production*; Nizhny Novgorod State Technical University: Novgorod, 2004; p 253.
4. Patent holder(s): Akzo N.V. Nobel (nl). Continuous Dosage of Initiators with a Very Short Half-Life Value During the Polymerization. Russian Federation Patent No. 2249014.
5. Ulyanov, V. M.; Rybkin, E. P.; Gutkovich, A. D.; Pishin, G. A. *Polyvinylchloride*; Khimia ("Chemistry", in Rus.) Publishing House: Moscow, 1992; p 288.
6. Mironov, A. A.; Gutkovich, A. D.; Shebyrev, V. V.; Rybkin, E. P.; Olnev, N. N. Vinyl Chloride Suspension Polymerization in Reactors with Reflux Condenser. *Plastic Masses*, **1989**, *12*, 9–12.
7. Izikaev, A. F.; Shkenyova, N. A.; Asfandiyarov, R. N.; Fatkullin, R. N.; Kantor, E. A. Use of Antifoam Suspension in Vinyl Chloride Suspension Polymerization Process. Proceedings of the 1st International Conference Theory and Practice of New Technologies and Materials Implementation in Manufacturing and Building Industries (in Rus.). Moscow, 2012.

CHAPTER 22

LIQUID PHASE FILLING OF BUTADIENE–STYRENE POLYMERS

I. V. PANKOV[1], V. P. YUDIN[2], and V. N. VEREZHNIKOV[1]

[1]Voronezh State University, Voronezh 394006, Universitetskaya pl No.1, Russian Federation
[2]Voronezh Affiliated Societies of Science Researching Synthetic Rubber, Voronezh 394014, Mendeleeva Str. No.3, Building B, Russian Federation, E-mail: skilful25@mail.ru

CONTENTS

ABSTRACT

Development of the master batch production process is bond to necessity of simplification of tread rubber manufacturing process used for "green" tires as well as elimination of toxic carbon white "dusting." An insertion of nitrous fragments to the polymer chain, which interact with carbon white surface, was an evolution of research work over Li-PBSR rubber. Carbon white, modified by various nitrous and silicon additives, can be produced on an industrial scale as master batch with Li-PBSR. In this case, carbon white "dusting" is prevented while vulcanizate remain its properties. Thus, it can be assumed that the developed compositions can be applied to the production of tire rubbers using "green" tire technology.

22.1 INTRODUCTION

The use of silica extenders (white carbon) in recipe of rubber mixture used for tire treads is one of the greatest developments in the range of tire production technology. The use of silica allows to improve tire traction with a road surface (with a wet, as well with icing) to decrease rolling losses and to increase wear resistance. In domestic rubber industry, the use of high-dispersed silica extenders is not wide spreaded. But it is possible to say today that, in the near future, an essential increase of the high-dispersed grades of carbon white in Russian Federation is predicted. The single, widely used in the industry, method of silica extended rubbers is their preparation in the internal mixer or in mills.[1]

The above-mentioned method has some drawbacks, particularly—a low mixing quality and high energy and working expenses. These drawbacks are specified by a low compatibility of the hydrophobic besieged silica filler (BSF) and as a result—a number and time of mixing stages is increased. The alternative version to the «dry» mixing is rubber extending with silica extenders (masterbatch) at the liquid phase. The world and domestic experience generalization in the range of mixtures production, using liquid phase method demonstrated the advantage of the liquid phase process in comparison with the traditional:

1. the uniform ingredients distribution in extended rubber;
2. the energy consumption decrease till 50% in case of rubber mixtures production on the base of rubber, extended at liquid phase;
3. the improvement of quality characteristics of the rubber mixtures and vulcanized rubber.

In comparison with the classical conditions of the rubber extending with BSF in the industry, which is carried out in mixers at high temperatures (at 160 °C), the liquid phase technology does not require such extremal conditions.[2]

During the masterbatches preparation, the silane coupling agent Si-69 (bis-(triethoxysilylpropyl)-tetrasulfide—TESPT) is used, which has some disadvantages, connected with its structure:

1. evolution during its reaction with carbon white a byproduct as—ethanol, which is not preferred at production;
2. Si-69 is high priced at its use in a quite great quantities in standard recipes of rubber mixture preparation (10 mass parts/100 mass parts of carbon white).

The above-mentioned disadvantages are eliminated via azeotrope during the organic solvents to remove ethyl alcohol molecules and to prevent side reactions. Particularly, replacement of Si-69, the use of the low molecular nitrogenated PB-N in amount of 10 mass parts/100 mass parts of extender was proposed and as a result of this the silane coupling agent content was decreased till 3 mass parts.[3] The raw material expences were decreased till 600 USD of finished product.

For the silica extender use, it is necessary to solve following tasks:

- to provide stability of the prepared suspension;
- to obtain the uniform suspension distribution in the rubber solution;
- to provide effective isolation of the prepared system «rubber-silica» without rubber or silica losses.

The development of the masterbatches production process is also connected with a necessity of the tread rubber process, used for «green» tires manufacturing, simplification as well as elimination of the toxic carbon white «dusting». This makes production of the extended compositions ecology friendly.[4]

22.2 EXPERIMENTAL PART

Two butadiene–styrene rubbers were used:

1. Li-PBSR—non-branched.
2. Li-PBSR-SSBR—block copolymer.

These rubbers were obtained via anionic polymerization, with a narrow and corrected molecular mass distribution ($M_w/M_n = 1.2$–1.6), in this case, the vulcanization net has a more regulated structure during vulcanization process.

The base technology of such rubber production process is developed at the scientific-research institute of synthetic rubber Voronezh branch (Voronezh branch of the Federal state unitary enterprise NIISK).

Li-PBSR rubber has following advantages:

1. The presence of the short end blocks of polystyrene in macromolecules gives an increased cohesive resistance to the raw rubber mixture and in tread rubber, tire traction with a wet road surface increased.

2. The functional compounds contained N–, OH– groups may be introduced in the Li-PBR molecule, which increase rubber interaction with
   extender.

The method of polydiene synthesis via anionic polymerization, using Li-
organic compounds was used. The n-BuLi as well as a product of the 2-stage
styrene and butadiene copolymerization, namely, polybutadienyl–polystyryl
lithium (PB–PS-Li)—35% solution in Nefras M_w = 20,000, active Li-content
Li^+ 0.0021 N were used. Butadiene was charged batchwise, polymerization was
carried out at the temperature of 60–90 °C. Nefras was used as a solvent.

During Li-PBSR-SSBR-N block copolymers synthesis, polymerization product of butadiene with n-BuLi as initiator as well as proton-donor additive M-11
tetraoxypropylene-diamine (modified lapramol) were used. In this case, active
polymer was deactivated via introduction of α-pyrrolidone or ε-caprolactame in
the reaction mixture as a solution in toluene. The rubbers also contain antioxidant Agidol-2 (2,2'-methylene-bis(4-methyl-6-tretbuthylphenol). The obtained
polymerizate and their characteristics are presented in the Table 22.1.

TABLE 22.1 Composition and Properties of the Synthesized Rubbers.

Parameter	Sample no.			
	1	2	3	4
Butadiene, mL	1400	2500	1350	1200
Styrene, mL	250	250	80	20
Dry residue, (%)	17	23	15	15
M_w	100,000	100,000	185,000	320,000
Content polystyrene, (%)	26	19.5	8	8
1,2-links, (%)	54	64.7	11.2	10
Mooney viscosity, arbitrary units	76	87	47	94

While the best results were obtained during Li-PBSR (non-branched structure), namely, No. 4 sample use, which has the best microstructure and Mooney
viscosity values, further the rubber compounds preparation on the basis of one
of the recipes was carried out, using sample No. 4.

The SSBR-N samples preparation was carried out via anionic polymerization, using 35% PB-PS-Li (active Li^+ content 0.0021 N) solution in Nefras as-
iniator, as a modifying agent the proton-donor additive M-11 was used. The
process was deactivated via introduction of the compound in polymerization
mixture on the basis of carbonylic nitrogenated compound of α-pyrrolidone or

ε-caprolactame. The obtained polymerizate and their characteristics are presented in the table.

BSF of «Rhodia» company Zeosil-1165, as well as silane coupling agent Si-69 were used in this work.

The silica extender preparation was carried out via azeotropic drying in organic solvents in the mixture of Si-69 with acetic acid as catalyst for interaction of hydrophobizators and silica surface.

The «carbon white» dispersion was treated with a low molecular polybutadiene (M_w = 2000–2500), modified with nitrogenated end groups on the «carbon white» surface. The low molecular polybutadiene was synthesized via anionic polymerization by deactivation with carbonylic nitrogenated compounds as α-pyrrolidone, ε-caprolactame, and N-methylpyrrolidone.

As this preparation stage, a dry modified silica as extender is used with the trade grades of rubbers, used in standard recipes of rubber mixture preparation.

The carbon white content was changed from 50 mass parts/100 mass parts of Li-PBSR. Then silica filler was mixed with Li-PBSR-N solution and modified DSSK-N with following isolation. The extended rubber isolation was carried out via two methods:

1.   using water–steam degassing;
2.   using waterless degassing.

The degassing and isolation methods of the prepared compositions are very important in this case, and it is necessary to pay great attention to these methods. Because of hydrophility of carbon white, the use of traditional methods of degassing is very difficult. As a result, it is necessary to modify surface of carbon white and/or use agglomerating agents. But using generally accepted degassing and isolation flow diagrams a new technology will be low price and quickly realized.

The use of water–steam degassing allows to minimize capital expenditures, which are necessary for introduction of developed technology in industry. Data, obtained for PB-N, used as modifying agent for carbon white demonstrate that extender losses at such isolation method are low—in the range of 0.5–1.5%, and this is agreed with all norms.

One of the advantages of waterless degassing is quite simple equipment design and as a result of this production of the finished products, which is not require additional drying, as well as decrease of energy losses.

Masterbatch, isolated use in one of the indicated methods, was mixed on mills with other ingredients of rubber mixture. Optimal composition of masterbatch is presented in Table 22.2.

TABLE 22.2 Optimal Composition of Masterbatch.

Rubber	Amount, mass parts
SSBR-N[1]	80
Li-PBSR-N[2]	20
SiO_2[3]	50

Note:

[1] styrene content 30%; 1,2-butadiene units in diene part 70%;

[2] with end NH–CH$_3$ groups;

[3] treatment by 3% Si-69 and 10% polybutadiene with NH-R end groups ($M_w = 2500$).

Three versions of rubber mixture were prepared on mills at 70–80 °C during 30 min. Vulcanization was carried out at 150 °C during 25–50 min. The obtained results show that the use of a more active vulcanization group is resulted in improvement of all indices of tread rubber (in sample B—amount of diphenylguanidine and sulfenamide-T is increased 1.5 times).

22.3 RESULTS AND DISCUSSION

Composition of rubber compounds and results of testing of rubber compounds are presented in Tables 22.3 and 22.4.

TABLE 22.3 Composition of Rubber Compounds.

Composition of mixture, (g)	Standard	Sample no.	
		A	B
Li-PBSR + SSBR-25-60	100	–	–
Zeosil 1165	50	–	–
Masterbatch		150	150
TU N-330	15	15	15
Sulfur	1.3	1.3	1.3
Diphenylguanidine	1.5	1.5	2.25
ZnO	3	3	3
Stearic acid	1	1	1
Oil PN-6	6	6	6
OP deafen	1	1	1
Si-69	8	–	–
Sulfenamide-T	1.7	1.7	2.55

TABLE 22.4 Results of Testing of Rubber Compounds.

Results	Standard	Sample no.					
		A				B	
Mooney viscosity, arbitrary units	82	90				75	
Optimum cure time, min	15.5	25	35	50	25	35	50
M 300%, MPa	9.6	9.5	9.7	10.7	11.7	12.2	11.4
σ, MPa	16.9	17.1	18.1	18.9	20.1	21.9	20.1
L, (%)	440	540	550	530	510	500	520
L, (%)	19.6	35	34	32	24	22	22
Resilience, (%) (20/100 °C)	18 = 32 − 14	–	20 = 34 − 14	–	–	21 = 34 − 13	–
Tear resistance, kN m^{-1}	62	35				107	

From experimental data it is clear that the use of the developed compositions allows, in whole, to improve physical–mechanical and hysteresis properties of vulcanizates in comparison with the well-known production technology of rubber compounds, using trade grades of rubbers. During the obtained compositions use, the process energy was decreased more than one-third.

On the basis of the obtained results, it is possible to advise the above-mentioned compositions for manufacturing of various tire types: lorry and car tires, as well as farm service and industrial tires.

22.4 CONCLUSION

In accordance with a forecast the world consumption in the nearest time will be near 100,000 ton/year and will increase, because of ecology problems.

As it was estimated before, the economic benefit during extended compositions production is more than 600 USD per ton of the finished product, taking into consideration only expenses for raw materials. The energy expenses during this production process carrying out are decreased double.

The economic process indices may be evaluated after organization of trial production, which will be realized on the basis of Voronezh Affiliated Societies of Science Researching Synthetic Rubber (NIISK) in 2015. The development shops of enterprise have possibility to produce more than 2000 ton/year extended compositions for «green tires».

KEYWORDS

- rubbers
- silica filler
- solution polymerization
- master batch, green tires

REFERENCES

1. Rakhmatullin, A. I. Liquid Filling with Rubber Solution Polymerization of the Silica Filler. Thesis PhD Rakhmatullin, A. I., Candidate of Technology Sciences, Kazan, 2010, p. 152.
2. Pankov, I. V.; Verezhnikov, V. N.; Yudin, V. P. A Development of Masterbatch Mixture Based on Lithium Polybutadiene Li-PBR for "green" Tire Production. In *Synthesis, Properties Study, Modification and Recycling of Macromolecular Compounds*, All-Russian Youth Conference, Ufa, 11–14 September 2012.
3. Pankov, I. V.; Yudin, V. P.; Verezhnikov, V. N. Liquid Phase Filling of Butadiene–Styrene Rubbers, Obtained by Solution Polymerization, with Carbon White, Modified with Silicon and Nitrous Compounds. In *Compilation of Reports Theses in 2 Sections. Section One*, 6th All-Russian Kargin Conference "Polymers 2014," Moscow, January 2014, Vol. II, 27–31.
4. Grishin, B. S. *Materials Rubber Industry*; KSTU: Kazan, Vol. 1, 2010, p 506.

CHAPTER 23

A STUDY ON COMPOSITION OF STEREOISOMER OF AMINO ACID COMPLEXES

T. V. BERESTOVA, G. V. MIFTAKHOVA, G. YU. AMANTAEVA, L. G. KUZINA, N. A. AMINEVA, and I. A. MASSALIMOV

Bashkir State University, Faculty of Chemistry, Chair of Inorganic Chemisry, 32 Zaki Validi Str., Ufa 450043, Russian Federation, E-mail: berestovatv@gmail.com

CONTENTS

ABSTRACT

The *bis*-(amino acidato)copper(II) **1–3** were synthesized under standard conditions according to the published method[1] and characterized by IR spectroscopy and XRD.

IR and XRD data show that *bis*-(glycinato)copper(II) (**1**) turned *cis-isomer* (**1a**), *bis*-(*L*-)alaninato)copper(II) (**2**) was obtained as *cis*-(**2a**) and *trans*-(**2b**) isomers in ratio 1:2 and *bis*-(*DL*-valinato)copper(II) (**3**) represented only as *trans*-(**3b**) geometric isomer.

23.1 INTRODUCTION

In recent years the importance of metal ions for smooth operation of plant and animal organisms was evidenced by the publication of numerous papers in the field of biophysics and biochemistry. Latest researches in these areas focus on the synthesis and characterization of biological compounds containing metal ions, due to their applicability in pharmacy, medicine, and agronomy.[2,3]

Synthesis and study of the structure of the amino acid complexes of biometals (Cu, Ni, Co, Zn, and other), including optically active, are summarized in some reviews of last years.[4–8] As a rule, analysis of the *bis*-(amino acidato) metal(II) complexes is carried out by IR, UV spectroscopy, and EPR.[9–13]

In recent years, an increased focus on studies carried out on the complexation of transition metal ions, including Cu(II) with different bioligands. In literature,[14] various applications of metal complexes with amino acids—in medicine as antibacterial activity (complexes with Cu-*alanine*, *arginine*, *histidine*, and *lysine*) on bacteria like *Staphylococcus aureus*, *Streptococcus pyogenes*, and *Escherichia coli* in regulating gastric (Cu-*tryptophan*, *phenylalanine*) are also reported.[2,3]

Thus, study composition of stereoisomer of amino acid complexes is an important and urgent task.

23.2 EXPERIMENTAL PART

All used glasswares were washed with a detergent, rinsed with distilled water, and dried in an oven before use. Amino acids (glycine, *L*-alanine, (*DL*)-valine) were purchased from Avilon-companychem. IR-spectra were recorded on spectrometer FTIR-8400S (Shimadzu). The X-ray diffraction analysis was made on the diffractometer DRON-4.

23.1.1 SYNTHESIS OF *BIS*-(GLYCINATO)COPPER(II) (1A)

According to the published method,[1] in the 50 mL flask equipped with a magnetic stirrer, we mixed 0.82 g $CuSO_4 \cdot 5H_2O$ (3.3 mmol) with glycine (0.5 g, 6.6 mmol). The reaction mixture was stirred for 1 h at room temperature. Further, in the mixture $Ba(OH)_2$ water solution (3.3 mmol) was added. The reaction mixture was left for 1 h at room temperature and blue solution was filtered and dried. The blue powder of *cis-bis*-(glycinato)copper(II) **(1)** was obtained, weighed (yield 70% based on glycine), and analyzed by IR and XRD.

IR (powder) cm^{-1}: 3267, 3165, 2976, 2885, 1605, 1580, 1593, 1396, 1404.

23.1.2 SYNTHESIS OF *BIS*-(*L*-)ALANINATO)COPPER(II) (2A,B)

According to the published method,[1] we mixed in the 50 mL flask equipped with a magnetic stirrer we mixed 0.82 g. $CuSO_4 \cdot 5H_2O$ (3.3 mmol) with *L*-alanine (0.5 g, 5.6 mmol). The reaction mixture was stirred for 1 h at room temperature. Further, in the mixture $Ba(OH)_2$ water solution (3.3 mmol) was added. The reaction mixture was left for 1 h at room temperature and blue solution was filtered and dried. The lilac powder of *cis*-**(a)** and *trans*-**(b)** isomers *bis*-(*L*-)alaninato) copper(II) **(2a,b)** in ratio 1:2 was obtained, weighed (yield 87% based on alanine) and analyzed by IR and XRD.

IR (powder) cm^{-1}: *major (trans-isomer):* 3278, 3290, 2982, 2885, 1618, 1358.

minor (cis- isomer): 3240, 3141, 2985, 2970, 1576, 1394.

23.1.3 SYNTHESIS OF *BIS*-(*DL*-VALINATO)COPPER(II) (3B)

According to the published method,[1] we mixed in the 50 mL flask equipped with a magnetic stirrer we mixed 0.82 g. $CuSO_4 \cdot 5H_2O$ (3.3 mmol) with (*DL*)-valine (0.5 g, 4.3 mmol). The reaction mixture was stirred for 1 h at room temperature. Further, in the mixture, $Ba(OH)_2$ water solution (3.3 mmol) was added. The reaction mixture was left for 1 h at room temperature and blue solution was filtered and dried. The violet powder of only *trans-bis*-(*DL*-valinato)copper(II) **(3)** was obtained, weighed (yield 48% based on valine) and analyzed by IR and XRD.

IR (powder) cm^{-1}: 3302, 3255, 2962, 2908, 1621, 1390.

23.3 RESULTS AND DISCUSSION

We have synthesized and studied the *bis*-(amino acidato)copper(II) complexes: *cis*-[Cu(gly)$_2$] (**1**), *cis*-[Cu(*L*-ala)$_2$] (**2a**), *trans*-[Cu(*L*-ala)$_2$] (**2b**), *trans*-[Cu((*DL*)-val)$_2$] (**3**) (where gly-glycinato, ala-*L*-alaninato, val-*DL*-valinato) by IR spectroscopy and XRD (Scheme 23.1, Table 23.1).

Scheme 1

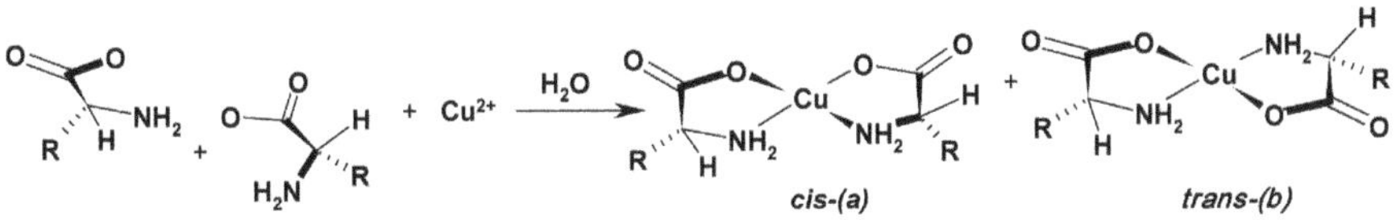

R = H (1), CH$_3$ (2), CH(CH$_3$)$_2$ (3)

TABLE 23.1 Infrared Spectral Data of Glycine, (*DL*)-Valine, *L*-Alanine, and Complexes of **1–3** (cm^{-1})

Complex	Color	Yield, (%)	Characteristic IR spectra						
			v(–COO⁻) as	v(–COO⁻) s	Δv (COO⁻)	v(–C(CO)–) as	v(–C(CO)–) s	v(–CH, CH$_3$)	v(–NH$_2$) (as, s)
Glycine (gly)		–	1608	1412	196	1506	1334	2923	3125, 3180
(*DL*)-valine (val)		–	1595	1418	177	1504	1325	2961	3126
L-alanine (ala)		–	1593	1410	183	–	–	2854 2924	3078
cis-[Cu(gly)$_2$] (**1**)	Blue	70	1605 1580 1593	1396 1404	201	–	–	2976 2885	3165, 3267
cis-[Cu(*L*-ala)$_2$] (**2a**)			1576	1394	182	–	–	2985 2970	3141, 3240
trans-[Cu(*L*-ala)$_2$] (**2b**)	Lilac	87	1618	1358	260	–	–	2982	3278, 3290
trans-[Cu((*DL*)-val)$_2$] (**3**)	Violet	48	1621	1390	231	–	–	2962 2908	3255, 3302

It was established that the ratio of *cis*- and *trans*-isomers of 1–3 determined by the nature of the ligand. Thus, the *bis*-(glycinato)copper(II) 1 isolated in solid form, represented the *cis*-isomer exclusively. The use of the sterically hindered ligands (valine, alanine) resulted in the formation of mainly *trans*-isomers 2, 3.

In IR spectra, the characteristic absorption bands (cm^{-1}) of the complexes **1–3** were found of the –NH_2– group and COO^- anion. It was found that spectra (–NH_2) group appears at 3078–3302 cm^{-1} for symmetric v_s (–NH_2) and asymmetric v_{as} (–NH_2) stretching vibrations. For the carboxylate, anion is characteristic of the region 1576–1621 cm^{-1} for v_{as} (COO^-) and 1358–1404 cm^{-1} for v_s (COO^-). Absorption band of 2885–2985 cm^{-1} corresponds to the stretching vibration –$CH(CH_3)$ groups.

It is revealed that the *trans*-isomers tend to present two characteristic lines of absorption bands in the region 3200–3300 cm^{-1}, unlike the *cis*-isomers which were three groups NH_2– absorption peak in a wider range of 3100–3350 cm^{-1}.

Furthermore, it was found that $\Delta v_{as,s}$(COO^-) for the geometric isomers **1–3** are different. The *cis*-isomer (**1,2**) and *trans*-isomer (**2,3**) exhibit $\Delta v_{as, s}$ (COO^-) at 182–201 cm^{-1} and 231–260 cm^{-1}, respectively. Thus, Δv for *cis*-isomer is substantially less than the *trans*-isomer.

The structure of the complexes **1–3** also studied by XRD. Thus, for a mixture of stereoisomers in ratio of 1:2 results in two crystalline phases, while the *trans*-isomer is characterized by only one crystalline phase.

23.4 CONCLUSION

Finally, some *bis*-(amino acidato)copper(II) were characterized by IR spectroscopy and XRD and found that the ratio obtained by *cis*- and *trans*-isomers of **1–3** was determined by the nature of the ligand. For the identification of the geometrical isomers by IR spectroscopy we can use the value $\Delta v_{as,s}$(COO^-), which is different for the geometric isomers **1–3**.

KEYWORDS

- copper
- amino acids complexes
- *cis*-, *trans*-**isomers**
- glycine
- *L*-alanine
- (*DL*)-valine
- IR spectroscopy
- XRD

REFERENCES

1. Malinin, V. V.; Pushkarev, A. N.; Chromov, A. N. Patent Russian Federation 2430733.
2. Mariana, U. Structural Studies of Metal Complexes with Amino Acids and Biomarkers for use in Diagnostic. Doctoral Thesis Summary, Cluj-Napoca, 2012, p 19.
3. Arish, D.; Nair, M. S. Synthesis, Characterization and Biological Studies of Co(II), Ni(II), and Zn(II) Complexes with Pyrral-L-histidinate. *Arab. J. Chem.* **2012**, *5*, 179–186.
4. Chaturvedi, K.S.; Hung, C. S.; Crowley, J. R.; Stapleton, A. E.; Henderson, J. P. The Sidero-phore Yersiniabactin Binds Copper to Protect Pathogens during Infection. *J. Nature Chem. Biol.* **2012**, *8*, 731–736.
5. Corradi, A. B. Structures and Stabilities of Metal(II) (Co(II), Ni(II), Cu(II), Zn(II), Pd(II), Cd(II)) Compounds of N-Protected Amino Acids. *Coord. Chem. Rev.* **1992**, *117*, 45–98.
6. Llobet, I.; Álvarez, M.; Albericio, F. Amino Acid-Protecting Groups. *Chem. Rev.* **2009**, *109* (*6*), 2455–2504.
7. Santini, C.; Pellei, M.; Gandin, V.; Porchia, M.; Tisato, F.; Marzano, C. Advances in Copper Complexes as Anticancer Agents. *Chem. Rev.* **2014**, *114*, 815–862.
8. Gielen, M.; Tiekink, E. R. T. *Metallotherapeutic Drugs and Metal-Based Diagnostic Agents the Use of Metals in Medicine*; John Wiley and Sons, Ltd.: USA, 2005; p 638.
9. Herlinger, A. W.; Wenhold, S. L.; Long, T. V. Infrared Spectra of Amino Acids and their Metal Complex. II. Geometrical Isomerism in bis(amino acidato)copper(II) Complexes. *JACS* **1970**, *92* (*22*), 6474–6481.
10. Fernandes, M. C. M. M.; Paniago, E. B.; Carvalho, S. Copper(II) Mixed Ligands Complexes of Hydroxamic Acids with Glycine, Histamine and Histidine. *J. Braz. Chem. Soc.* **1997**, *8* (*5*), 537–548.
11. Stradeit, H.; Strasdeit, H.; Büsching, I.; Behrends, S.; Saak, W.; Barklage W. Syntheses and Properties of Zinc and Calcium Complexes of Valinate and Isovalinate: Metal α-Amino Acid-ates as Possible Constituents of the Early Earth's Chemical Inventory. *Chem. Eur. J.* **2001**, *7* (*5*), 1133–1142.
12. Reddy, P. R., Radhika, M; Manjula, P. Synthesis and Characterization of Mixed Ligand Complexes of Zn(II) and Co(II) with Amino Acids: Relevance to Zinc Binding Sites in Zinc Fingers *J. Chem. Sci.* **2005**, *117*, 239–246.
13. Shukla, S.; Kashyap, A.; Kashyap, A. Spectroscopic and Thermal Studies of Copper(II) with Different Amino Acid. *J. Chem. Pharm. Res.* **2013**, *5* (*9*), 142–145.
14. Cuevas, A.; Viera, I.; Torre, M. H.; Kremer, E.; Etcheverry, S.B.; Baran, E. J. Infrared Spectra of the Copper(II) Complexes of Amino Acids with Hydrophobic Residues. *Acta Farm. Bon.* **1998**, *17* (*3*), 213–218.

CHAPTER 24

MICROTUBES FROM A CHITOSAN SOLUTION IN GLYCOLIC ACID USING DRY SPINNING METHOD

T. S. BABICHEVA, N. O. GEGEL, and A. B. SHIPOVSKAYA

Educational and Research Institute of Nanostructures and Biosystems, Saratov State University, 83 Astrakhanskaya Str., Saratov 410012, Russian Federation, E-mail: Tatyana.babicheva.1993@mail.ru, GegelNO@yandex.ru, ShipovskayaAB@rambler.ru

CONTENTS

ABSTRACT

Microtubes were prepared from chitosan solutions in glycolic acids by dry molding methods with salting-out agents of different nature: NaOH (an organic base), $N(C_2H_4OH)_3$ (an inorganic base), and $C_{12}H_{25}C_6H_4SO_3Na$ (an anionic surfactant). Their morphology, biocompatibility, elastic deformation, and physicomechanical properties were examined. The influence of the salting-out agent nature on the mechanism of the chemical reaction proceeding during the microtube wall formation was estimated. The usage of sodium dodecylbenzenesulfonate (SDBS) as a salting-out agent allows obtaining microtubes with high-strength properties. High adhesion and high proliferative activity of the epithelial-like MA-104 cellular culture on the surface of our microtubular substrates in model in vitro experiments were revealed. Certain physicochemical and biochemical parameters are comparable with the similar characteristics of human blood vessels.

24.1 INTRODUCTION

Nowadays, implants made of biocompatible biodegradable polymers are widely used as pins, screws, plates, membranes, and meshes in reconstructive surgery (orthopedics, maxillofacial surgery, dentistry) to stimulate osteosynthesis and to improve the healing of bones and soft tissues. In recent years, materials made of biodegradable polymers have begun to be used as prostheses, stents, and grafts in cardiovascular, urological, and ophthalmic surgery. The design and usage of such medical materials are based on the idea of replacing the damaged sections of hollow tubular organs (blood vessels, ureter, tear ducts, etc.) by their bioartificial analogs. Over time, such a biodegradable tube-frame becomes covered with the human connective and endothelial tissues and is excreted from the body. After the final bioresorption of the polymer framework, a new living organ (or part thereof) is formed and the functional quality of the old tissue completely restores.

In this connection, the design of preparation methods of hollow cylindrical structures with the width of their walls in the micron range (microtubes) from biodegradable polymers for the use as prostheses of biomedical purposes, particularly, as analogs of blood vessels, is urgent. It is known from the literature of the preparation of such materials from biodegradable synthetic and natural polymers. For example, approaches are known to the preparation of tissue-engineering hollow cylindrical structures by electrospinning of poly(L-lactide-co-glycolide) and poly (ε-caprolactone),[1,2] a blend of polydioxanone and soluble elastin,[3] poly(diol citrate) as a biodegradable polyester elastomer,[4] and self-assembled hydrogel plates based on poly(ethylene glycol) diacrylate.[5] Blood vessel analogs are also made of heteropolysaccharides, such as alginates, gellan gum, Kappa carrageenan, hyaluronan, and chitosan.[6–9] However, the methods and techniques of obtaining microtubular structures described in Refs.1–9 are multistage and often involve aggressive solvents. The latter circumstance, in turn, would adversely affect the biocompatibility and other biochemical properties of the finished material, and would limit its use in clinical practice. In addition, the functional transformations of a bioprosthesis related to its resorption in the human body and, consequently, to cell division on it and tissue repairing, determine specific requirements to the physicomechanical properties of the material. For example, the strength and elasticity indicators of the original structures must exceed those of vascular prostheses and patches made of non-biodegradable polymers. In this connection, the search for new approaches to designing hollow cylindrical frame structures (blood vessel analogs) with desired parameters is topical.

The aminopolysaccharide chitosan is a promising polymer for making biodegradable materials of biomedical purposes. Due to its properties such as biocompatibility with human tissues, nontoxicity, nonallergic nature, the ability to regenerate tissues, and bioresorbability in the metabolic environment, chitosan can be used for making biodegradable prostheses for reconstructive surgery.[10–12] In Refs. 13 and 14, we describe the production of 3D microtubes from chitosan solutions by the wet and dry molding processes. These methods are based on phase separation processes and the polymer-analogous transformation polysalt → polybase. Aqueous solutions of NaOH and triethanolamine (TEA) were used as transfer agents (salting-out agents) to convert chitosan in microtubes from its polysalt form into the water-insoluble polybasic form. Citric and lactic acids were used as solvents in forming the chitosan solutions. The microtubes obtained from chitosan solutions in citric acid are found to have a fragile porous inner layer. For the microtubes obtained from chitosan solutions in lactic acid, their morphology, biocompatibility, elastic deformation, and physicomechanical properties were evaluated. The strength of the microtubes obtained by the dry method is much higher than in the case of the wet one. It is suggested that chitosan microtubes are promising as vascular prostheses. To continue our research in this direction, glycolic acid was used to prepare forming chitosan solutions in this work. Its choice is due to the fact that it falls into the class of pharmacopoeial acids,[15] and the polyesters and copolymers based thereon are not only actively studied to produce materials for surgery, tissue engineering, and bioartificial organs,[16–18] but are already widely used in regenerative medicine.[19,20]

The aim of this work was to obtain microtubes from a chitosan solution in glycolic acid by the dry spinning method with the usage of salting-out agents of different chemical nature to study their morphology, mechanical and biochemical properties.

24.2 EXPERIMENTAL

24.2.1 MATERIALS AND REAGENTS

Chitosan with a molecular weight of 700 KDa and a deacetylation degree of 80 mol.% was received from Bioprogress Ltd. (Russian Federation). Glycolic acid (GA) was used as an organic solvent, $C_{GA} = 1.5\%$. The physicochemical characteristics of glycolic acid are as follows: mp 79 °C, the dissociation constant in water $K = 1.48 \times 10^{-4}$ (25 °C). Chitosan solutions with $C_{CH} = 3.5–4.0$ wt.% were prepared by dissolving a sample of air-dried powder of the polymer in the presence of the acid on a magnetic stirrer at room temperature during 5–7 h. The resulting solution was left to remove air bubbles for 24 h and used to form microtubes by both wet and dry methods. To convert the chitosan in microtubes from its polysalt form to the water-insoluble polybasic one, several salting-out agents were used (in various combinations): an organic base—50% TEA ($K = 4.5 \times 10^{-7}$, 20 °C), an inorganic base—5% NaOH ($K = 5.9$, 25 °C), and an anionic micelle-forming surfactant—0.1 M SDBS (CMC = 1.3 mM), capable of forming ionic bonds with protonated amino groups.[21] Bidistilled water was taken to prepare solutions of the acid, alkalis, triethanolamine, and SDBS. Xenopericardial biomaterial was gotten from bovine pericardium, the manufacturer Cardioplant Ltd. (Russian Federation).

24.2.2 MICROTUBE PREPARATION TECHNIQUE

Figure 24.1 presents a diagram of microtube preparation from chitosan by the dry molding method. The microtube formation process was performed as follows. A glass rod with a circular cross-section (the diameter of 5 mm) was vertically immersed into a chitosan solution and kept for 1 min during its surface to be coated with a uniform layer of the solution. The rod coated with a chitosan solution layer was then immersed in an aqueous solution of the salting-out agent for 1 min to form a first layer of the water-insoluble polymer (FLWIP), removed, and dried in an oven at 45–50 °C until complete evaporation of the solvent (4–5 h). A condensed phase of chitosan glycolate was formed. Then, the rod was again immersed in the aqueous solution of the salting-out agent to form a second layer of the water-insoluble polymer (SLWIP) and, correspondingly, to form the wall of the microtube as a hollow cylinder. Samples of the microtubes obtained by both dry methods were removed from the rod, washed with distilled water until neutral reaction, and stored in a swollen state in distilled water. The volume fraction of the solid phase in swollen samples was ~80 wt.%.

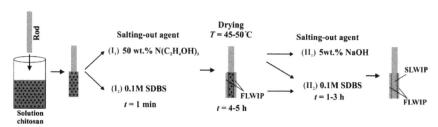

FIGURE 24.1   A scheme of making microtubes from chitosan solutions in glycolic acid by the dry molding method, the salting-out agent being 5% NaOH, 50% TEA, or 0.1 M SDBS.

24.3   METHODS OF EXAMINATION

The morphology of sample chitosan microtubes was evaluated visually for color, transparency, and the uniformity of the inner and outer surfaces. The samples were photographed with a digital camera Canon EOS 650D Kit (China).

SEM photos were obtained on a MIRA LMU scanning electron microscope (Tescan, Czech Republic), a 15 kV accelerating voltage, a 2 pA current. Cuts of the samples were prepared for microscopic examination as follows: a microtube was taken out from distilled water and put into a special holder; a ring of a 3-mm length was cut with a sharp blade and dried in air for 24 h. The prepared sample was then removed from the holder and used in experiments. A 5-nm thick layer of gold was deposited on the obtained sample with a K450X Carbon Coater device (Deutschland).

The wall thickness (d, mm) of our microtubes was measured with an electronic digital outside micrometer CT 200-521 (China) with a scale division of 10 μm. Measurements were carried out several times at different sites of every microtube, and the thickness was then averaged.

The humidity of the samples was determined on a moisture analyzer AND MX-50 (Japan), the weighing accuracy being ±0.0001

Elastoplastic properties were examined on a tensile testing machine of uniaxial tension Tinius Olsen H1KS (Deutschland) with a 50 N load cell. Microtubes were cut along, unfolded, and fixed in the terminals of the machine as plates. Every sample was tested at least five times. The breaking force and elongation at break were calculated according to standard procedures. The data obtained were plotted as stress–strain curves $\sigma = f(\varepsilon)$. The tensile stress (σ, MPa) was calculated on the basis of the cross-sectional area of original samples. The elongation at break (ε, %) was calculated using the original length of sample microtubes.

To assess the biocompatibility of our microtubes, the MA-104 culture of epithelial-like cells (rhesus monkey embryonic kidney) was used. Cells were cultivated under sterile conditions—in a laboratory specially equipped with a complex of "clean" laminates (second class protection, Nuaire, USA). A microtube sample was sterilized in a 70% ethanol solution for 20 min, placed into sterile Petri dishes (FalconBD), filled with the DMEM culture medium (Biolot, RF) supplemented with 10% fetal bovine serum (Fetal bovine serum, Hyclone UK) and an antibiotic–antimycotic mixture, and a suspension of the cell culture was introduced at a concentration of 1×10^4 cells/cm^3. Culturing was performed in a SanyoMCO-18 M CO$_2$ incubator (Sanyo, Japan) in a 5% CO$_2$ atmosphere at 37 °C. The culture medium was replaced on changing the color of the medium indicator. Cells were cultivated during 7 days. The adhesion and proliferation of the cell culture on our microtubes were evaluated daily on an inverted Biolam P microscope (Russian Federation) with a digital CCD camera DMS 300 Scopotek (China): a 10× lens, 3 Mp× resolution.

24.4   RESULTS AND DISCUSSION

To form the FLWIP–SLWIP of the chitosan microtube walls by the dry spinning process, several combinations of the salting-out reagents were used: TEA–NaOH ($I_1 - II_1$), TEA–0.1 M DDBSN ($I_1 - II_2$), and DDBSN–DDBSN ($I_2 - II_2$) (Fig. 24.1). The interaction mechanism of chitosan macromolecules with the molecules of the salting-out agent is as follows.

In the initial aqueous acidic solution, chitosan exists in its polysalt form ($-NH_3^+$) due to protonation of its side NH$_2$ groups:

$$-NH_2 + HOCH_2COOH \rightarrow -NH_3^{+ -}OOCCH_2OH.$$

When organic or inorganic bases are used to form microtube walls, in the thin layer of the chitosan solution in glycolic acid deposited on the glass rod, the polymer-analogous polysalt $\rightarrow$ polybase conversion proceeds accompanied by phase separation. When FLWIP of the microtube wall is being formed, the chemical reaction proceeds at the interface of the two liquid phases: chitosan solution–salting-out agent solution. When SLWIP of the microtube wall is being formed, the reaction occurs at the solid–liquid interphase: condensed chitosan glycolate layer–salting-out agent solution. Note that the solid (condensed) polysalt phase is formed at the drying stage (Fig. 24.1) as a result of evaporation of water molecules, and probably some of the molecules of glycolic acid, unbound ionically with chitosan amino groups.

In an alkaline medium (NaOH), the salt groups are converted to amine ones to form a polybase (–NH2), and the polymer loses its solubility ($\downarrow$). The conversion scheme is shown below:

$$-NH_3^{+\,-}OOCCH_2OH + Na^{+\,-}OH \rightarrow -NH_2(\downarrow) + HOCH_2COONa + H_2O.$$

The polymer-analogous reaction of polysalt $\rightarrow$ polybase conversion proceeding in the solid polymer phase is due to the directed diffusion of low molecular weight ions (OH$^-$, Na$^+$) to the condensed polymer layer due to a concentration gradient at the –NH$_3^+$|NaOH interface, resulting in the equalization of the chemical potential of the low molecular weight component over the system.

Similar processes also proceed when an aqueous TEA solution is used as the salting-out agent. Glycolic-triethanolamine ester is one of the products of this chemical interaction:

$$(HOCH_2CH_2)_3-N + H_2O \rightarrow [(HOCH_2CH_2)_3-NH]^{+\,-}OH$$

$$-NH_3^{+\,-}OOCCH_2OH + [(HOCH_2CH_2)_3-NH]^{+} + {}^-OH \rightarrow$$

$$-NH_2(\rightarrow) + (HOCH_2CH_2)_2-N-CH_2CH_2OOCCH_2OH + 2H_2O.$$

As TEA is a weak base ($K = 4.5 \times 10^{-7}$), the formation rate of the polybase is significantly lower than for a stronger base, NaOH. The possibility of such a reaction is evidenced by Ref. 22, which shows the possibility of the formation of a similar ester by reacting TEA with acetic acid.

When SDBS is taken, the microtube wall is formed by the ionotropic gelation mechanism as a result of frontal diffusion of the surfactant through the interphase. When FLWIP or SLWIP is being formed, the chitosan glycolate macromolecules interact with the surfactant molecules in the liquid phase or the solid (condensed) polysalt phase, respectively. In both cases, a polyelectrolyte–surfactant complex is formed according to the following reaction:

$$-NH_3^{+\,-}OOCCH_2OH + C_{12}H_{25}C_6H_4SO_2O^{-\,+}Na$$

$$\rightarrow -NH_3^{+\,-}OSO_2C_6H_4C_{12}H_{25} + HOCH_2COONa.$$

Figure 24.2a–c shows photos of our samples of chitosan microtubes obtained using several salting-out agents. The microtubes have no seams and are characterized by smooth external and internal wall surfaces without visual defects.

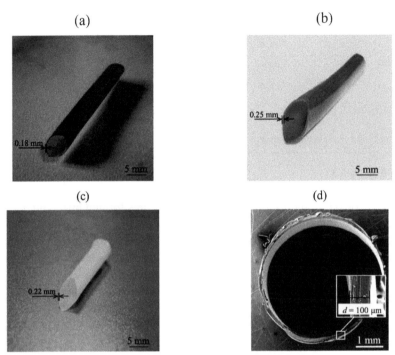

FIGURE 24.2 Photos of our microtubes obtained from a 4.0 wt.% chitosan solution in glycolic acid using the following combinations of salting-out agents: (a) 50% TEA–5% NaOH, (b) 50% TEA–0.1 M SDBS, (c) SDBS 0.1M–0.1 M SDBS, and (d) a SEM photo of a cut of the dried tube shown in Figure 24.2a.

All samples are characterized by the almost uniform distribution of wall thicknesses along the length of the sample. The wall thickness of our microtubes depends on the nature of the salting-out agent and reaches its maximum value for SDBS (Fig. 24.2a–c; Table 24.1). This is probably due to some peculiarities of the chemical and physicochemical properties of the salting-out agents used.

TABLE 24.1 Physicomechanical Properties of our Chitosan Microtube Samples.

Chitosan concentration (C_{CH}), wt.%	Salting-out agent		Microtube wall thickness (d), mm	Relative elongation (ε), %	Elongation at break (σ), MPa
	FLWIP	**SLWIP**			
	TEA	NaOH	0.09	22.0	3.5
3.5	TEA	SDBS	0.17	10.0	9.7
	SDBS	SDBS	0.18	28.0	4.0

TABLE 24.1   (*Continued*)

	TEA	NaOH	0.16	21.0	2.0
4.0	TEA	SDBS	0.25	29.0	5.7
	SDBS	SDBS	0.31	29.0	6.8
Xenopericardial biomaterial			0.28	20.0	5.7
Human blood vessels (vien, artery)			0.25–0.71	5–10	1.4–11.1*

*Data from Ref. 23.

Let us consider possible supramolecular structures formed by the interaction of chitosan glycolate with salting-out agents of different chemical nature. In all cases, that is, whether a base or a surfactant is used as a salting-out agent, the structure of the surface layers of the microtube wall is a semipermeable membrane that is permeable to the molecules of low molecular weight substances (water). The application of an organic or inorganic base to form SLWIR and FLWIR of the microtube wall leads to the formation of a relatively dense supramolecular structure of the polybasic condensed phase (Fig. 24.2d) stabilized with a network of intramolecular and intermolecular contacts (hydrogen bonding). When a micellar surfactant solution is used (with a SDBS concentration well above its CMC), the supramolecular structure of the microtube wall can be represented as a loose 3D network formed by micelle-like clusters of the surfactant–polyelectrolyte complexes. The size of the internal cavities in such a structure is known to be determined by the characteristic size of the polyelectrolyte–surfactant complex measured as the average distance between the micelle-like aggregates in the gel.[24,25] The size of the latter, during ionotropic chitosan gelation in the presence of anionic surfactants such as sodium dodecyl sulfate, can reach 3.5–7.5 nm.[26] Therefore, the differences in the wall thickness of the microtubes obtained with TEA, NaOH, or SDBS are logically explained by the different character of the supramolecular organization of the solid phase of polybase and polycomplex.

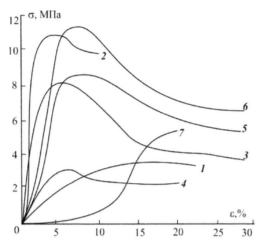

FIGURE 24.3 Stress–strain curves for microtube samples obtained from chitosan solution in glycolic acid at polymer concentrations $C = 3.5$ (curves 1–3) and 4.0 wt.% (curves 4–6) and using the following combinations of salting-out agents: 50% TEA–5% NaOH (1:4), 50% TEA–0.1 M SDBS (2, 5), and SDBS 0.1 M–0.1 M SDBS (3, 6); and for a xenopericardial sample, a conventional material for bioprostheses and patches of blood vessels (7).

Figure 24.3 shows the deformation curves $\sigma = f(\varepsilon)$ of chitosan microtube samples obtained by using various reagents to convert the protonated amino groups of the polysalt into the basic amino groups of the polybase.

Figure 24.3 (curve 1) shows the deformation curve of the sample obtained from a chitosan solution with $C = 3.5$ wt.% with the TEA–NaOH pair of salting-out agents to form FLWIP and SLWIP of the microtube wall (Fig. 24.1, $I_1 – II_1$). It is evident that the stress–strain dependence has an elastic deformation fragment, whose value is about 4–5%, and a plastic deformation one, this deformation occurring uniformly without yield. Such $\sigma = f(\varepsilon)$ curves are typical of viscoplastic materials. The stress–strain curves of the samples obtained from the same solution but using TEA–SDBS (Fig. 24.1, $I_1 – II_2$; Fig. 24.3, curve 2), SDBS–SDBS (Fig. 24.1, $I_2 – II_2$; Fig. 24.3, curve 3) to form FLWIP–SL-WIP, and the solutions with $C = 4.0$ wt.% (Fig. 24.3, curves 4–6) show forced elastic deformation characteristic of viscoelastic materials. These samples are deformed at tension to form a neck due to the flow effect.

It should be noted that the use of SDBS to form the microtube wall leads to significantly stronger samples (Fig. 24.3, curves 2, 3, 5, and 6; Table 24.1) as compared to using TEA and NaOH for the chemical polysalt → polybase reaction (curves 1 and 4). The elasticity of these samples does not change and remains sufficiently high, ~28–30% (curves 3, 5, and 6). The maximum values of $\sigma = 11.5$ MPa and $\varepsilon_p = 29.5\%$ were observed for the microtube obtained from the chitosan solution with $C = 4.0$ wt.% with the use of SDBS–SDBS (curve 6). Such high values of tensile strength are perhaps due to peculiarities of the supramolecular structure of the polymer system formed by the surfactant–polyelectrolyte complexes.[21,24,25] Moreover, hardening of the structure is facilitated by heat treatment at one of the stages of forming microtubes (Fig. 24.1), it enhances the interaction between surfactant molecules and chitosan macromolecules.[27]

It should be noted that the elastic deformation characteristics and strength values of microtubes are significantly higher than σ_p and ε_p of xenopericardial biomaterial samples traditionally used in reconstructive surgery of the heart and blood vessels (Fig. 24.3, curve 7) and are comparable with the same characteristics of human blood vessels.[23]

In model in vitro experiments, the biocompatibility of our chitosan microtubes was assessed with the epithelial cell culture MA-104 as an example. The morphology of the cells cultured for 1 week on the microtube surface is illustrated in Figure 24.4. The experimental results show good adhesion and high proliferative activity of epithelial cells on the surface of the microtubular substrates.

(a)   (b)

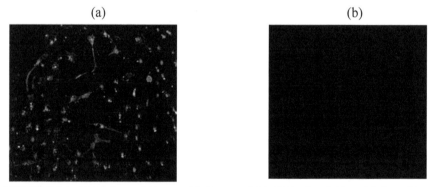

FIGURE 24.4 Fetal rhesus kidney epitheliocytes (MA-104) after 4 (a) and 7 days (b) of culturing on the surface of a microtube obtained from a 4.0 wt.% chitosan solution in glycolic acid using the TEA–NaOH pair of salting-out agents. Light microscopy. Magnification × 100.

24.5 CONCLUSION

The possibility of producing hollow cylindrical structures with a wall thickness in the micron range (microtubes) from chitosan solution in glycolic acid by the dry molding method is shown. The influence of the salting-out agent nature (a base or a surfactant) on the supramolecular ordering, elastic–plastic, and physicomechanical properties of the samples was evaluated. The usage of an organic or inorganic base as a salting-out agent leads to the formation of a densely packed supramolecular structure of the microtube wall formed like a physical network of entanglements and stabilized by hydrogen bonding. When SDBS is used as a salting-out agent, a network structure of the microtube wall is formed, consisting of water-insoluble electrostatic complexes between the oppositely charged macromolecules of the polyelectrolyte (chitosan) and the surfactant molecules. The use of SDBS, in comparison with other salting-out agents, significantly raises the strength of the samples. The mechanical properties of our microtubes are superior to those of xenopericard, a biomaterial successfully used in cardiac surgery for several decades, and are comparable to native blood vessels (veins and arteries). In vitro experiments have shown high biocompatibility of our chitosan microtubes with epithelial-like cells. The results obtained allow us to consider chitosan microtubes as promising bioartificial materials to repair damaged human blood vessels.

ACKNOWLEDGMENTS

The work was supported by the Russian Ministry of Education and Science, State Task No 4.1212.2014 / K.

KEYWORDS

- chitosan
- microtube
- molding
- polymer-analogs transformations
- salting-out agent
- biodegradable vascular prostheses

REFERENCES

1. Panseri, S.; Cunha, C.; Lowery, J.; Carro, U. D.; Taraballi, F.; Amadio, S.; Vescovi, A.; Gelain, F. Electrospun Micro- and Nanofiber Tubes for Functional Nervous Regeneration in Sciatic Nerve Transections. *BMC Biotechnol.* **2008**, *8*, 39.
2. Vaz, C. M.; Tuijl, S.; Bouten, C. V. C.; Baaijens, F. P. T. Design of Scaffolds for Blood Vessel Tissue Engineering using a Multi-Layering Electrospinning Technique. *Acta Biomater.* **2005**, *1*, (*5*), 575–582.
3. Smith, M. J.; McClure, M. J.; Sell, S. A.; Barnes, C. P.; Walpoth, B. H.; Simpson, D. G.; Bowlin, G. L. Suture-Reinforced Electrospun Polydioxanone–Elastin Small-Diameter Tubes for Use in Vascular Tissue Engineering: A Feasibility Study. *Acta Biomater.* **2008**, *4*, 58–66.

4. Yang, J.; Motlagh, D.; Webb, A. R.; Ameer, G. A. Novel Biphasic Elastomeric Scaffold for Small-Diameter Blood Vessel Tissue Engineering. *Tissue Eng.* **2005**, *11* (11–12), 1876–1886.

5. Baek, K.; Jeong, J. H.; Shkumatov, A.; Bashir, R.; Kong, H. In Situ Self-Folding Assembly of a Multi-Walled Hydrogel Tube for Uniaxial Sustained Molecular Release. *Adv. Mater.* **2013**, DOI: 10.1002/adma.201300951, 203–213.

6. Barros, A. A.; Duarte, A. R. C.; Pires, R. A.; Lima, A.; Mano, J. F.; Reis, R. L. Tailor Made Degradable Ureteral Stents from Natural Origin Polysaccharides. Materials Tenth Conference on Supercritical Fluids and Their Applications, 2013, p 1–6.

7. Lepidi, S.; Grego, F.; Vindigni, V.; Zavan, B.; Tonello, C.; Deriu, G. P.; Abatangelo, G.; Cortivo, R. Hyaluronan Biodegradable Scaffold for Small-Caliber Artery Grafting: Preliminary Results in an Animal Model. *Eur. J. Vasc. Endovasc. Surg.* **2006**, *32*, 411–417.

8. Kong, X.; Han, B.; Wang, H.; Li, H.; Xu, W.; Liu, W. Mechanical Properties of Biodegradable Small-Diameter Chitosan Artificial Vascular Prosthesis. *J. Biomed. Mater. Res. Part A.* **2012**, DOI: 10.1002/jbm.a.34136, 67–89.

9. Zhu, C.; Fan, D.; Duan, Z.; Xue, W.; Shang, L.; Chen, F.; Luo, Y. Initial Investigation of Novel Human-like Collagen/Chitosan Scaffold for Vascular Tissue Engineering. *J. Biomed. Mater. Res.* **2009**, *89A*, 829–840.

10. Rinaudo, M. Chitin and Chitosan: Properties and Applications. *Prog. Polym. Sci.* **2006**, *31* (7), 603–632.

11. Kim, I. Y.; Seo, S. J.; Moon, H. S.; Yoo, M. K.; Park, I. Y.; Kim, B. C.; Cho, C. S. Chitosan and its Derivatives for Tissue Engineering Applications. *Biotech. Adv.* **2008**, *26* (1), 1–21.

12. Huang, C.; Chen, R.; Ke, Q.; Morsi, Y.; Zhang, K.; Mo, X. Electrospun Collagen–Chitosan–TPU Nanofibrous Scaffolds for Tissue Engineered Tubular Grafts. *Colloids Surf. B: Biointerfaces* **2011**, *82* (2), 307–315.

13. Gegel, N. O.; Shipovskaya, A. B.; Vdovykh, L. S.; Babicheva, T. S. Preparation and Properties of 3D Chitosan Microtubes. *J. Soft Matter.* **2014**, *34*, Article ID 863096, 9.

14. Kuchanskaya, L. S.; Gegel, N. O.; Shipovskaya, A. B. Preparation of Chitosan-Based Microtubes. *Modern Perspectives in the* Study of chitin and Chitosan: Materials XIth International Conf. Murmansk, 2012, p 59–63.

15. General Pharmacopeia, Article № 42-0070-07. 16. Makadia, H. K.; Siegel, S. J. Poly Lactic-Co-Glycolic Acid (PLGA) as Biodegradable Controlled Drug Delivery Carrier. *Polymers* **2011**, *3* (3), 1377–1397.

17. Carletti, E.; Motta, A.; Migliaresi, C. Scaffolds for Tissue Engineering and 3D Cell Culture. *Methods Mol. Bio.* **2011**, *695*, 17–39.

18. Pan, Z.; Ding, J. Poly(Lactide-Co-Glycolide) Porous Scaffolds for Tissue Engineering and Regenerative Medicine. *Interface Focus* **2012**, *2* (3), 366–377.

19. *Biodegradable Systems in Tissue Engineering and Regenerative Medicine*; Reis, R. L., San, R. J., Ed.; CRC Press: Boca Raton, 2004; p 592.

20. Nair, L. S.; Laurenein, C. T. Biodegradable Polymers as Biomaterials. *Prog. Polym. Sci.* **2007**, *32* (8–9), 762–798.

21. Kildeeva, N. R.; Babak, V. G.; Vichoreva, G. A.; Ageev, E. P. et al. A New Approach to the Development of Materials with Controlled Release of the Drug Substance. *Her. Moscow Univ. Ser. 2. Chem. (in. Rus.)* **2000**, *41* (6), 423–425.

22. Stepanova, T. J. Modification of Surface Properties of Warp Yarns from Natural and Synthetic Fibers. Apriori. Series: Natural and Technical Sciences: UK, 2014; Vol. 3, p 17.

23. Kumar, V. A.; Caves, J. M.; Haller, C. A.; Dai, E.; Liu, L.; Grainger, S.; Chaikof, E. L. Acellular Vascular Grafts Generated from Collagen and Elastin Analogs. *Acta Biomater.* **2013**, *9* (9), 8067–8074.

24. *Chitin and Chitosan*; Skriabin, K. G., Vikhoreva, G. A., Varlamov, V. P., Ed.; 2002; p 211–216 [in Rus.].

25. Rinaudo, M.; Kildeyeva, N. R.; Babak, V. G. Surfactant-Polyelectrolyte Complexes Based on Chitin Derivatives. *Russ. Chem. J.* **2008**, *LII* (*1*), 84–90.

26. Babak, V. G.; Merkovich, E. A.; Galbraikh, L. S.; Shtykova, E. V.; Rinaudo, M. Kinetics of Diffusionally Induced Gelation and Ordered Nanostructure Formation in Surfactant–Polyelectrolyte Complexes Formed at Water/Water Emulsion Type Interfaces. *Mendeleev Commun.* **2000**, *10* (*3*), 94–95.

27. Masubon, T.; McClements, D. J. Influence of pH, Ionic Strength, and Temperature on Self-Association and Interactions of Sodium Dodecyl Sulfate in the Absence and Presence of Chitosan. *Langmuir* **2005**, *21* (*1*), 79–86.

CHAPTER 25

PROPAGATION RATE CONSTANT IN THE METHYL METHACRILATE BULK RADICAL POLYMERIZATION

V. M. YANBORISOV[1], A. A. SULTANOVA[1], and S. V. KOLESOV[2]

[1]Ufa State University of Economics and Service, Chernyshevsky Str. 145, 450078 Ufa, Russia, E-mail: yanborisovvm@mail.ru, tyc92@yandex.ru
[2]Institute of Organic Chemistry of Ufa Science, Center of the Russian Academy of Sciences, Prospect Oktyabrya Str. 71, 450054, Ufa, Russia, E-mail: kolesovservic@rambler.ru

CONTENTS

ABSTRACT

Propagation rate coefficient was defined for the process of methyl methacrylate bulk radical polymerization in the temperature range of 40–90 °C in quasi-stationary approximation. Also, it was defined by Monte Carlo simulation.

The theoretical model of radical polymerization was developed with regard to the gel effect. The software which can imitate the process of radical polymerization considering the gel effect with eligible costs of computer time was implemented. The software has been tested at the asymptotic examples and used for the simulation of methyl methacrylate bulk radical polymerization. The created program allows calculating monomer conversion, molecular weight change, molecular weight distribution, etc.

25.1 INTRODUCTION

High interest toward the process of radical polymerization is supported not only by its great practical significance and the necessity of the controlling methods elaboration, but also by a rapid development of viable option research of a process—pseudo living, complex-radical, and radical-coordinated polymerization, which helps, side by side with elaboration of experimental methods research, to develop mathematical modeling methods of polymerization processes. Via modern computer technologies the effective forecasting of kinetics of polymerization, molecular structure, and molecular weight characteristics of the products; also forecasting the practically important physicochemical properties of polymers are realized.[1–4] The usage of calculating methods is especially relevant for the analysis of complex mechanisms of radical polymerization. Definitely a positive experience of application of mathematical modeling to the study of other types of catalytic polymerization[2–3] proves the potential of this approach also in the case of radical processes. At the same time, even the classical bulk radical polymerization of typical monomers modeling is not a trivial task. In the late 40s, it was found that the bulk radical polymerization of methyl methacrylate (MMA) is accompanied with the gel effect appearance,[5] that is, the sharp increase in the polymerization rate and molecular mass (MM) at relatively high conversion. Thus, acceleration in the polymerization of MMA observed at conversion of 0.15–0.2[6] or 0.6.[7] Therefore, the study of kinetics and determination of the rate constants of polymerization should be carried out with the gel effect considering one of the models which is considering the gel effect in the case of the styrene polymerization is a Hui–Hamielec model. The model is based on the empirical dependency of the termination rate constant (k_t) from the monomer conversion (x): $k_t = k_{t0} \cdot \exp(-E/RT) \cdot \exp(-2(A_1 x + A_2 x^2 + A_3 x^3))$, where $k_{t0} \cdot \exp(-E/RT)$ is a termination rate constant excluding gel effect.[8]

For the numerical simulation of the MMA polymerization process and quantitative agreement between the calculated and experimental dependences it is necessary to know the exact actual values of the propagation rate constant (k_p) at various temperatures.

There are many known methods for determination of a k_p value. The values found by different researchers are given in the Table 25.1. In Ref. 9, the propagation rate constant was found by the interpretation of the experimental data of various researchers. However, as a rule, researchers determine the propagation rate constant only by interpretation of their experimental data.

TABLE 25.1 The Propagation Rate Constant in the Form of an Arrhenius Equation $k_p = k_{p0} \cdot \exp(-E/RT)$.

$k_{p0} \cdot 10^5$ L/(mol·s)	E/R, K	Determination method	Ref.
6.6	2367	Experimental data interpretation	9
4.92	2198	SIP	10
4.9	2190	SIP	11
25	2766	Electron spin resonance	12

For a long time, the method of "rotating sector"[13] was the main determination method of the propagation rate constant. A spatially discontinuous polymerization (SIP) is described in Ref. 14. The technique of SIP is similar to a technique of the rotating sector. Solution of monomer and the initiator flows through the dark tubular reactor by regularly located cracks through which light passes. The propagation rate constant found by SIP (Table 25.1) is published in Refs. 10–11. The received constant was used in Refs. 15–16 for the theoretical description of the MMA bulk radical polymerization.

The method of "spin traps" became widely used to obtain information about the propagation rate constant. The method is based on the reaction of specially introduced system non-paramagnetic molecule (trap) with a short-lived radical. As a result, a stable radical is formed by the reaction having a characteristic signal electron paramagnetic resonance. The use of spin traps allows the identification of short-lived radical products.[17] In Ref. 12, the propagation rate constant which was obtained by this method is published (Table 25.1).

The most reliable method of determining the propagation rate constant is the method of pulsed laser polymerization (PLP).[18] The essence of a method consists in irradiation of a vessel with monomer the pulsing laser radiation of big power as a result of which the huge number of radicals arises, earlier formed radicals destroying practically all. The survived new radicals initiate emergence of the new macroradicals growing to the following impulse.[19] Works with the published value of k_p, determined by this method for the MMA polymerization were not found.

This research aims to determine the propagation rate constant in quasi-stationary approximation, as well as by modeling the bulk polymerization of MMA by Monte Carlo method (MC) with consideration of the gel effect.

25.2   KINETIC MODEL

We considered the following reactions in modeling the polymerization of MMA:

initiation $\qquad\qquad I \xrightarrow{k_d} 2R_0^\bullet$

propagation $\qquad\qquad R_p^\bullet + M \xrightarrow{k_p} R_{p+1}^\bullet$

monomer transfer $\qquad\qquad R_p^\bullet + M \xrightarrow{k_m} R_1^\bullet + P_p$

termination by combination $\qquad\qquad R_p^\bullet + R_j^\bullet \xrightarrow{k_{com}} P_{p+j}$

termination by disproportionation $\qquad\qquad R_p^\bullet + R_j^\bullet \xrightarrow{k_{disp}} P_p + P_j$,

where I, M, R*, P—initiator, monomer, growing radical, "dead" macromolecule; index—the degree of polymerization, k_d, k_p, k_m, k_{disp}, k_{com}—rate constants of the reactions: initiation, chain propagation, chain transfer to monomer, disproportionation, and combination of macroradicals, respectively.

The possibility of thermal initiation is neglected due to the simulation of the synthesis of PMMA at relatively low temperatures. Also, unlikely chain transfer reaction to the initiator and the polymer are not considered.

This process of polymerization of MMA is described by a system of differential equations:

$$\frac{d[I]}{dt} = -k_d \cdot [I] \qquad\qquad (25.1)$$

$$\frac{d[M]}{dt} = -[R^\bullet] \cdot [M] \cdot (k_p + k_m) \qquad\qquad (25.2)$$

$$\frac{d[R^\bullet]}{dt} = 2 \cdot f \cdot k_d \cdot [I] - [R^\bullet]^2 \cdot (k_{disp} + k_{com}), \qquad\qquad (25.3)$$

where [R*]—total concentration (macro) radicals any length, f—efficiency of the chemical initiation.

Literary values of the rate constants of reactions were used (Table 25.2) in the simulation of the MMA radical polymerization in the presence of azobisisobutyronitrile (AIBN) and benzoyl peroxides (BPO).

TABLE 25.2 Parameters Used for MMA Bulk Radical Polymerization Modeling.

Parameters	Values	Refs.
$k_{t0} = k_{rec} + k_{disp}$	$9.8 \times 10^7 \cdot \exp(-353/T)$, L/(mol·s)	10
k_t	$k_{t0} \cdot \exp(-2(A_1x + A_2x^2 + A_3x^3))$	8
$n = k_{rec}/k_{disp}$	$3.956 \times 10^{-4} \cdot \exp(-2065/T)$	20
k_{disp}	$k_t \cdot n/(n + 1)$, L/(mol·s)	21
k_{com}	$k_t/(n + 1)$, L/(mol·s)	
$f(AIBN)$	0.5	
$k_d(AIBN)$	$1.053 \times 10^{15} \cdot \exp(-15440/T)$, s^{-1}	22
$f(BPO)$	1	9
$k_d(BPO)$	$1.18 \times 10^{14} \cdot \exp(-15097/T)$, s^{-1}	

25.3 EXPERIMENTAL

MMA (Fluka) was distilled in vacuo twice. The fraction with $T_b = 39$ °C at $p = 100$ mmHg, $n_D^{20} = 1.4130$, $d_4^{20} = 0.936$ g/mL for the MMA polymerization were used. The purity of the monomer was monitored by NMR ^{1}H- and ^{13}C. The initiators (AIBN, BPO) were recrystallized twice from methanol and dried in vacuo to constant weight at room temperature.

Conversion dependences were received by the dilatometric method.[6] For the bulk polymerization of MMA the reaction mixture was poured into a vial and solution was degassed at triplicate cycles of freezing–thawing at a residual pressure of 1.3 Pa. The vial was then sealed, placed in a thermostat (temperature accuracy ±0.1 °C), and maintained until the necessary degree of conversion. Conversion was calculated according to the formula: $x = \Delta V/(V_0 \times k)$, where V_0— the initial volume of the monomer, ΔV—its change, $k = (V_M - V_n)/V_M$—coefficient of contraction, V_M and V_n—the specific volume of the monomer and polymer.[23] After polymerization, the vials were cooled and opened. The resulting polymeric product was dissolved in acetone and then precipitated with a 10–15-fold excess of ethanol. From the remnants of the initiator polymer was purified by triple procedure of dissolution—precipitation. The purified polymer was dried in vacuum ($T = 40$ °C) to constant weight.

MMD and average MM of the polymer was determined by gel permeation chromatography. Analyses were carried out on a liquid chromatograph Waters GPC 2000 System (eluent: chloroform, flow rate: 0.5 ml/min). System columns were calibrated with polystyrene standards with $\overline{M}_w/\overline{M}_n \leq 1.2$.

25.4 DETERMINATION OF THE PROPAGATION RATE CONSTANT IN THE QUASI-STATIONARY APPROXIMATION

Historically, the first method of calculation of the propagation rate constant is a method based on the quasi-stationary approximation. It was suggested that the reaction rate constant of chain termination does not depend on the conversion: $k_t = k_{t0}$ = constant. Quasi-stationary macroradicals concentration was calculated (eq 25.4) using the approximation $d[R^\bullet]/d = 0$

$$[R^\bullet] \approx \sqrt{\frac{2 \cdot f \cdot k_d}{k_{t0}}[I]_0} \tag{25.4}$$

The propagation rate constant was evaluated from the initial rate of polymerization:

$$k_p \approx \frac{V_{i\dot{a}\dagger}}{[M]_0 \cdot [R^\bullet]}, \tag{25.5}$$

where $[M]_0 = 9.39$ mol/L—the initial monomer concentration.

The initial rate of polymerization was determined by fitting the experimental conversion dependences. We used data of this research, as well as data published in Refs. 24–29 (in the initial part of the conversion curve $x \leq 0.15$) (Table 25.3). The initial segment of the conversion curves was approximated by a linear dependence with a high correlation coefficient $r \geq 0.9964$ (Fig. 25.1).

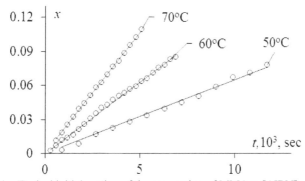

FIGURE 25.1   Typical initial section of the conversion of MMA at $[AIBN]_0 = 1$ mmol/L.

TABLE 25.3 Kinetic Parameters for the Bulk Radical Polymerization of MMA.

T, (°C)	Initiator	$[И]_0$, ммоль/л	Refs.	$V_0 \cdot 10^5$, моль/(л·с)			$[R^\cdot] \cdot 10^8$, моль/л	k_p, л/ (моль·с)
	AIBN	6.25		5.90	±	0.20	1.3	480
	AIBN	12.5		10.0	±	1.0	1.9	560
45	AIBN	25	24	12.5	±	8.7	2.6	509
	AIBN	50		16.7	±	1.4	3.7	480
	AIBN	100		22.0	±	5.0	5.2	427
	AIBN	200		29.52	±	0.15	7.4	425
								$480 \pm 51^{**}$
	AIBN	1	*	4.11	±	0.02	0.8	580
	AIBN	3	*	7.3	±	0.08	1.3	595
	AIBN	15.48	25	14.2	±	3.7	3	485
50	AIBN	15.5	25	15.1	±	2.9	4.3	354
	AIBN	25.8	26	18.8	±	3.6	3	674
	AIBN	25.8	26	18.19	±	0.93	3.8	506
	BPO	43	27	15.73	±	0.31	4	421
								$516 \pm 111^{**}$
	AIBN	1	*	11.38	±	0.04	1.5	797
	AIBN	3	*	19.57	±	0.19	2.9	708
	AIBN	5.6	28	22.5	±	1.0	3.6	688
	AIBN	11.2	28	30.5	±	0.66	5.1	617
60	AIBN	28	28	46.38	±	0.41	8	659
	AIBN	56	28	63.04	±	0.99	11.4	590
	AIBN	100	29	83.7	±	1.8	15.2	586
	BPO	1	*	8.610	±	0.01	1.2	761
	BPO	21.4	27	26.86	±	0.29	5.6	513
								$658 \pm 78^{**}$

TABLE 25.3   (*Continued*)

	AIBN	1	*	20.43	±	0.08	2.3	740
	AIBN	3	*	40.4	±	1.2	5.1	844
	AIBN	15.48	25	94.9	±	1.8	11.6	874
70	AIBN	15.5	26	86	±	37	11.6	794
	AIBN	25.8	25	110	±	16	15	788
	AIBN	25.8	26	119.8	±	7.1	14.9	854
	BPO	1	*	17.81	±	0.07	2.3	827
								$817 \pm 46^{**}$
80	AIBN	3	*	85.89	±	0.17	9.5	964
	BPO	1	*	37.81	±	0.17	4.2	955
								$959 \pm 6^{**}$
	AIBN	15.5	26	445	±	47	38.9	1220
90	AIBN	15.48	25	391	±	37	38.9	1071
	AIBN	25.8	25	460	±	181	50.1	978
								$1146 \pm 122^{**}$

* **Dates** of this work. ** Average values.

The value of k_p is determined with a rather large error, which has two components: a statistical error in the determination of initial rate and systematic error arising as a result of the quasi-stationary approximation. The first one we can calculate exactly when fitting the experimental data—the average relative error is 15%, which is comparable to the average statistical error in the determination of k_p from the experimental data of various researchers—8% (Table 25.1).

The second component—a systematic relative error in the determination of the reaction rate constant of chain propagation in quasi-stationary approximation was evaluated as follows. Differentiating eqs 25.4–25.5, following was obtained:

$$\frac{\Delta k_p}{k_p} = \frac{\Delta V}{V} + \frac{\Delta [R^\bullet]}{[R^\bullet]} + \frac{\Delta [M]_0}{[M]_0}, \quad \frac{\Delta [R^\bullet]}{[R^\bullet]} = \frac{1}{2}\left(\frac{\Delta [I]_0}{[I]_0} + \frac{\Delta k_{t_0}}{k_{t_0}}\right),$$

where $\Delta V/V$ is the statistical error. Let us estimate the systematic error of the concentration of monomer and initiator. For calculating eqs 25.4–25.5, it is necessary that concentration of monomer remains invariable. Actually, on a site of conversion curve 0–20% concentration decreases by 20%. For accuracy let us take the median value of $\Delta[M]_0/[M]_0 = 10\%$. Similarly, the initiator concentration decreases by $\Delta[I]_0/[I]_0 = 10\%$ (AIBN, 60 °C). Neglecting these errors leads to an overestimation of the macroradicals concentration and, consequently, to an underestimation of the constant of chain propagation.

Error arising due to the slowing down of the termination reactions was estimated using the Hui–Hamielec model.[8] In accordance with the eq 25.5, termination rate constant decreases by 71% in the area of 0–20% (60 °C). This systematic error leads to an overestimation of k_p in the quasi-stationary approximation and partially compensates for the error in the concentrations of monomer and initiator.

Thus, it is possible to believe that the systematic error of introduction of quasi-stationary approach in case of bulk MMA polymerization leads to overestimate k_p approximately for 50%.

Approximating the average values of the propagation rate constants (Table 25.3) by Arrhenius equation, $k_p = 7.26 \cdot 10^5 \cdot \exp(-2333/T)$ is received. The obtained value corresponds rather well with previously published values of k_p, found by SIP[10–11] (Fig. 25.2).

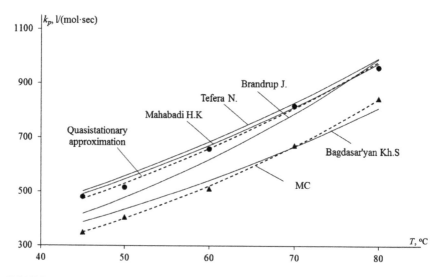

FIGURE 25.2 Propagation rate constant for MMA polymerization.

25.5   MODELING OF POLYMERIZATION OF MMA WITH MONTE CARLO METHOD

25.5.1   DESCRIPTION OF THE ALGORITHM

A MC method is an effective statistical method for solving systems of equations. Classical algorithm of the MC method is to draw events for each element of the ensemble using a pseudorandom number generator. Time taken for the calculation depends on the number of accessing the generator. However, there are cases where nothing happens with the selected element. Therefore, time spent on the calculation at this approach of modeling is sufficiently large.

In the proposed Monte Carlo simulation algorithm, "idle" accessing to random number generator are eliminated, which significantly increases the efficiency of the algorithm. Note that this algorithm is an evolution of the idea of the algorithm,[2-3] developed for modeling ion coordination polymerization of dienes in the presence of Ziegler–Natta catalysts. Difference following. In case of ion-coordination polymerization initiation is considered instant. In radical polymerization initiation happens during all polymerization. Therefore, in the simulation assumed that at the time $t = 0$, the size of the ensemble is the initial number of initiator molecules. The entire polymerization process is divided into small time intervals dt. The essence of the algorithm is that, at each time step random number element of the ensemble is played for events. The event for this element happens with the probability equal to unit. Number of random draws acts in step corresponds to the rate of reactions. This approach allows modeling to either significantly reduce the time of computer calculation or to increase the statistical ensemble.

One act of propagation happens, increasing number of units of a chain by unit. The number of molecules of monomer is reduced by one.

Act of chain transfer to monomer was modeled as follows. The item number of the ensemble (macroradical) was played, which undergoes a reaction of chain transfer to monomer. At the time of the transfer, there arises a "dead" macromolecule; its degree of polymerization is equal to the length of the growing molecule. At this moment, a new growing chain arises. This chain length is equal to one.

Chain termination is carried out by one of the two mechanisms such as disproportionation or combination. Two macroradicals was randomly played. The disproportionation of macroradicals formed two "dead" macromolecules with the relevant chain lengths. The macroradicals combination reaction formed a "dead" macromolecule with the chain length equal to the sum of the lengths of the chains of faced macroradicals. Number of macroradicals after chain termination was reduced by two units.

During simulation, the quantity of "dead" macromolecules with defined chain lengths was counted, which allowed to calculate number average and weight average MM PMMA and a molecular mass distribution (MMD) at different polymerization times.

The program was implemented in Borland Delphi 7, calculations were made on the computer AMD Athlon (2.81 GHz, 2 GB). Computation time is directly proportional to the magnitude of the ensemble. And the ensemble of 50–100 million molecules of the initiator is quite sufficient for correct playback of MMD. For example, when the number of the initiator molecules of the ensemble equals to 108, which corresponds to the ensemble of the monomer molecules 1012, the simulation of 1 h MMA polymerization was 25 min, which is significantly faster than in Ref. 4, where the simulation of 1 h of polymerization required 2 h of computing time and on a much more powerful computer: Intel Core i5 (2 cores, 3.46 GHz, 16 GB) at the less ensemble with 1011 monomer molecules. Thus, we believe that our program is 50 times more effective by the calculating speed.

25.5.2 DEFINING THE PARAMETERS OF THE HUI–HAMIELEC MODEL

In Ref. 8, values of the Hui–Hamielec coefficients A_1, A_2, and A_3 from eq 25.1 for the bulk radical polymerization of styrene are shown. An indispensable condition for this is a monotonic decrease of the termination rate coefficient with increasing conversion. Values of Hui–Hamielec parameters for a case of radical polymerization of MMA are defined in work[30] as a solution of the inverse task at a simultaneous variation of Hui–Hamielec coefficients and the reaction rate constant of chain propagation. Agreement between the experimental and calculated conversion of MMA was sought in the presence of an initiator AIBN in the temperature range 50–80 °C at different initial concentrations of the initiator. It was found that the inverse problem has several solutions (sets), describes the experimental dependence of the conversion from almost the same correlation coefficient. Firstly, some sets at high conversions lead to an increase in k_t. Such sets were excluded from the decision as not having physical sense. Secondly, it has been found that the dependence $k_t(x)$ has a point of intersection at different temperatures (Fig. 25.3b). Therefore, sets for which the intersection point lies in the conversion $x < 1$ were also excluded. As a result, the following solution was identified:

$$A_1 = 31.25 - 0.074 \cdot T, \quad A_2 = 14.92 - 0.059 \cdot T, \quad A_3 = -38.66 + 0.132 \cdot T. \quad (25.6)$$

Hui–Hamielec model coefficients (eq 25.6) applied to consider the gel effect in the Monte Carlo simulation of bulk radical MMA polymerization.

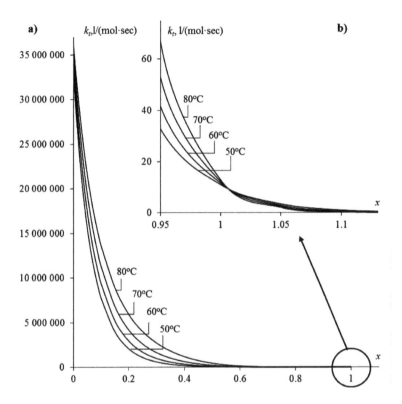

FIGURE 25.3   Termination rate constant versus conversion at different MMA polymerization temperatures.

25.5.3 PROGRAM TESTING

In order to test the simulation program of radical polymerization by MC method the computational experiments were carried out for some asymptotic cases.

Modeling growth of macromolecules in the absence of any transfer reactions and chain termination, expectedly obtained MMP described by the function of the Poisson distribution (Fig. 25.4):

$$q_w(p) = \frac{p}{\overline{P}_n} \frac{(\overline{P}_n)^p}{p!} \exp(-\overline{P}_n), \tag{25.7}$$

where p is the degree of polymerization, $\overline{P}_n$ is the number average degree of polymerization.

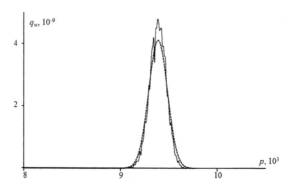

FIGURE 25.4 MMD obtained by modeling the propagation reaction: dashed line—the Poisson distribution (25.7), solid line—obtained with the MC simulation with $t = 10$ s, $k_p = 100$ L/(mol·s), the magnitude of the ensemble 100 million, $T = 70$ °C, $[AIBN]_0 = 1$ mmol/L.

It is generally known that chain transfer to monomer and/or a disproportionation reaction of the macroradicals leads to the Flory distribution function:

$$q_w(p) = \frac{p}{\overline{P}_n^2} \exp(-\frac{p}{\overline{P}_n}).$$

(25.8)

Indeed, MMP resulting from modeling reactions of chain propagation and chain transfer to monomer is Flory distribution (Fig. 25.5, curve 1). The same result is obtained by modeling the chain propagation reactions, chain transfer to monomer, and disproportionation macroradicals (Fig. 25.5, curve 2).

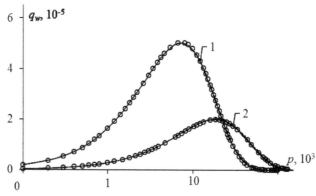

FIGURE 25.5 MMD PMMA: Flory distribution (25.8), solid line—MC (50 million quantity ensemble, $T = 70$ °C, $[BPO]_0 = 1$ mmol/L, $k_p = 670$ L/(mol·s), $k_m = 0.035$ L/(mol·s)). (1) Modeling chain propagation and chain transfer to the monomer ($t = 190$ s); (2) simulation chain propagation, chain transfer to monomer and disproportionation macroradicals ($t = 6600$ s, k_{disp} (Table 25.2)).

If during the polymerization, only chain growth reaction and combination of macroradicals proceed or macroradical combination predominates over reaction chain transfer to the monomer, then MMD should occur as the result described by Schulz distribution:

$$q_w(p) = \left(\frac{2}{P_n}\right)^2 \frac{p^2}{P_n} \exp(-\frac{2p}{P_n}). \qquad (25.9)$$

Program testing in this asymptotic case as well was successful (Fig. 25.6).

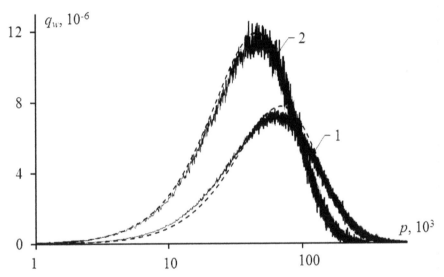

FIGURE 25.6   MMD PMMA, dashed line—Schulz distribution (25.9), the solid lines—MC (50 million quantity ensemble, $T = 70$ C, $[BPO]_0 = 1$ mmol/L, $k_p = 670$ L/(mol·s), $t = 1$ h, k_{com} (Table 25.2)): (1) chain propagation and combination macroradicals; (2) chain propagation, chain transfer to the monomer ($k_m = 10^{-4}$ L/(mol·s)) and combination of macroradicals.

In order to verify the results of the MC the system of differential eqs 1–3 have also been solved by the Runge–Kutta method of the fourth order. For the case of the polymerization of MMA in the presence of an initial concentration of BPO 1 mmol/L at 60 °C, results obtained after solving the system conducted by the two methods coincided (Fig. 25.7).

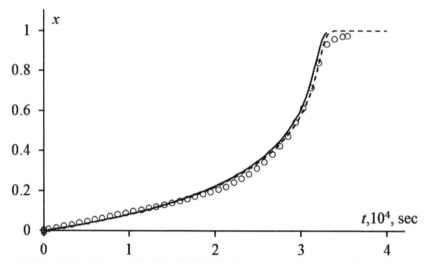

FIGURE 25.7 Conversion histories of MMA: solid lines—MC method, dashed line—Runge–Kutta method, ○—experimental data ([BPO]$_0$ = 1 mmol/L, 60 °C).

Thus, the tests results showed that the developed simulation software of chemical-initiated radical polymerization by MC considering gel effect Hui–Hamielec model correctly reproduces the change in monomer conversion and MMD during polymerization.

25.5.4 CONVERSIONAL DEPENDENCES

Search strategy for the propagation rate constant was in solving the inverse problem by approximating the experimental dependences of the conversion of MMA in the presence of an initiator AIBN at different initial concentrations of initiator. That is, the function to be minimized: $F(k_p) = \sum_{i=1}^{i_{max}} (x_i^{exp} - x_i^{calc})^2$, where i is the number of experimental point. On the choice of the value i_{max}. Experimental conversion dependences of MMA never reach the theoretical maximum possible value of $x = 1$, and strive only for some limit value $x_{limit} < 1$ (Fig. 25.8). It is explained by vitrification of PMMA in the field of high conversion[11,31–32] in which all reactions pass into the diffusive mode. The theoretical description of glass area at polymerization of MMA is possible at introduction to model of additional kinetic parameters[32–33] that complicates model. Therefore, we limited i_{max} not to consider glass region.

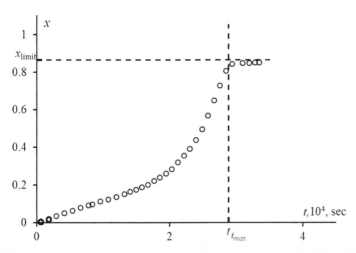

FIGURE 25.8 Conversion versus time for the bulk polymerization of MMA ($[AIBN]_0 = 1$ mmol/L, $T = 60\ °C$).

Propagation rate constant was determined on an array of experimental data of the MMA bulk radical polymerization in the presence of an initiator AIBN, and then checked for the initiator BPO. The fairly good coincidence between the calculated and experimental conversion curves was obtained (Fig. 25.9). Found $k_p = 2.5 \cdot 10^6 \cdot \exp(-2823/T)$, L/(mol·s) (Fig. 25.2) corresponds to the range of values of k_p^{12} (Table 25.1).

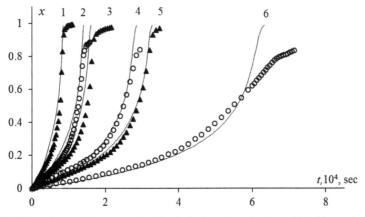

FIGURE 25.9 Conversion versus time for the bulk polymerization of MMA: ○—$[AIBN]_0 =$ 1 mmol/L, ▲—$[BPO]_0 = 1$ mmol/L; (1) 80 °C; (2,3) 70 °C; (5,4) 60 °C; (6) 50 °C; symbols—experimental data, lines—MC simulation.

25.5.5 MACRORADICALS CONCENTRATION

Only in Ref. 29, the concentration of macroradicals was experimentally measured for MMA polymerization conducted for 60 °C with the presence of AIBN
at initial concentration of 100 mmol/L. In this research, the dependence $y(t) =$
$[R^{\bullet}]/M$, is shown, which is almost identical to the concentration of radicals at
coordinates $[R^{\bullet}](t)$ in the process of modeling the MC (Fig. 25.10, curves 1–2),
which once again confirms adequacy of the Hui–Hamielec model in case of the
MMA polymerization.

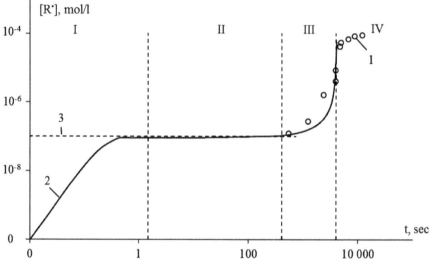

FIGURE 25.10 Macroradicals concentration versus time of the polymerization of MMA:
(1) experimental data[29], (2) MC simulation (60 °C, [AIBN]$_0$ = 100 mmol/L, k_p = 520 L/
(mol·s)), (3) quasi-stationary approximation at k_p = 660 L/(mol·s).

There are four areas on the kinetic curve: I—the initial portion of increasing
concentrations for stationary value, II—section with a constant concentration
macroradicals, III—an area on which there is a sharp increase in the concentration of macroradicals, IV—glass transition region. Note that the duration of the
first two stages is very small compared with the time of polymerization. Practically, all the monomer polymerizes in the area III, at the final stage IV because
of the glass transition of the polymer all reactions are slowed down.

25.5.6 MMD PMMA

Thus, the propagation rate constant was found at the first step of solving the inverse problem. The rate constant of chain transfer to the monomer are determined at the second stage of solution of the inverse problem by coincidence of the calculated and experimental average MM and MMD (Fig. 25.11). The following values were obtained: $k_m = 2.79 \times 10^5 \cdot \exp(-5450/T)$, l/(mol·s).

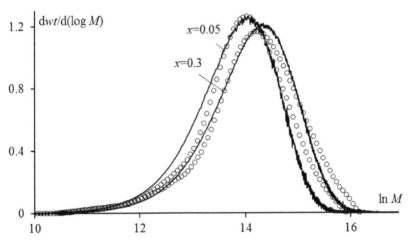

FIGURE 25.11　Typical MMD PMMA: symbols—experimental data ($T = 70$ °C, $[BPO]_0 = 1$ mmol/L), lines—MC simulation at $k_p = 670$ L/(mol·s), $k_m = 0.035$ L/(mol·s).

25.6　CONCLUSION

The propagation rate constant was determined in quasi-stationary approximation and by Monte Carlo simulation of bulk radical polymerization of MMA in presence AIBN and BPO for temperature range 50–90 °C. The use of quasi-stationary approximation leads to a significant overestimation of this constant. Therefore, it is necessary to consider the gel effect, which was held in MC modeling. The propagation rate constant value was obtained in the simulation appears to be more correct.

The theoretical model of the bulk radical polymerization of MMA taking into account gel effect in Hughie–Hamelek's form is constructed.

An original algorithm for the simulation of the radical polymerization by MC method was developed. The highly effective software is created. Concentration of reactants, speed of the imitated reactions, MMR polymer, and the value of an average MM may be calculated at various moments of polymerization with the acceptable computer time expenses.

An important advantage of the algorithm of modeling the process of radical polymerization is uncomplicated opportunity to add the program blocks to simulate other reactions.

KEYWORDS

- **propagation rate constant**
- **gel effect**
- **methyl methacrylate**
- **Monte Carlo method**
- **bulk radical polymerization**

REFERENCES

1. Bain, E. D.; Turgman-Cohen, S.; Genzer, J. Progress in Computer Simulation of Bulk, Confined, and Surface-Initiated Polymerizations. *Macromol. Theory Simul.* **2013**, *22*, 8–30.
2. Yanborisov, V. M.; Yanborisov, E. V.; Spivak, S. I. Algorithm of Modeling Ziegler–Natta Polymerization Taking Account of Change of Catalyst Activity. *Mathem. Mod.* **2010**, *22* (*3*), 15–25.
3. Yanborisov, V. M.; Yanborisov, E. V.; Spivak, S. I.; Ziganshina, A. S.; Ulitin, N. V. The Evolution of the Molecular Weight Distribution of Polydienes During Polymerization on Polycenter Catalysts of the Ziegler–Natta. *Herald Kazan State Technol. Univ.* **2014**, *17* (*4*), 155–158.
4. Drache, M.; Drache, G. Simulating Controlled Radical Polymerizations with mcPolymer—A Monte Carlo Approach. *Polymers* **2012**, *4*, 1416–1442.
5. Trommsdorff, E.; Kohle, H.; Lagally, P. ZurPolymerisation des Methacrylsiiuremethylesters. *J. Macromol. Chem. Phys.* **1948**, *1* (*3*), 169–196.
6. Gladyshev, G. P.; Popov, V. A. Radical Polymerization at High Degrees of Conversion. Nauka (Science, in Rus.): Moscow, 1974; 244.
7. Brooks, B. W. Viscosity Effects in the Free-Radical Polymerization of Methyl Methacrylate. *Proc. R. Soc. Lond.* **1977**, *357* (*A*), 183–192.
8. Hui, A. W.; Hamielec, A. E. Thermal Polymerization of Styrene at High Conversions and Temperatures. An Experimental Study. *J. Appl. Polym. Sci.* **1972**, *16*, 749–469.
9. Bagdasaryan, Kh. S. *Theory of Free-Radical Polymerization*. Davey, D., Ed.; Jerusalem: Israel Program for Scientific Translations, Hartford, Conn., 1968; p 328.
10. Mahabadi, H. K.; O'Driscoll, K. F. Absolute Rate Constants in Free-Radical Polymerization. *J. Macromol. Sci. Chem.* **1977**, *11A* (*5*), 967–976.
11. Tefera, N.; Weickert, G.; Westerterp, K. R. Modeling of Free Radical Polymerization up to High Conversion: 2. Development of a Mathematical Model. *J. Appl. Polym. Sci.* **1997**, *63* (*12*), 1663–1680.

12. Brandrup, J.; Immergut, E. H.; Grulke, E. A. *Polymer Handbook*; 4th ed., John Wiley and Sons. Inc.: USA, 1999; p. 2366.

13. Nagy, A.; Foldes-Berezsnich, T.; Tudos, F. Kinetics of Radical Polymerization-XLIII Investigation of the Polymerization of Methyl Acrylate in Solution by the Rotating Sector Method. *J. Eur. Polym.* **1984**, *20 (1)*, 25–29.

14. Mahabadi, H. K.; O'Driscoll, K. F. Spatially Intermittent Polymerization. *J. Polym. Sci.* **1976**, *14*, 869–881.

15. Chiu, W. Y; Carratt, G. M.; Soong, D. S. A Computer Model for the Gel Effect in Free-Radical Polymerization. *Macromolecules* **1983**, *16*, 348–357.

16. Kiparissides, C.; Achilias, D. Modeling of Diffusion-Controlled Free Radical Polymerization Reactions. *J. Appl. Polym. Sci.* **1988**, *35*, 1303–1323.

17. Emanuel, N. M.; Kuzimina, M. G. *Experimental Methods of Chemical Kinetics;* University of Moscow: Moscow, 1985; p 384.

18. Willemse, R. X. E. New Insights into Free-Radical (Co)Polymerization Kinetics; Technische Universiteit Eindhoven: Eindhoven, 2005; p 160.

19. Hutchinson, R. A.; Aronson, M. T.; Richards, J. R. Analysis of Pulsed-Laser-Generated Molecular Weight Distributions for the Determination of Propagation Rate Coefficients. *J. Macromol.* **1993**, *26*, 6410–6415.

20. Fenouillot, F.; Terrisse, J.; Rimlinger, T. Polymerization of Methyl Methacrylate at High Temperature with 1-Butanethiol as Chain Transfer Agent. *J. Appl. Polym. Sci.* **1999**, *72 (12)*, 1589–1599.

21. Bamford, C. H.; Barb, W. G.; Jenkins, A. D.; Onyon, P. F. *The Kinetics of Vinyl Polymerization by Radical Mechanisms*; Butterworths Scientific Publication: London, 1958; pp 103–117.

22. Tobolsky, A. V.; Baysal, B. A Review of Rates of Initiation in Vinyl Polymerization: Styrene and Methyl Methacrylate. *Polym. Sci.* **1953**, *11 (5)*, 471–486.

23. Gladyshev, G. P.; Gibov, K. M. *Polymerization at High Degrees of Conversion and Methods of Research*; Alma-Ata: Nauka (Science, in Rus.), 1968; p 142.

24. Ito, K. Estimation of Molecular Weight in Terms of the Gel Effect in Radical Polymerization. *J. Polym.* **1980**, *12*, 499–506.

25. Marten, F. L.; Hamielec, A. E. High Conversion Diffusion-Controlled Polymerization. *ACS Symp. Ser.* **1979**, *104*, 43–70.

26. Balke, S. T.; Hamielec, A. E. Bulk Polymerization of Methyl Methacrylate. *J. Appl. Polym. Sci.* **1973**, *17*, 905–949.

27. O'Neil, G. A.; Wisnudel, M. B.; Torkelson, J. M. An Evaluation of Free Volume Approaches to Describe the Gel Effect in Free Radical Polymerization. *Macromolecules* **1998**, *31 (3)*, 4537–4545.

28. Stickler, M.; Panke, D. Polymerization of Methyl Methacrylate up to High Degrees of Conversion: Experimental Investigation of the Diffusion-Controlled Polymerization. *J. Polym. Sci.* **1984**, *22*, 2243–2253.

29. Carswell, T. G.; Hill, D. J. T.; Londero, D. I.; O'Donnell, J. H.; Pomery, P. J.; Winzor, C. L. Kinetic Parameters for Polymerization of Methyl Methacrylate at 60 °C. *Polymer* **1992**, *33 (1)*, 137–140.

30. Kolesov, S. V.; Shiyan, D. A.; Yanborisov, V. M.; Burakova, A. O.; Tereshchenco, K. A.; Ulitin, N. V.; Zaikov, G. E. Description of Radical Polymerization Kinetics of Methyl Methacrylate with Consideration of Autoacceleration. *J. Inform., Intelligence Knowled.* 25, 34–60.

31. Vrentas, J. S. Review of the Applications of Free Volume Theory for Diffusion in Polymers. *Polym. Rev.* **1981**, *15*, 136–138.

32. Fleury, P. A.; Meyer, Th.; Renken, A. Solution Polymerization of Methyl-Methacrylate at High Conversion in a Recycle Tubular Reactor. *Chem. Eng. Sci.* **1992**, *47 (9)*, 2597–602.

33. Nising, P.; Meyer, T. Mathematical-Modeling of Polymerization Kinetics at High Monomer Conversions a Critical-Review. *Plaste Kautsch* **1986**, *33 (8)*, 281–285.

CHAPTER 26

ENZYMATIC DEGRADATION OF CHITOSAN IN ACETIC ACID SOLUTION

I. F. TUKTAROVA, V. V. CHERNOVA, and E. I. KULISH

Bashkir State University, Republic of Bashkortostan, Ufa 450074, ul. Zaki Validi 32, Russia, Tel.: +7 (347) 229 96 14, E-mail: alenakulish@rambler.ru

CONTENTS

26.1 INTRODUCTION

Polymer of natural origin chitosan attracts attention of researchers because it can be used to produce bioactive protecting coatings for treatment of surgical wounds and burns. In this case, the biodegradability (enzymatic degradation) of the chitosan material will be affected under the action of nonspecific enzymes of the human body (e.g., hyaluronidase, which is present at the wound surface, etc.). The speed of this process will determine the life of the polymer material on the wound surface. Thus, the study of the process of enzymatic degradation of chitosan under the action of nonspecific enzyme preparations will be important, both from scientific and practical points of view. In this paper, we determine the kinetic parameters of the process of enzymatic degradation of chitosan by the enzyme hyaluronidase.

26.2 EXPERIMENTAL

The object of investigation was a CHT specimen produced by the company «Bioprogress» (Russia) and obtained by acetic deacetylation of crab chitin with a molecular weight of M_{sn}=11,3000. As an enzyme preparation, hyaluronidase enzyme was used («Liraze»), production of «Microgen» (Moscow, Russia). The concentration of the enzyme preparation was 0.1, 0.2, and 0.3 g/L. Acetic acid of 1 g/dL concentration was used as a solvent. CHT concentration in solution ranged from 0.1 to 5 g/dL.

The process of enzymatic degradation was judged by the falling of the intrinsic viscosity [η] of CHT. The intrinsic viscosity in a solution of acetic acid was determined at 25 °C, using the method of Irzhak and Baranov.[1] To determine the initial values of the intrinsic viscosity of CHT solution, $[η]_0$ was used at a concentration of $c = 0.15$ g/dL. To determine the values of intrinsic viscosity during the enzymatic digestion [η], CHT was dissolved in acetic acid to which a solution of the enzyme preparation maintained for a certain time was added. Then, a process of enzymatic degradation was quenched by boiling the original solution for 30 min in a water bath. Next, from the initial concentration of the solution c_{ed}, a solution to determine the inherent viscosity at a concentration of $c = 0.15$ g/dL was made. The process of enzymatic degradation was performed at 36 °C.

The initial rate of enzymatic degradation of chitosan V_0 evaluated on the linear part by the drop in its intrinsic viscosity [η] and calculated by the formula[2]:

$$V_0 = \frac{c_{ed} \, K^{1/a} ([η]^{-1/a} - [η]_0^{-1/a})}{t}$$ (26.1),

where c_{ed}—the concentration of chitosan in the solution was subjected to enzymatic degradation, g/dL; t—time of degradation, min; K and α—constants of the equation Mark–Houwink $[\eta] = KM^{\alpha}$; M—molecular weight of chitosan.

To determine the constants in Mark–Houwink equation, it is necessary to calculate values of the initial velocity of the enzymatic degradation of the eq 26.1, CHT sample was fractionated into 10 fractions of molecular weight ranging from 20,000 to 150,000 Daltons. The absolute value of the molecular weight fractions of CHT were determined by a combination methods sedimentation velocity and viscosimetry.

The molecular weight of the fractions was determined by the formula:

$$M_{s\eta} = \left(\frac{S_0 \eta_0 [\eta]^{1/3} N_A}{A_{hi}(1-\bar{v}\rho_0)} \right)^{3/2}$$
(26.2),

where S_0—sedimentation constant; η_0—dynamic viscosity of the solvent, equal to 1.2269×10^{-2} PP; $[\eta]$—intrinsic viscosity dL/g; N_A—Avogadro number, equal to 6023×10^{23} mol^{-1}; $(1-v\rho_0)$—Archimedes factor or buoyancy factor; v—the partial specific volume, cm^3/g, ρ_0—density of the solvent g/cm^3; A_{hi}—hydrodynamic invariant, equal to 2.71×10^6.

26.3   RESULTS AND DISCUSSION

Figure 26.1 shows the dependence of the intrinsic viscosity of the CHT solution on the time of standing with the enzyme preparation.

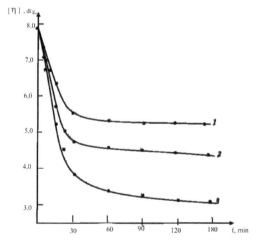

FIGURE 26.1   Dependence of the intrinsic viscosity on the time of standing 1% CHT solution with the enzyme solution with concentration 0.1 (1), 0.2 (2), and 0.3 g/L (3).

It can be seen that when increasing time of CHT exposure to the enzyme solution, the viscosity decreases regularly, indicating a reduction in molecular weight of CHT. The most significant drop in viscosity occurs in the initial period. Further exposure of CHT to enzymatic solution causes the degree of viscosity fall significantly to a lesser extent, that is, the reaction rate decreases with time. This course of the kinetic curve is typical for most enzymatic reactions.[3] Increasing the concentration of the enzyme preparation leads to a natural increase in the rate of fall of the intrinsic viscosity.

As established in the course of the work, for all the studied solutions for small CHT times degradation decrease the intrinsic viscosity curves are linear. On this plot, the value of the initial rate of enzymatic degradation of V_0 was determined, serving as a measure of enzymatic activity of the enzyme with respect to CHT.

To determine the rate of enzymatic degradation of the eq 26.1, it is necessary to determine the value of the constants in the equation of Mark–Houwink. For the sample, we used their significance, which was defined as $\alpha = 1.02$ and $R = 5.57 \times 10^{-5}$.

As studies have shown, the observed dependence of the initial rate of enzymatic degradation of the substrate concentration can be described within the Michaelis–Menten scheme. Figure 26.2 shows the dependence of the initial rate of CHT concentration in solution.

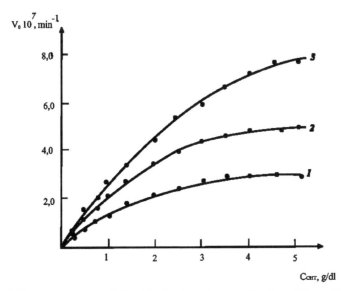

FIGURE 26.2 Dependence of the initial rate of enzymatic degradation of CHT at a concentration of enzyme preparation 0.1 (1), 0.2 (2), and 0.3 g/L (3).

It is clear that it has a classic look and occurs as a part of a rectangular hyperbola. Submission to double check the coordinates (a graphical method of Lineweaver–Burk) can accurately determine the value of the Michaelis constant K_m (Table 26.1). As can be seen from the data, its value is constant, independent of the concentration of the enzyme and, in fact, characterizes the affinity of the enzyme to the substrate value $V_{max} = k_2 C_e$, where k_2, the decay constant of the enzyme–substrate complex, gives a description of the catalytic activity of the enzyme, that is, defines the maximum possible formation of the reaction product at a given concentration of enzyme in an excess of substrate. However, the reaction substrate in excess of the maximum rate of reaction depends linearly on the concentration of the enzyme (Fig. 26.3).

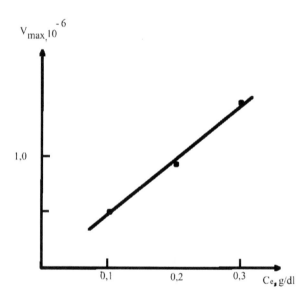

FIGURE 26.3 The dependence of the maximum reaction rate of enzymatic degradation of CHT depending on the concentration of the enzyme preparation in the solution.

TABLE 26.1 The Values of the Constants in the Equation of Michaelis–Menten for CHT Solution in 1% Acetic Acid.

C_e, g/L	K_m, g/dL	V_{max}, 10^6
0.1	3.37	0.50
0.2	3.47	0.92
0.3	3.42	1.51

26.4   CONCLUSION

For the first time, the kinetic characteristics of the enzyme activity of hyaluronidase for the enzymatic degradation of a CHT solution of 1% acetic acid was determined. Certain value of the Michaelis constant $K_m = 3.4$ g/dL, significantly higher than the $K_m = 0.03$ g/dL, as defined in Ref. 1 for the carboxymethyl cellulose under the cellulase system of *Geotrilium candidum*. This is due to the fact that the enzyme is used in the nonspecific with respect to CHT and the conditions of the enzymatic degradation did not meet the temperature and pH optimum of the enzyme hyaluronidase.

KEYWORDS

- **chitosan**
- **viscosity**
- **enzymatic degradation**
- **hyaluronidase**

REFERENCES

1. Baranov, V; Brestkin, Y; Agranova, S. A.; Pinkevich, V. N. The Behavior of Macromolecules of Polystyrene in the "Thickened" Good Solvent. *Polymer.* **1986,** *T. 28B* (10), S. 841–843.
2. Rabinovich, M. L.; Klesov, A. A.; Berezin, I. V. The Kinetics of Enzyme Action Tsellyuliticheskih Geotrilium Candidum. Viscometric Analysis of the Kinetics of Hydrolysis of CMC. *Bioorg. Chem.* **1977,** *3* (3), S. 405–414.
3. *Biochemistry;* Severin, E. S., Ed.; Geotar Med: Moscow, 2004; p 779.

CHAPTER 27

MEDICINAL FILM COATINGS ON THE BASIS OF CHITOSAN

A. S. SHURSHINA and E. I. KULISH

Bashkir State University, Republic of Bashkortostan, Ufa 450074, ul. Zaki Validi 32, Russia, Tel.: +7 (347) 229 96 14, E-mail: alenakulish@rambler.ru

CONTENTS

The medicinal film coatings on the basis of chitosan and antibiotics, both of cephalosporin series and of aminoglycoside one, have been considered. It has been shown that the antibiotics release from a film will be determined by the amount of antibiotics connected with chitosan by hydrogen bonds, on the one hand, and by the state of the polymer matrix, on the other.

27.1 INTRODUCTION

Decrease in effectiveness of therapy by the antibiotics, being observed recently, is caused, generally distribution of strains of bacteria steady against them. Polymeric derivatives of antibiotics can help with the solution of this task.[1] Advantages of use of polymeric derivative antibiotics are obvious in that case when polymer carrier of medicinal substance is in a soluble form. However, it is not less important to consider the case when polymer carries out a matrix role—the carrier of medicinal substance. In this work, some approaches to creation of sheet having antibacterial coverings on the basis of chitosan (ChT) of the prolonged action suitable for treatment of surgical, burn, and slow wounds of a various etiology are considered. The choice as the ChT carrier is not casual as this polymer possesses the whole range of the unique properties doing it by irreplaceable polymer[2] for medicine.

27.2 EXPERIMENTALS

The objects of investigation were a ChT specimen produced by the company "Bioprogress" (Russia) and obtained by acetic deacetylation of crab chitin and antibiotics, both of cephalosporin series—cefazolin sodium salt (CPhZ) and cefotaxime sodium salt (CPhT), and of aminoglycoside series—amikacin sulfate (AMS) and gentamicin sulfate (GMS). The investigation of the interaction of medicinal preparations with ChT was carried out according to the techniques described in Refs. 2 and 3.

ChT films were obtained by means of casting of the polymer solution in acetic acid onto the glass surface with the formation of chitosan acetate (ChTA). The polymer mass concentration in the initial solution was 2 g/dL. The acetic acid concentration in the solution was 1, 10, and 70 g/dL. Aqueous antibiotic solution was added to the ChT solution immediately before films' formation. The content of the medicinal preparation in the films was 0.1 mol/mol ChT. The film's thickness in all the experiments was maintained constant and equal to 0.1 mm. The kinetics of antibiotics release from ChT film specimens into aqueous medium was studied spectrophotometrically at the wavelength corresponding to the maximum absorption of the medicinal preparation.

In order to regulate the ChT ability to be dissolved in water, the anion nature varied while obtaining ChT salt forms. So, a ChT–CPhZ film is completely soluble in water. The addition of aqueous sodium sulfate solution in the amount of 0.2 mol/mol ChT to the ChT–CPhZ solution makes it possible to obtain an insoluble ChT–CPhZ–Na$_2$SO$_4$ film. On the contrary, a ChT–AMS film being formed at the components ratio used in the process of work is not soluble in water. Obtaining a water-soluble film is possible if amikacin sulfate is transformed into amikacin chloride (AMCh). In this case, the obtained ChT–AMCh film will be completely soluble in water. Thus, the following film specimens have been analyzed in the investigation: ChT–CPhZ and ChT–CPhT (soluble forms), ChT–CPhZ–Na$_2$SO$_4$ (insoluble-in-water form), ChT–AMCh (soluble form), ChT–AMS, and ChT–GMS (insoluble-in-water forms).

With the aim of determining the amount of medicinal preparation held by the polymer matrix, synthesis of adducts of the ChT–antibiotic interaction in the mole ratio 1:1 in acetic acid solution was carried out. The synthesized adducts were isolated by double re-precipitation of the reaction solution in NaOH solution followed with the washing of precipitated complex residue with isopropyl alcohol. Then, the residue was dried in vacuum up to constant mass. The amount of preparation strongly held by chitosan matrix was determined according to the data of the element analysis on the analyzer EUKOEA-3000.

27.3   RESULTS AND DISCUSSION

On the basis of the chemical structure of the studied medicinal compounds,[4] one can suggest that they are able to combine with ChT forming polymer adducts of two types—ChT-antibiotics complexes and polymer salts produced due to exchange interaction. As a result, some quantity of medicinal substance will be held in the polymer chain. The interaction taking place between the studied medicinal compounds and ChT was demonstrated by UV- and IR-spectroscopy data. The interaction energies evaluated by the shift in UV-spectra are about 7–12 kJ/mole, which allows us to speak about the formation of complex ChT-antibiotic compounds by means of hydrogen bonds.

Table 27.1 gives the data on the amount of antibiotics determined in polymer adducts obtained from acetic acid solution.

TABLE 27.1 The Amount of Antibiotics Determined in Reaction Adducts.

C_{CH_3COOH} g/dL in the initial solution	The antibiotics used	The amount of antibiotics in reaction adduct, mass%
1	CPhZ	10,1
	CPhT	15,9
	AMS	61,5
	GMS	59,4
10	CPhZ	5,88
	CPhT	57,5
	AMS	55,8
	GMS	31,3
70	CPhZ	3,03
	CPhT	3,7
	AMS	41,3
	GMS	40,1

Attention should be paid to the fact that the amount of medicinal preparation in the adduct of the ChT-medicinal preparation reaction is considerably higher in the case of antibiotics of aminoglycoside series than in the case of antibiotics of cephalosporin series. This can be connected with the fact that CPhZ and CPhT anions interact with ChT polycation forming salts that are readily soluble in water. In the case of using AM and GM sulfates because of two-base character of sulfuric acid, one may anticipate the formation of water-insoluble "double" salts—ChT–AM or ChT–GM sulfates—due to which additional quantity of antibiotics is held on the polymer chain.

Table 27.2 gives the data on the value of the rate of AM and GM release from film specimens formed from acetic acid solutions of different concentrations. The rate was evaluated only for water-insoluble films because by using soluble films, the antibiotic release was determined not by medicinal preparation diffusion from swollen matrix but by film dissolving.

TABLE 27.2 Transport Properties of Chitosan Films in Relation to Medicinal Preparation Release.

Acetic acid concentration g/dL	The antibiotics used	Release, mass%/h for chitosan specimens
1	AMS	0.5
	GMS	0.4

TABLE 27.2 (*Continued*)

10	AMS	0.8
	GMS	0.5
70	AMS	1.5
	GMS	1.3

Attention must be given to the fact of interaction between the rate of anti-biotics release from chitosan films and their amount, which is strongly held in ChT chain. For example, on increasing the concentration of acetic acid used as a solvent, the amount of medicinal preparation connected with the polymer chain decreases in all the cases considered by us. Correspondingly, the rate of antibiot-ics released from films insoluble in water increases.

The influence of the amount of medicinal preparation strongly held in ChT matrix on the rate of medicinal substance released from the film must be most pronounced at comparing the rates of release of antibiotics of aminoglycoside series and cephalosporin one. However, ChT–CPhZ and ChT–CPhT films are soluble in water while ChT–AMS and ChT–GMS ones do not dissolve in water and it is not correct to compare them. At ChT transition into insoluble form (by adding sodium sulfate), the rate of release of antibiotics of cephalosporin series decreases considerably (Fig. 27.1, curve 1) as compared with a soluble form, but still it is higher than that in the case of antibiotics of aminoglycoside series (Fig. 27.1, curve 2). It should be also noted that the rate of antibiotics release from soluble ChT–CPhZ film (Fig. 27.1, curve 3) is also higher than in the case of ChT–AMCh film (Fig. 27.1, curve 4). Thus, considerable difference between the rate of release of aminoglycoside series antibiotics and that of cephalospo-rin series antibiotics is evidently explained by the difference in the amount of ChT–antibiotics adduct.

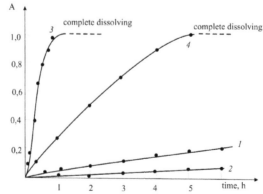

FIGURE 27.1 The kinetic curve of the release of CPhZ (1,3) and AM (2,4) from insoluble (1,2) and soluble (3,4) films.

Thus, while forming film coatings, one should proceed from the fact that a medicinal preparation can be distributed in the polymer matrix in two ways. One part of it connected with polymer chain, for example, by complex formation is rather strongly held in that polymer chain. The rest of it is concentrated in polymer-free volume (in polymer pores). The rate of release of antibiotics from the film will be determined by the amount of antibiotics connected with ChT by hydrogen bonds, on the one hand, and by the state of the polymer matrix including its ability to dissolve in water, on the other hand.

The work has been carried out due to the financial support of the RFFR and the republic of Bashkortostan (grant_r_povolzhye_a No 11-03-97016)

KEYWORDS

- **chitosan**
- **modification**
- **medicinal preparation**
- **the state of polymer matrix**

REFERENCES

1. Skryabin K. G.; Vikhoreva G. A.; Varlamov V. P. *Chitin and Chitosan: Obtaining, Properties and Application*; Nauka: Moscow, 2002; p 365.
2. Mudarisova R. Kh.; Kulish E. I.; Kolesov S. V.; Monakov Yu. B. Investigation of Chitosan Interaction with Cephazolin. *JACh* **2009**, *82* (*5*), 347–349.
3. Mudarisova R. Kh.; Kulish E. I.; Ershova N. R.; Kolesov S. V.; Monakov Yu. B. The Study of Complex Formation of Chitosan with Antibiotics Amicacin and Hentamicin. *JACh* **2010**, *83* (*6*), 1006–1008.
4. Mashkovsky M. D. *Medicinal Preparations*; Torsing: Kharkov, 1997; Vol. 2, p 278.

INDEX

Milton Keynes UK
Ingram Content Group UK Ltd.
UKHW031139141024
449569UK00024B/1199